ADVANCES IN
MOLECULAR AND CELL BIOLOGY
VOLUME 36

Caveolae and Lipid Rafts: Roles in Signal Transduction and the Pathogenesis of Human Disease

ADVANCES IN
MOLECULAR AND CELL BIOLOGY
VOLUME 36

Caveolae and Lipid Rafts: Roles in Signal Transduction and the Pathogenesis of Human Disease

Series Editor

E. Edward Bittar
University of Wisconsin – Madison
Madison, Wisconsin

Volume Editors

Philippe G. Frank
Departments of Molecular Pharmacology and Urology
Albert Einstein College of Medicine
Bronx, New York

Michael P. Lisanti
Department of Molecular Pharmacology
Albert Einstein College of Medicine
Bronx, New York

2005

ELSEVIER
ACADEMIC
PRESS

AMSTERDAM • BOSTON • HEIDELBERG • LONDON
NEW YORK • OXFORD • PARIS • SAN DIEGO
SAN FRANCISCO • SINGAPORE • SYDNEY • TOKYO

Elsevier Academic Press
525 B Street, Suite 1900, San Diego, California 92101-4495, USA
84 Theobald's Road, London WC1X 8RR, UK

This book is printed on acid-free paper. ♾

For all information on all Elsevier Academic Press publications
visit our Web site at www.books.elsevier.com

ISBN-13: 978-0-444-51500-3
ISBN-10: 0-444-51500-3
ISSN: 1569-2558

PRINTED IN THE UNITED STATES OF AMERICA
05 06 07 08 09 9 8 7 6 5 4 3 2 1

Contents

Section IV. Caveolae and Pathologies

Contributors

Soman N. Abraham *Department of Pathology, Duke University Medical Center, Durham, North Carolina 27710*

Barbara P. Atshaves *Department of Physiology and Pharmacology, Texas A & M University, College Station, Texas 77843-4466*

Judith M. Ball *Department of Pathobiology, Texas A & M University, College Station, Texas 77843-4467*

Ken L. Chambliss *Department of Pediatrics, University of Texas Southwestern Medical Center at Dallas, Dallas, Texas 75390*

Matthew J. Duncan *Department of Molecular Genetics and Microbiology, Duke University Medical Center, Durham, North Carolina 27710*

William V. Everson *University of Kentucky, Sanders Brown Center on Aging, Lexington, Kentucky 40536-0230*

Olivier Feron *University of Louvain Medical School, Unit of Pharmacology and Therapeutics, B-1200 Brussels, Belgium*

Philippe G. Frank *Departments of Molecular Pharmacology and Urology, Albert Einstein College of Medicine, Bronx, New York 10461*

Adalberto M. Gallegos *Department of Pathobiology, Texas A & M University, College Station, Texas 77843-4467*

Ghada S. Hassan *Department of Molecular Pharmacology, Albert Einstein College of Medicine, Bronx, New York 10461*

Huan Huang *Department of Physiology and Pharmacology, Texas A & M University, College Station, Texas 77843-4466*

Junsuke Igarashi *Department of Cardiovascular Physiology, Kagawa University Faculty of Medicine, 761-0793 Kagawa, Japan*

Jean-François Jasmin *Department of Molecular Pharmacology, Albert Einstein College of Medicine, Bronx, New York 10461*

Ann B. Kier *Department of Pathobiology, Texas A & M University, College Station, Texas 77843-4467*

Michael P. Lisanti *Department of Molecular Pharmacology, Albert Einstein College of Medicine, Bronx, New York 10461*

J. C. Steve Liu *Computer Science Department, Texas A & M University, College Station, Texas 77843-3312*

Avery L. McIntosh *Department of Physiology and Pharmacology, Texas A & M University, College Station, Texas 77843-4466*

M. A. McNiven *Department of Biochemistry and Molecular Biology, and Center for Basic Research in Digestive Diseases, Mayo Clinic College of Medicine, Rochester, Minnesota 55905*

L. S. Orlichenko *Department of Biochemistry and Molecular Biology, and Center for Basic Research in Digestive Diseases, Mayo Clinic College of Medicine, Rochester, Minnesota 55905*

Friedhelm Schroeder *Department of Physiology and Pharmacology, Texas A & M University, College Station, Texas 77843-4466*

Philip W. Shaul *Department of Pediatrics, University of Texas Southwestern Medical Center at Dallas, Dallas, Texas 75390*

Eric J. Smart *University of Kentucky, Sanders Brown Center on Aging, Lexington, Kentucky 40536-0230*

Peter Strålfors *Department of Cell Biology and Diabetes Research Centre, Linköping University, SE58185 Linköping, Sweden*

Matthias Vorgerd *Department of Neurology, Neuromuscular Center Ruhrgebiet, Ruhr-University Bochum, D-44789 Bochum, Germany*

Preface

Caveolae are 50–100 nm plasma membrane invaginations found at the surfaces of most terminally differentiated cell types. Palade (1953) and Yamada (1955) first morphologically identified caveolae (a.k.a., plasmalemmal vesicles) about fifty years ago–in both endothelial and epithelial cells. Their organization and cellular localization indicated that these organelles might play an important role in the transport of molecules between different peripheral compartments. However, at that time, their molecular composition remained completely unknown.

In 1992, almost 40 years later, Glenney & Soppet (1992) and Rothberg *et al.* (1992) were the first to identify the most important structural protein of caveolae, namely caveolin. This protein was re-named caveolin–1 after the discovery of two other caveolin genes, caveolin–2 and caveolin–3. Subsequent studies clearly demonstrated the essential role of caveolin–1 in the formation of caveolae, as well as in the regulation of cell signaling.

The existence of caveolae was initially considered controversial (Severs, 1988); they were thought to represent an EM fixation artifact. However, several groups have now biochemically purified caveolae, using caveolin–1 as a marker protein for the organelle (Lisanti *et al.*, 1994a; Schnitzer *et al.*, 1995; Smart *et al.*, 1995; Anderson, 1998). In addition, the absence of caveolae in caveolin–1 deficient mice dramatically demonstrates their existence in a normal in vivo environment (Drab *et al.*, 2001; Razani *et al.*, 2001; Zhao *et al.*, 2002; Cao *et al.*, 2003).

Simons and collaborators were among the first to propose the concept and the existence of lipid raft domains (Simons and Ikonen, 1997); it is now well accepted that caveolae represent a subset of these lipid rafts. The specific lipid composition of caveolae (sphingolipid and cholesterol-rich) led to the idea that these microdomains may play an important role in the regulation of cellular cholesterol homeostasis (Brown and London, 1998; Liscum and Munn, 1999; Fielding and Fielding, 2001) (see Section 1 of the book).

Besides the important role of caveolae in the regulation of endocytosis (see Section 2 of the book), the enrichment of signaling molecules in caveolae suggested that these domains play a key role in the regulation and organization of various cell-signaling cascades—forming "pre-assembled signaling complexes" (Lisanti *et al.*, 1994b) (see Section 3 of the book).

Finally, more recent studies using caveolin-deficient mice have confirmed many of the "controversial hypotheses" put forth in earlier publications (Drab *et al.*, 2001; Razani *et al.*, 2001; Zhao *et al.*, 2002; Cao *et al.*, 2003). However, even if caveolin proteins are not essential for life, they still play a critical role in the pathogenesis of a number of human diseases, as highlighted in studies using caveolin-deficient mouse animal models.

These studies have shown that caveolin proteins play an important role in regulating muscular, cardiovascular, and pulmonary patho-physiology (see Section 4 of the book). Furthermore, the negative regulatory role of caveolin–1 in various signaling pathways has suggested that caveolin–1 may indeed function as a tumor suppressor gene. Clear genetic links between the caveolin proteins and human disease have now been established in patients suffering from muscular dystrophy, cardiomyopathy, and breast cancer.

Overall, this book summarizes the essential features and functions of caveolae in cells, mouse animal models, and in human disease. We thank all the excellent contributors who extensively participated in making this book possible.

Philippe G. Frank and Michael P. Lisanti

REFERENCES

Anderson, R. G. W. (1998) The caveolae membrane system. *Ann. Rev. Biochem.* **67,** 199–225.

Brown, D. A. and London, E. (1998) Functions of lipid rafts in biological membranes. *Ann. Rev. Cell Dev. Biol.* **14,** 111–136.

Cao, G., Yang, G., Timme, T. L., Saika, T., Truong, L. D., Satoh, T., Goltsov, A., Park, S. H., Men, T., Kusaka, N., Tian, W., Ren, C., Wang, H., Kadmon, D., Cai, W. W., Chinault, A. C., Boone, T. B., Bradley, A., and Thompson, T. C. (2003) Disruption of the caveolin–1 gene impairs renal calcium reabsorption and leads to hypercalciuria and urolithiasis. *Am. J. Pathol.* **162,** 1241–1248.

Drab, M., Verkade, P., Elger, M., Kasper, M., Lohn, M., Lauterbach, B., Menne, J., Lindschau, C., Mende, F., Luft, F. C., Schedl, A., Haller, H., and Kurzchalia, T. V. (2001) Loss of caveolae, vascular dysfunction, and pulmonary defects in caveolin–1 gene-disrupted mice. *Science* **293,** 2449–2452.

Fielding, C. J. and Fielding, P. E. (2001) Caveolae and intracellular trafficking of cholesterol. *Adv. Drug. Deliv. Rev.* **49,** 251–264.

Glenney, J. R. J. and Soppet, D. (1992) Sequence and expression of caveolin, a protein component of caveolae plasma membrane domains phosphorylated on tyrosine in Rous sarcoma virus-transformed fibroblasts. *Proc. Natl. Acad. Sci. USA* **89,** 10517–10521.

Lisanti, M. P., Scherer, P., Tang, Z.-L., and Sargiacomo, M. (1994a) Caveolae, caveolin and caveolin-rich membrane domains: A signalling hypothesis. *Trends Cell Biol.* **4,** 231–235.

Lisanti, M. P., Scherer, P. E., Vidugiriene, J., Tang, Z., Hermanowski-Vosatka, A., Tu, Y. H., Cook, R. F., and Sargiacomo, M. (1994b) Characterization of caveolin-rich membrane domains isolated from an endothelial-rich source: Implications for human disease. *J. Cell Biol.* **126,** 111–126.

Liscum, L. and Munn, N. J. (1999) Intracellular cholesterol transport. *Biochim. Biophys. Acta* **1438,** 19–37.

Palade, G. E. (1953) Fine structure of blood capillaries. *J. Appl. Phy.* **24,** 1424.

Razani, B., Engelman, J. A., Wang, X. B., Schubert, W., Zhang, X. L., Marks, C. B., Macaluso, F., Russell, R. G., Li, M., Pestell, R. G., Di Vizio, D., Hou, H. Jr., Kneitz, B., Lagaud, G., Christ, G. J., Edelmann, W., and Lisanti, M. P. (2001) Caveolin–1 null mice are viable but show evidence of hyperproliferative and vascular abnormalities. *J. Biol. Chem.* **276,** 38121–38138.

Rothberg, K. G., Heuser, J. E., Donzell, W. C., Ying, Y. S., Glenney, J. R., and Anderson, R. G. (1992) Caveolin, a protein component of caveolae membrane coats. *Cell* **68,** 673–682.

Schnitzer, J. E., Oh, P., Jacobson, B. S., and Dvorak, A. M. (1995) Caveolae from luminal plasmalemma of rat lung endothelium: Microdomains enriched in caveolin, Ca(2+)-ATPase, and inositol trisphosphate receptor. *Proc. Natl. Acad. Sci. USA* **92,** 1759–1763.

Severs, N. J. (1988) Caveolae: Static inpocketings of the plasma membrane, dynamic vesicles or plain artifact. *J. Cell Sci.* **90,** 341–348.

Simons, K. and Ikonen, E. (1997) Functional rafts in cell membranes. *Nature* **387,** 569–572.

Smart, E. J., Ying, Y., Mineo, C., and Anderson, R. G. (1995) A detergent-free method for purifying caveolae membrane from tissue culture cells. *Proc. Natl. Acad. Sci. USA* **92,** 10104–10108.

Yamada, E. (1955) The fine structure of the gall bladder epithelium of the mouse. *J. Biophysoc. Biochem. Cytol.* **1,** 445–457.

Zhao, Y. Y., Liu, Y., Stan, R. V., Fan, L., Gu, Y., Dalton, N., Chu, P. H., Peterson, K., Ross, J., and Chien, K. R. (2002) Defects in caveolin–1 cause dilated cardiomyopathy and pulmonary hypertension in knockout mice. *Proc. Natl. Acad. Sci. USA* **99,** 11375–11380.

Section I

Caveolae Organization and Role in Lipid Cholesterol Metabolism

Chapter 1

Lipid Rafts and Caveolae Organization

Friedhelm Schroeder,[1] Barbara P. Atshaves,[1] Adalberto M. Gallegos,[2] Avery L. McIntosh,[1] J. C. Steve Liu,[3] Ann B. Kier,[2] Huan Huang,[1] and Judith M. Ball[2]

[1]*Department of Physiology and Pharmacology, Texas A & M University, College Station, Texas 77843-4466*
[2]*Department of Pathobiology, Texas A & M University, College Station, Texas 77843-4467*
[3]*Computer Science Department, Texas A & M University, College Station, Texas 77843-3312*

I. INTRODUCTION

The advent and almost universal acceptance of the Singer fluid-mosaic model over 30 years ago (Singer and Nicolson, 1972) led to the concept that membrane lipids spontaneously form a bilayer structure composed of randomly assembled lipids wherein

Advances in Molecular and Cell Biology, Vol. 36

ISBN: 0-444-51500-3

DOI: 10.1016/S1569-6603(05)36001-6

proteins are inserted. In this view, lipids do not contribute to organizing proteins within the membrane. However, subsequent studies from many laboratories indicate that within model and biological membrane bilayers both lipids and proteins are non-randomly organized across (reviewed in Thompson *et al.*, 1974; Schroeder and Nemecz, 1990; Schroeder *et al.*, 1996) and within the lateral plane (reviewed in Thompson *et al.*, 1974; Edidin, 1990; Schroeder *et al.*, 1991a; Anderson, 1993; Bretscher and Munro, 1993; Glaser, 1993; Lisanti *et al.*, 1995; Smart *et al.*, 1995; Schroeder *et al.*, 1996; Brown and London, 1998b; Hooper, 1999; Edidin, 2001; Schroeder *et al.*, 2001a; Anderson and Jacobson, 2002; Lin and Tian, 2003). It is now believed that lipids such as cholesterol and sphingolipids spontaneously form cholesterol-rich, sphingolipid-rich lipid domains (reviewed in Bretscher and Munro, 1993; Brown, 1998; Brown and London, 1998b) (Fig. 1). Although cholesterol is not required for sphingolipids to form phase-segregated domains, cholesterol stabilizes such domains (reviewed in Bretscher and Munro, 1993; Brown, 1998; Brown and London, 1998b). Thus, by spontaneously organizing into domains, lipids such as cholesterol have been postulated to provide the driving force for selective recruitment and organization of proteins into domains (reviewed in Bretscher and Munro, 1993; Brown, 1998; Brown and London, 1998b).

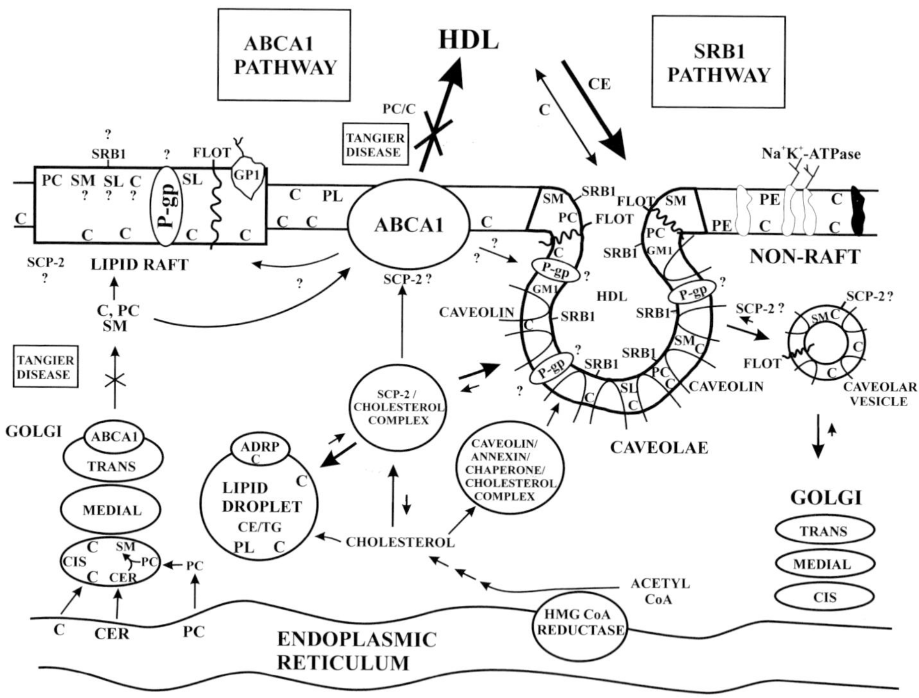

Figure 1. Lipid rafts and caveolae mediate HDL-cholesterol trafficking through the plasma membrane of tissues that efflux cholesterol: role of sterol carrier protein-2. Caveolin (∩), flotillin (Flot), high-density lipoprotein (HDL), scavenger receptor B1 (SRB1), P-glycoprotein (P-gp), ABCA1 transporter (ABCA1), sterol carrier protein-2 (SCP-2), cholesterol (C), sphingolipid (SL), sphingomyelin (SM), ceramide (CER), ganglioside GM1 (GM1), phosphatidylcholine (PC), phosphatidylethanolamine (PE), phospholipids (PL).

Cholesterol-rich microdomains, termed lipid rafts (reviewed in Marx, 2001), are present in almost all cell plasma membranes studied (Fig. 1), while caveolae, a specialized subclass of flask-shaped lipid rafts (Fig. 1), are present only in cells expressing caveolin protein (reviewed in Masserini and Ravasi, 2001). The literature contains more than 2000 publications from the past decade that ascribe a growing variety of functions to plasma membrane caveolae/lipid rafts. Many signaling pathways (reviewed in Lavie and Liscovitch, 2000) are organized in non-caveolar lipid rafts (Fig. 1) that individually appear to be relatively short lived, but that exhibit a stable overall pattern (reviewed in Edidin, 2001). Diverse plasma membrane processes such as diffusional cholesterol uptake/efflux to HDL (Fig. 1), SRB1-facilitated cholesterol uptake/efflux to HDL (Fig. 1), SRB1-facilitated selective cholesteryl ester uptake from HDL (Fig. 1), receptor-effector coupling (insulin receptor), cell signaling (eNOS), immune function, transcytosis, and cell recognition appear organized within cholesterol-rich microdomains such as caveolae (reviewed in Lavie and Liscovitch, 2000). Caveolae, in contrast to other lipid rafts, appear to be much more stable and longer lived (reviewed in Sheets *et al.*, 1997; Pralle *et al.*, 2000; Edidin, 2001; Anderson and Jacobson, 2002). While HDL-mediated cholesterol uptake/efflux via caveolae/lipid rafts is very rapid (reviewed in Smart and van der Westhuyzen, 1998), the mechanism(s) whereby cholesterol moves in and out of these highly cholesterol-rich domains is only beginning to be understood ("?" in Fig. 1). Growing evidence also indicates that many potential bioterror agents, including bacterial (anthrax toxin, cholera toxin, Shiga toxin, enterohemorrhagic *E. coli* Shiga-like toxin, NSP4 enterotoxic peptides) and plant (ricin) toxins (Sandvig and van Deurs, 1999; Abrami *et al.*, 2003), viruses (Ebola, Marburg, Echovirus, influenza) (Scheiffele *et al.*, 1997; Marsh and Pelchen-Matthews, 2000; Bavari *et al.*, 2002; Empig and Goldsmith, 2002; Marjomaki *et al.*, 2002; Sieczkarski and Whittaker, 2002), and parasites (malaria) (reviewed in Shin and Abraham, 2002), utilize lipid rafts/caveolae as cell entry portals (Norkin, 2001). With the exception of cholera toxin (binds GM1 in caveolae), however, the mechanisms whereby other toxins such as the rotaviral and retroviral enterotoxic peptides (Huang *et al.*, 2001, 2004; Swaggerty *et al.*, 2004) and other organisms are recruited to or influence lipid rafts/caveolae are only beginning to be resolved.

The importance of cholesterol to the structure and protein organization of plasma membrane caveolae/lipid rafts is evidenced by the fact that cholesterol depletion or disruption abolishes many functions associated with caveolae/lipid rafts (reviewed in Smart and van der Westhuyzen, 1998). Nevertheless, very little is known regarding the lipid composition, distribution, and structure of caveolae/lipid rafts either *in vitro* or in intact cells. Part of the problem is that the majority of biochemical studies of caveolae/lipid rafts utilize detergents to isolate these microdomains from whole cells and, in a few cases, from purified plasma membranes. Interpretation of plasma membrane caveolar/lipid raft lipid composition and structure from whole cell detergent-based preparations is complicated by the presence of lipid domains from intracellular membranes (reviewed in Pike *et al.*, 2002; Eckert *et al.*, 2003). Even when they are isolated from purified plasma membranes, it is not possible to directly correlate the properties of detergent-resistant membrane domains with those of caveolae/lipid rafts isolated without detergents or in the physiologically intact membrane of living cells (reviewed in Pike *et al.*, 2002; Eckert *et al.*, 2003). For example, estimates of the amount of plasma membrane composed of caveolae/lipid rafts range from a few percent (determined by non-detergent isolation) to a majority (determined from detergent-resistant isolation) of membrane lipids, depending on the amount of detergent, type of detergent, and conditions used (Brown, 1998; Masserini and Ravasi, 2001). Finally, there is considerable discussion regarding the size of lipid rafts/caveolae, which, depending on the specific marker (GPI-anchored protein, GM1, phosphatidylethanolamine) and method (fluorescence photobleaching recovery, single particle tracking, laser trap, electron microscopy morphology, etc.) used, range from 26 to 2000 nm (reviewed in Sheets *et al.*, 1997; Pralle *et al.*, 2000; Edidin, 2001). Despite the fact that lipid rafts/caveolae are highly enriched in cholesterol, almost nothing is known about the size of lipid rafts/caveolae sensed by cholesterol itself (McIntosh *et al.*, 2003).

The advent of non-detergent methods for isolating caveolae/lipid rafts has permitted biochemical and structural characterizations of caveolae/lipid raft lipids in a more physiological context. Increasingly, lipid compositional analyses of caveolae/lipid rafts isolated without detergents have begun to appear (Smart *et al.*, 1995; Smart and van der Westhuyzen, 1998; Pike *et al.*, 2002; Atshaves *et al.*, 2003; Eckert *et al.*, 2003). Recent reports provided insights on lipid distribution (Pitto *et al.*, 2000; Atshaves *et al.*, 2003), cholesterol organization (Atshaves *et al.*, 2003; McIntosh *et al.*, 2003; Gallegos *et al.*, 2004), and structure/fluidity (Gallegos *et al.*, 2004) of caveolae/lipid rafts purified without the use of detergents. Fluorescent sterol and multiphoton imaging of living cells are providing answers to the question of whether cholesterol-rich domains are actually present in plasma membranes of living cells or whether they are simply artifacts of the methods used to isolate caveolae/lipid rafts (McIntosh *et al.*, 2003). With the exception of caveolin, relatively little is known about the mechanisms whereby other intracellular lipid transfer proteins may contribute to the distribution of cholesterol and signaling lipid precursors such as phosphatidylinositol to lipid rafts/caveolae. Studies with N-terminal peptides as well as intact sterol carrier protein-2 (SCP-2) yielded insights to understanding the role of SCP-2 in transporting and targeting cholesterol and phosphatidylinositol to lipid rafts/caveolae (Huang *et al.*, 1999a,b, 2002; Schroeder *et al.*, 2003). Similarly, with the exception of caveolin, relatively little is known about other structural features that influence the targeting of intracellular peptides and proteins to lipid rafts/caveolae. Recent studies with enterotoxic peptides derived from rotaviral proteins have shed new light on the molecular basis for targeting of such enterotoxins to lipid rafts/caveolae (Tian *et al.*, 1996; Huang *et al.*, 2001, 2004; Swaggerty *et al.*, 2004). This review focuses on these recent advances in our understanding of lipid raft/caveolae lipid distribution, cholesterol dynamics, structure, and peptide interactions.

II. A NON-DETERGENT METHOD OF SIMULTANEOUSLY ISOLATING CAVEOLAE/LIPID RAFTS AND NON-RAFT DOMAINS

Neither detergent-based (Ahmed *et al.*, 1997; London and Brown, 2000) nor non-detergent (Smart *et al.*, 1995; Pike *et al.*, 2002; Eckert *et al.*, 2003; McIntosh *et al.*, 2003; Gallegos *et al.*, 2004) methods of isolating lipid rafts/caveolae recover the non-raft domains or differentiate between lipid rafts and caveolar lipid rafts. While non-detergent, anti-caveolin immunoisolation has been used to purify a caveolae-enriched fraction (Stan *et al.*, 1997; Oh and Schnitzer, 1999; Waugh *et al.*, 1999; Riddell *et al.*, 2002), these methods do not recover or resolve non-rafts from non-caveolar lipid rafts. Since none of these procedures allow biochemical or structural characterization of the non-lipid raft domain fraction of the plasma membrane, the properties of non-lipid raft domains in such studies could only be inferred. These difficulties were recently overcome by the use of concanavalin A affinity chromatography to simultaneously resolve and recover both lipid raft/caveolae and non-lipid raft domains from plasma membrane vesicles isolated from a cultured cell line, L-cell fibroblasts (Atshaves *et al.*, 2003).

The concanavalin-A affinity method is based on the discovery over 30 years ago that concanavalin-A sepharose affinity chromatography resolves plasma membrane vesicles (from nucleated cells) into two subfractions, an adherent and a non-adherent fraction (reviewed in Brunner *et al.*, 1976; Resch *et al.*, 1978; Schroeder *et al.*, 1982). While these early studies did not correlate these fractions with specific markers for lipid domains/caveolae, nevertheless, biochemical characterization revealed that the concanavalin-A adherent and non-adherent fractions represent different "patches" or "membrane domains" of the plasma membrane (reviewed in Brunner *et al.*, 1976; Resch *et al.*, 1978; Schroeder *et al.*, 1982). Concanavalin-A is known to specifically interact with select carbohydrate structures (reviewed in Baenziger and Fiete, 1979), and such carbohydrate structures are present in glycopeptides such as SR-B1 or glycolipids such as glucosylceramide (reviewed in Atshaves *et al.*, 2003). To determine if concanavalin-A affinity chromatography separates lipid rafts/caveolae from non-raft domains of the plasma membrane, advantage was taken of the fact that SR-B1 and glucosylceramide are both present in L-cell plasma membranes (Frolov *et al.*, 2000; J. Jefferson, Luther College, Decorah, IA, personal communication). Concanavalin-A affinity chromatography of plasma membrane vesicles purified from cultured L-cells resolves an adherent and a non-adherent fraction. Both SR-B1 and glucosylceramide are enriched in the lipid rafts/caveolae (reviewed in Atshaves *et al.*, 2003). Further confirmation that the concanavalin-A adherent fraction of plasma membrane vesicles represents lipid rafts/caveolae is based on the fact that this fraction also contains high levels of other markers for lipid rafts/caveolae (flotillin, caveolin, ganglioside GM1, cholesterol, phospholipid, sphingomyelin, and specific phospholipids classes), while being concomitantly deficient in markers for non-raft domains (Na^+K^+-ATPase, cholesterol poor, lipid poor, GM1 poor, sphingomyelin poor) (Atshaves *et al.*, 2003).

The fact that concanavalin-A chromatography results in nearly complete recovery of lipid rafts/caveolae and non-raft domains without the use of detergents (Atshaves *et al.*, 2003) allowed the resolution of a long-standing question in the lipid raft/caveolae field: do lipid rafts/caveolae represent a major or a minor constituent of the plasma membrane? Earlier non-detergent methods indicated that lipid rafts/caveolae comprise only a very small portion (a few percent) of the plasma membrane, possibly due to low recoveries (reviewed in Smart *et al.*, 1995; Atshaves *et al.*, 2003). In

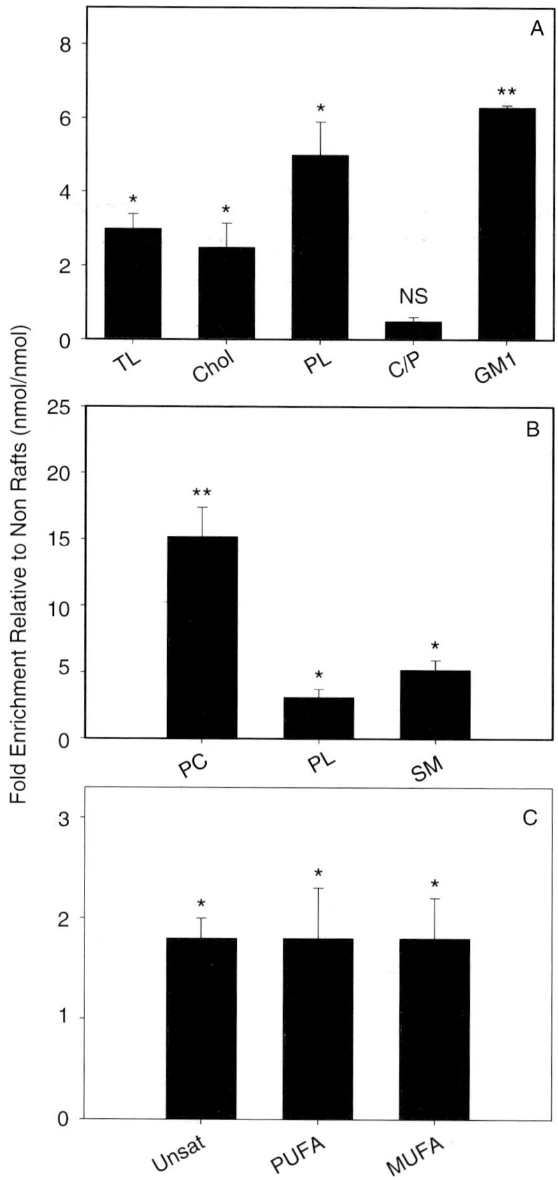

Figure 2. Relative enrichment of lipids in lipid rafts/caveolae versus non-raft domains isolated from L-cell plasma membranes by concanavalin-A affinity chromatography. Plasma membranes were isolated from L-cell fibroblasts; plasma membrane vesicles were resolved into lipid rafts/caveolae and non-raft fractions by concanavalin-A affinity chromatography without the use of detergents; lipids were extracted, resolved into lipid classes, and quantitated; and relative enrichment of each lipid class was calculated from the ratio [mass (nmol/mg protein) in lipid rafts/caveolae]/[mass (nmol/mg protein) in non-rafts] using data appearing in Atshaves *et al.* (2003). Values represent the mean ± SEM with (*) and (**) indicating significance $P < 0.05$

contrast, detergent-based methods for isolating lipid rafts/caveolae yield proportions varying from almost none to essentially the entire membrane, depending on the temperature used during the isolation procedure (Brown and London, 1998b, 2000). This issue was resolved by near-complete recovery of lipid rafts/caveolae and non-raft domains upon concanavalin-A affinity chromatography. The concanavalin-A adherent fraction, enriched in lipid raft/caveolae markers, contains 31.9 ± 5.7% of plasma membrane protein and 59.3 ± 10.1% of plasma membrane total lipid (Atshaves *et al.*, 2003). Similarly, early studies of lymphocyte plasma membrane vesicles separated by concanavalin-A affinity chromatography also showed that about one-third of the plasma membrane vesicle protein was in the concanavalin A-adherent fraction (Brunner *et al.*, 1976; Resch *et al.*, 1978). Thus, lipid rafts/caveolae represent the majority of plasma membrane lipids and contain a substantial portion of plasma membrane proteins. In contrast, non-raft domains are more highly enriched in proteins, but contain only about 40% of plasma membrane total lipid.

III. LIPID COMPOSITION OF LIPID RAFTS/CAVEOLAE AND NON-RAFT DOMAINS

Since almost nothing is known regarding the lipid composition of non-raft domains in plasma membranes, the concanavalin-A method provided the opportunity to (i) experimentally establish the lipid composition of non-lipid rafts, (ii) concomitantly isolate lipid rafts/caveolae from the same plasma membrane vesicle preparation, and (iii) directly compare the lipid composition of lipid rafts/caveolae with that of non-lipid raft domains of the same plasma membrane (Atshaves *et al.*, 2003). When expressed as a ratio of (nmol/mg)/(nmol/mg), the lipids of the concanavalin-A non-adherent fraction, composed of non-lipid rafts, are relatively poor in (i) total lipid, ganglioside GM1, cholesterol, and total phospholipid (Fig. 2A), (ii) specific phospholipid classes such as phosphatidylcholine, sphingomyelin, and phosphatidylinositol (Fig. 2B), and (iii) phospholipid unsaturated fatty acids (Fig. 2C). Concomitantly, non-lipid rafts are relatively enriched in total protein, phosphatidylethanolamine, and saturated fatty acids. Thus, the composition of lipid rafts/caveolae differs significantly from that of non-lipid rafts.

---

and $p < 0.01$ (n = 3–4) as compared to the non-lipid raft fraction isolated from L cells. (A) Individual bars show relative enrichment of total lipid (TL), cholesterol (Chol), total phospholipids (PL), and ganglioside GM1 (GM1) in lipid rafts/caveolae versus non-rafts. Cholesterol/phospholipid (C/P) refers to the molar ratio of cholesterol/phospholipid in lipid rafts/caveolae divided by the molar ratio of cholesterol/phospholipid in non-rafts. (B) The phospholipid fractions from lipid rafts/caveolae and non-rafts were resolved into individual phospholipids classes. Individual bars show relative enrichment of phosphatidylcholine (PC), phosphatidylinositol (PI), and sphingomyelin (SM) in lipid rafts/caveolae versus non-rafts. (C) The fatty acid content of the phospholipid fraction from lipid rafts/caveolae and non-rafts was determined. Individual bars show relative enrichment of total unsaturated (Unsat), total polyunsaturated fatty acids (PUFA), and total monounsaturated fatty acids (MUFA) in the phospholipids of lipid rafts/caveolae versus non-rafts.

As compared to non-lipid raft domains, when expressed as a ratio of (nmol/mg)/(nmol/mg), the lipids of the lipid raft/caveolae fraction isolated from the same plasma membrane vesicle preparations are enriched in (i) total lipid, ganglioside GM1, cholesterol, and total phospholipid (Fig. 2A); (ii) phosphatidylcholine, sphingomyelin, and phosphatidylinositol (Fig. 2B); and (iii) unsaturated fatty acids (especially 20:4, 20:2, 18:2, and monounsaturated fatty acids) (Fig. 2C). With some exceptions (phosphatidylcholine), many of these lipid compositional properties of lipid rafts/caveolae isolated by the concanavalin-A method are shared with those of lipid rafts/caveolae prepared by other non-detergent methods (Smart *et al.*, 1995; Pike *et al.*, 2002; Eckert *et al.*, 2003). For example, as compared to plasma membranes, lipid rafts/caveolae isolated by other non-detergent methods are also enriched in total lipid, ganglioside GM1, arachidonic acid (20:4), total phospholipid, sphingomyelin, cholesterol, and total phosphatidylinositides (albeit not phosphatidylinositol) (Smart *et al.*, 1995; Pike *et al.*, 2002; Eckert *et al.*, 2003). Since phospholipid and cholesterol are both enriched in lipid rafts/caveolae as compared to non-lipid raft domains, the sterol/phospholipid ratio does not significantly differ (Fig. 2A) between lipid rafts/caveolae and non-rafts (Atshaves *et al.*, 2003). The latter finding was consistent with early studies of lymphocyte plasma membrane vesicles separated by concanavalin-A affinity chromatography, which also showed essentially identical cholesterol/phospholipids molar ratios in the two fractions separated by concanavalin A affinity chromatography (Brunner *et al.*, 1976; Resch *et al.*, 1978). Similar findings are shown by comparison of non-detergent preparations of lipid rafts/caveolae versus plasma membranes from brain synaptosomes (Wood *et al.*, 1990; Eckert *et al.*, 2003) and by comparison of lipid rafts/caveolae from cultured kidney KB cells (Pike *et al.*, 2002) with plasma membranes from many other cell types (reviewed in Schroeder *et al.*, 1996, 2001b). In contrast, the lipid composition of lipid rafts/caveolae isolated by detergent-based methods differs significantly from the composition of those isolated by non-detergent based methods. For example, arachidonic acid (20:4) is not enriched in lipid rafts/caveolae isolated by use of detergents (Pike *et al.*, 2002). In contrast to lipid rafts/caveolae isolated by non-detergent methods, the content of anionic phospholipids is significantly depleted in lipid rafts/caveolae isolated by use of detergents (Pike *et al.*, 2002). Furthermore, the cholesterol/phospholipid ratio is significantly higher in lipid rafts/caveolae isolated by detergent-based methods as compared to plasma membranes and to caveolae/lipid rafts isolated without detergents from the same plasma membranes (Pike *et al.*, 2002; Eckert *et al.*, 2003). These data suggest that detergents elicit preferential loss of phospholipids and/or result in purification of a population of caveolae/lipid rafts different from those isolated by non-detergent methods (Smart *et al.*, 1995; Pike *et al.*, 2002; Eckert *et al.*, 2003). However, the basic similarity of enrichment of total lipid and select lipid classes in lipid rafts/caveolae isolated by the high-yield, concanavalin-A affinity chromatography method (Atshaves *et al.*, 2003) compared to the other non-detergent, but low-yield, methods (Smart *et al.*, 1995; Pike *et al.*, 2002; Eckert *et al.*, 2003) for isolating lipid rafts/caveolae suggests the former possibility.

IV. FLUIDITY AND STRUCTURE OF LIPID RAFTS/CAVEOLAE AND NON-RAFTS

Almost nothing is known regarding the physical structure of lipids in lipid rafts/caveolae or non-raft domains of the plasma membrane. Lipid rafts/caveolae isolated by use of detergents (i.e., detergent-resistant membranes [DRMs]) (reviewed in Brown and London, 1998b; Smart and van der Westhuyzen, 1998) or non-detergent methods (Pike *et al.*, 2002; Atshaves *et al.*, 2003; Eckert *et al.*, 2003) are enriched in cholesterol and sphingolipids (GM1, sphingomyelin). Model membrane studies indicate that membranes become more rigid with increasing cholesterol and sphingolipid content (reviewed in Brown, 1998; Brown and London, 1998b, 2000). Furthermore, membrane functions are generally slowed with increasing membrane rigidity, rather than enhanced (reviewed in Schroeder and Sweet, 1988; Sweet and Schroeder, 1988). Thus, it is difficult to reconcile predictions of lipid raft/caveolae structure based solely on cholesterol and sphingolipid enrichment with the observation that lipid rafts/caveolae are highly dynamic structures involved in multiple transmembrane processes. Several recent studies offer insights into our understanding of this conundrum.

First, although lipid rafts/caveolae are enriched several-fold in cholesterol, they are also enriched in phospholipids to the extent that the molar ratio of cholesterol/phospholipids in lipid rafts/caveolae does not differ significantly from that in the parent plasma membrane or in non-lipid raft domains (Fig. 2B). This would suggest that lipid rafts/caveolae might not be as "rigid" as was previously predicted.

Second, studies with caveolar microdomains isolated by use of detergents (i.e., DRMs) suggest that their "rigidity" may be significantly less than that of "rigid" lipids (Brown and London, 1998b; Pike *et al.*, 2002). When DRM lipid fluidity is probed by measurement of diphenylhexatriene (DPH) polarization, the fluorescent polarization of DPH in DRM lipids is intermediate between that in the fluid liquid-crystalline phase and that in the solid-ordered phase, a characteristic representative of the liquid-ordered phase (Schroeder *et al.*, 1994). However, it is unclear whether detergents used to isolate DRMs influence formation of this intermediate liquid-ordered phase.

Third, a recent study demonstrated that the lipid rafts/caveolae isolated by a non-detergent method are significantly more rigid (i.e., less fluid) than non-raft domains (Gallegos *et al.*, 2004). The higher fluorescence polarization of DPH, DiI-C18, *cis*-parinaric acid (*cis*-PnA), *trans*-parinaric acid (*trans*-PnA), and NBD-stearic acid (NBD-C18) in lipid rafts/caveolae as compared to plasma membranes (Fig. 3A) indicates that acyl chains in lipid rafts/caveolae are significantly less fluid (i.e., more rigid) than those in the parent plasma membrane (Gallegos *et al.*, 2004). While this had been predicted based on DPH polarization in lipid rafts/caveolae isolated by detergent-based methods, it is unclear whether detergents can result in formation of the less fluid DRMs. Thus, the finding of a lower fluidity in lipid rafts/caveolae than in plasma membranes directly showed that lipids of lipid rafts/caveolae are more rigid, independent of whether detergents are used to isolate the lipid rafts/caveolae (Gallegos *et al.*, 2004).

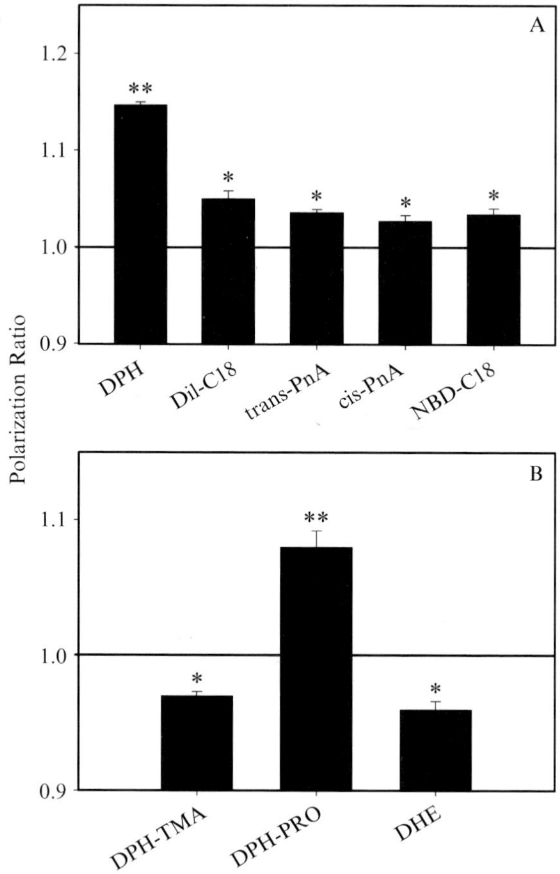

Figure 3. Structure/fluidity of lipids in lipid rafts/caveolae and the parent plasma membranes from L-cell fibroblasts. Plasma membranes were purified from L-cell fibroblasts; lipid rafts/caveolae were isolated by a non-detergent method; fluorescent lipid probes were incorporated into lipid rafts/caveolae and plasma membranes; fluorescence polarization of each probe was determined; and the ratio [polarization in lipid rafts/caveolae]/[polarization in plasma membranes] was calculated using values reported in Gallegos *et al.* (2004). Values represent the mean ± SEM with (*) and (**) indicating significance $P < 0.05$ and $p < 0.01$ ($n = 7$) as compared to plasma membranes isolated from L-cells. (A) Lipid probes reporting on lipid mobility in the center of lipid bilayers (diphenylhexatriene, DPH), phospholipids (DiI-C18), phospholipid acyl chains near the center of the bilayer (trans-parinaric acid [trans-PnA], cis-parinaric acid [cis-PnA]), phospholipid acyl chains near the surface of the bilayer (NBD-stearic acid, NBD-C18). (B) Lipid probes preferentially distributing to the exofacial leaflet (diphenylhexatriene-trimethylammonium, DPH-TMA) and cytofacial leaflet (diphenylhexatriene-propionic acid, DPH-PRO; dehydroergosterol, DHE).

Fourth, the fluidity of non-detergent lipid rafts/caveolae as measured by DPH polarization is intermediate between the fluidity in the fluid liquid-crystalline phase and that in the solid-ordered phase, again characteristic of the liquid-ordered phase (Gallegos *et al.*, 2004). The DPH polarization in lipid rafts/caveolae isolated without detergents is 0.3085 ± 0.0008, very similar to the polarization of DPH in the liquid-ordered phase of model membranes, 0.281, and falling between the DPH polarization in the non-fluid gel phase, 0.381, and in the liquid crystalline phase, 0.095 (Schroeder *et al.*, 1994).

Fifth, to determine if the fluidity of lipids of the exofacial and cytofacial leaflets of the lipid rafts/caveolae differed, the fluorescence polarization of DPH analogues (DPH-TMA, DPH-PRO) that selectively localize in the exofacial or cytofacial leaflet of the membrane, respectively, was determined in lipid rafts/caveolae (Gallegos *et al.*, 2004). The fluorescence polarization of DPH-propionic acid (DPH-PRO) in lipid rafts/caveolae ($P = 0.2896 \pm 0.0032$) is significantly higher than that of DPH-TMA in lipid rafts/caveolae ($P = 0.2577 \pm 0.0007$), $p < 0.01$ (Gallegos *et al.*, 2004). Thus, there exists a transbilayer fluidity gradient, indicating that the exofacial leaflet is more fluid and the cytofacial leaflet is more rigid in lipid rafts/caveolae. Such transmembrane fluidity gradients are thought to be essential for maintaining the functions of plasma membrane receptors and transporters (reviewed in Schroeder and Sweet, 1988; Sweet and Schroeder, 1988; Schroeder *et al.*, 1990b). Loss of transbilayer fluidity gradients (Sweet and Schroeder, 1988) is associated with the action of anesthetics (reviewed in Sweet *et al.*, 1987; Sweet and Schroeder, 1988; Wood *et al.*, 1996) and several diseases (reviewed in Wood *et al.*, 1999, 2002, 2003).

Sixth, to compare the transbilayer fluidity gradient in lipid rafts/caveolae with that of the parent plasma membrane, the preceding experiment was repeated with purified plasma membranes. As indicated by the higher ratio of DPH-PRO fluorescence polarization in lipid rafts/caveolae versus plasma membranes (Fig. 3B), the cytofacial leaflet of the lipid rafts/caveolae is more rigid than that of the bulk plasma membrane. In contrast, the lower ratio of DPH-TMA fluorescence polarization in lipid rafts/caveolae versus plasma membranes (Fig. 3B) indicates that the exofacial leaflet of the lipid rafts/caveolae is more fluid than that of the bulk plasma membrane. Quantitative comparison reveals that the transbilayer "fluidity gradient" in lipid rafts/caveolae is 4.4-fold greater than in plasma membranes (Gallegos *et al.*, 2004). Since transmembrane fluidity gradients appear important to the function of transmembrane coupled systems such as receptor/effectors and transporters (reviewed in Schroeder and Sweet, 1988; Sweet and Schroeder, 1988; Schroeder *et al.*, 1990b), these data suggest that lipid rafts/caveolae may be structurally optimized for such tasks but are concomitantly more sensitive to the action of agents that perturb transbilayer fluidity gradients (reviewed in Sweet *et al.*, 1987; Sweet and Schroeder, 1988; Wood *et al.*, 1996) or diseases that disrupt the transbilayer fluidity gradient (reviewed in Wood *et al.*, 1999, 2002, 2003).

V. CHOLESTEROL ORGANIZATION IN LIPID RAFTS/CAVEOLAE AND NON-RAFTS

Despite the importance of cholesterol in formation of lipid rafts/caveolae (reviewed in Brown and London, 1998b, 2000) and the role of lipid rafts/caveolae in cellular cholesterol efflux (reviewed in Fielding and Fielding, 2000, 2001a,b; Yokoyama, 2000), relatively little is known about either the structure or the molecular dynamics of cholesterol in lipid rafts/caveolae.

The transbilayer distribution of cholesterol is a major determinant giving rise to a gradient of fluidity in plasma membranes wherein the exofacial leaflet is cholesterol poor and more fluid than the cytofacial leaflet (reviewed in Kier *et al.*, 1986; Sweet and Schroeder, 1988; Schroeder *et al.*, 1991a,b). The observation that the exofacial leaflet of lipid rafts/caveolae is much more fluid than the cytofacial leaflet (Fig. 3A) suggests that the majority of cholesterol should localize in the cytofacial leaflet (Gallegos *et al.*, 2004). The available literature ascertaining the transbilayer distribution of cholesterol in lipid rafts/caveolae is not completely consistent. For example, the fact that only caveolar cholesterol is reactive to cholesterol oxidase has led some investigators to conclude that cholesterol is enriched in the exofacial leaflet of caveolae (reviewed in Fielding and Fielding, 2000). However, this conclusion does not take into account the rapid transbilayer migration rate ($t_{1/2}$ of 2–17 min) of cholesterol across a variety of plasma membranes (Schroeder and Nemecz, 1990; Wood *et al.*, 1990; Schroeder *et al.*, 1991b) and is inconsistent with the finding that the exofacial leaflet of lipid rafts/caveolae is more fluid than the cytofacial leaflet (Gallegos *et al.*, 2004). In contrast, a recent study utilized a photoactivatable ganglioside GM1 to assess the lipid distribution in caveolae (Pitto *et al.*, 2000). Since the ganglioside GM1 localizes essentially completely in the exofacial leaflet, the only lipids crosslinked by the photoactivatable GM1 are localized in the exofacial leaflet. While the photoactivatable GM1 crosslinked phosphatidylcholine, sphingomyelin, and cholesterol in a plasma membrane-enriched fraction, only sphingomyelin was crosslinked in the detergent-resistant lipid raft/caveolae fraction (Pitto *et al.*, 2000). These data suggest that sphingomyelin is enriched in the exofacial leaflet along with the ganglioside GM1. However, the lack of cholesterol crosslinking in the detergent-resistant lipid raft/caveolae fraction indicates that very little, if any, cholesterol is distributed to the exofacial leaflet and that almost all the cholesterol is localized in the cytofacial leaflet (Pitto *et al.*, 2000). Caveolin is known to bind cholesterol in membranes and in the cytosol as a heat shock protein–chaperone complex (reviewed in Murata *et al.*, 1995; Smart and van der Westhuyzen, 1998). Since caveolin is localized primarily in the cytofacial leaflet but does not extend completely across the caveolar bilayer (reviewed in Fielding and Fielding, 2000, 2001a,b), caveolin binding cholesterol in the cytofacial leaflet may also contribute directly to enrich the cytofacial leaflet in cholesterol (Uittenbogaard *et al.*, 1998). However, because the number of cholesterol molecules far exceeds the number of caveolin molecules and binding stoichiometry is 1:1, this explanation is unlikely to account for nearly 100% of total sterol in the lipid rafts/caveolae being localized in the cytofacial leaflet. Instead, other as-yet-unresolved factors must account for this transbilayer sterol gradient. While cholesterol spontaneously organizes itself into multiple lateral domains, sterol is randomly distributed across the bilayer of model membrane vesicles unless the limiting radius of curvature is approached (reviewed in Schroeder *et al.*, 2001a). In contrast to the near-absolute transbilayer distribution of cholesterol in the lipid rafts/caveolae, about 20% of the cholesterol is localized in the exofacial leaflet, while 80% is cytofacial in the parent plasma membrane (reviewed in Kier *et al.*, 1986; Sweet and Schroeder, 1988; Schroeder *et al.*, 1991a,b). The greater transbilayer difference of cholesterol distribution in the lipid rafts/caveolae as compared to that in the parent plasma membrane is consistent with the greater transbilayer fluidity gradient across the bilayer of lipid rafts/caveolae as compared to that in the parent plasma membrane (Gallegos *et al.*, 2004). These data also suggest that the transbilayer cholesterol and fluidity gradient are likely to be much smaller in the non-lipid raft domains of the plasma membrane. Finally, the paucity of cholesterol in the exofacial leaflet of the lipid rafts/caveolae suggests that factors regulating the transbilayer distribution of cholesterol to the exofacial leaflet are likely to have a major impact on determining efflux of cholesterol to HDL mediated through SRB1 present in caveolae (reviewed in Fielding and Fielding, 2000, 2001a,b).

Regarding the lateral distribution of cholesterol, model membrane studies indicate that at high levels the cholesterol phase separates into a pure crystalline cholesterol phase (reviewed in McIntosh *et al.*, 2003; Gallegos *et al.*, 2004). Despite the several-fold enrichment of cholesterol in lipid rafts/caveolae, however, recent studies with a fluorescent sterol (dehydroergosterol) suggest that this is not the case in lipid rafts/caveolae (McIntosh *et al.*, 2003; Gallegos *et al.*, 2004). Dehydroergosterol is a naturally occurring fluorescent sterol that is readily taken up by cultured cells and that replaces as much as 90% of cholesterol in biological membranes without significantly altering structure, lipid composition, or function of the membrane (reviewed in Hale and Schroeder, 1982; Schroeder, 1984; McIntosh *et al.*, 2003). Since the emission spectral properties of crystalline dehydroergosterol differ significantly from those of monomeric dehydroergosterol, it is possible to distinguish the physical form of sterol in lipid rafts/caveolae (McIntosh *et al.*, 2003). Despite the high level of sterol in lipid rafts/caveolae, the dehydroergosterol is primarily monomeric with only a small amount (i.e., $\leq$0.3%) occurring in the crystalline phase in detergent-free lipid rafts/caveolae (McIntosh *et al.*, 2003). In fact, dehydroergosterol is localized in an environment significantly more fluid than that of the surrounding lipids. For example, the fluorescence polarization of dehydroergosterol in lipid rafts/caveolae ($P = 0.3255 \pm 0.0022$) is significantly lower (i.e., more fluid) than that of fluorescent phospholipid analogues (DiI C18, $P = 0.3871 \pm 0.0031$; DiI C1, $P = 0.3482 \pm 0.0013$) and fluorescent fatty acid analogues (cis-parinaric acid, $P = 0.3478 \pm 0.0021$; NBD-stearic acid, $P = 0.3299 \pm 0.0018$) in lipid rafts/caveolae (Gallegos *et al.*, 2004). Finally, dehydroergosterol polarization in lipid rafts/caveolae is lower than that in the parent plasma membrane, suggesting that cholesterol in lipid rafts/caveolae may be more dynamic/mobile and available for transport than other lipids within the lipid rafts/caveolae (Atshaves *et al.*, 2003; Gallegos *et al.*, 2004).

Since cholesterol is less ordered in lipid rafts/caveolae than in the parent plasma membranes, cholesterol movement in and out of lipid rafts/caveolae may be faster than from non-lipid raft domains or the parent plasma membranes. Indeed, spontaneous sterol transfer from lipid rafts/caveolae is faster than that from the plasma membrane (Gallegos *et al.*, 2004) or from the non-lipid raft domains (Atshaves *et al.*, 2003). The initial rate and half-life of molecular sterol transfer from lipid rafts/caveolae is 7.7-fold faster and 9.4-fold shorter, respectively, as compared to sterol transfer from non-lipid rafts (Atshaves *et al.*, 2003). Kinetic analysis of molecular sterol transfer shows that lipid rafts/caveolae contain a substantial proportion of rapidly transferable sterol (i.e.,

23%), while non-lipid raft domains contain very little rapidly exchangeable sterol pool (Atshaves *et al.*, 2003). These findings are consistent with the apparent higher fluidity of the microenvironment wherein sterol was localized in the caveolae as compared to the parent plasma membrane (Gallegos *et al.*, 2004). The significance of spontaneous sterol movement, i.e., simple diffusion, to sterol efflux via lipid rafts/caveolae must take into consideration other factors. For example, the half-life of spontaneous sterol transfer from isolated lipid rafts/caveolae (1.1–1.7 h) is faster than transfer from plasma membranes (2.5 h) or non-lipid raft domains (10.8 h) (Atshaves *et al.*, 2003; Gallegos *et al.*, 2004). Nevertheless, a half-life >1 h is still very slow as compared to the very rapid (minutes) cholesterol uptake or efflux via the HDL receptor pathway mediated by caveolae in intact cells (Smart and van der Westhuyzen, 1998; Atshaves *et al.*, 2000; Frolov *et al.*, 2000). The slow half-life of spontaneous sterol transfer is not limited by transbilayer migration of sterol, which is fast (min) across both model membranes and plasma membranes (Schroeder *et al.*, 1991b). Cholesterol-rich, sphingomyelin-rich model membranes exhibit slow cholesterol dynamics similar to lipid rafts/caveolae (reviewed in Schroeder *et al.*, 1991a). Thus, it is likely that factors in addition to spontaneous/diffusional transfer must contribute to mediating the very rapid transfer of cholesterol out of lipid rafts/caveolae. In support of this possibility, lipid rafts/caveolae are also enriched in P-glycoprotein (P-gp; Fig. 1), which translocates cholesterol from cytofacial to exofacial leaflets to facilitate availability of cholesterol for spontaneous/diffusional efflux to HDL- or SRB1-mediated sterol efflux at the exofacial leaflet of the membrane (reviewed in Lavie and Liscovitch, 2000; Garrigues *et al.*, 2002). The presence of these lipid raft/caveolae proteins (i.e., caveolin, SRB1, P-gp; Fig. 1) contributes significantly to spontaneous sterol transfer as shown by the very slow sterol transfer from cholesterol-rich, sphingomyelin-rich model membranes (Schroeder and Nemecz, 1989; Schroeder and Nemecz, 1990; Schroeder *et al.*, 1991a).

Finally, spontaneous/diffusional sterol transfer between lipid rafts/caveolae and non-lipid raft microdomains of the plasma membrane is up to 12-fold slower than transfer between lipid rafts/caveolae (Atshaves *et al.*, 2003; Gallegos *et al.*, 2004). These data indicate that cholesterol moving through the caveolae/lipid raft domain slowly equilibrates with the non-lipid raft sterol domain. Consistent with this finding, in intact cells lipid raft/caveolar cholesterol poorly equilibrates with bulk plasma membrane cholesterol, except when the lipid rafts/caveolae are disrupted (reviewed in Smart and van der Westhuyzen, 1998; Schroeder *et al.*, 2001b).

VI. STEROL CARRIER PROTEIN-2 SELECTIVELY MODULATES CHOLESTEROL DYNAMICS IN LIPID RAFTS/CAVEOLAE

Although the vesicular trafficking pathways that slowly move cholesterol to and from the plasma membrane have been studied extensively, much less is known regarding rapid, protein-mediated, non-vesicular cholesterol transfer within the cell and how this is targeted to lipid rafts/caveolae (Fig. 1) (Smart and van der Westhuyzen, 1998; Atshaves *et al.*, 2000; Frolov *et al.*, 2000). The importance of non-vesicular cholesterol transfer is underscored by a recent real-time fluorescence imaging study showing that the majority of cholesterol in the living polarized hepatocytes trafficks from the basolateral membrane by non-vesicular pathways to the bile canalicular region (Wustner *et al.*, 2002), consistent with earlier *in vitro* studies (Hafer *et al.*, 2000; Fuchs *et al.*, 2001). Although several soluble cholesterol-binding proteins (caveolin, SCP-2, L-FABP) contribute to nonvesicular cholesterol trafficking from plasma membrane lipid rafts/caveolae (reviewed in Schroeder *et al.*, 1998, 2001b; Sviridov, 1999), this review focuses on recent advances in understanding the interactions of SCP-2 with lipid rafts/caveolae and its role in modulating lipid raft/caveolae cholesterol trafficking.

SCP-2 is highly expressed in liver hepatic cells (reviewed in Gallegos *et al.*, 2001b). Moreover, SCP-2 has high affinity (i.e., nanomolar K_ds) for sterols (Colles *et al.*, 1995; Schroeder *et al.*, 1998, 2000; Stolowich *et al.*, 1999), stimulates the *in vitro* transfer of sterol from plasma membranes (Frolov *et al.*, 1996b,c), enhances sterol transfer from the plasma membrane to the endoplasmic reticulum (ER) for esterification in intact SCP-2 overexpressing cells (Moncecchi *et al.*, 1996; Murphy and Schroeder, 1997), and preferentially redistributes plasma membrane cholesterol to lipid droplets of living cells (Atshaves *et al.*, 2000; Gallegos *et al.*, 2001a). Most recently, SCP-2 was shown to selectively alter cholesterol dynamics from lipid rafts/caveolae, but not non-lipid raft domains, isolated without the use of detergents (Atshaves *et al.*, 2003; Gallegos *et al.*, 2004). SCP-2 selectively enhances sterol transfer from lipid rafts/caveolae isolated without the use of detergents, increasing the initial rate of sterol transfer up to 5-fold, decreasing the $t_{1/2}$ of sterol transfer up to 2-fold, and increasing the size of the exchangeable sterol pool up to 4.2-fold such that >98% of sterol becomes transferable. In contrast, SCP-2 does not significantly alter the initial rate of sterol transfer, the $t_{1/2}$ of sterol transfer, or the size of the exchangeable sterol domain in non-raft domains. Thus, non-detergent lipid rafts/caveolae display unique responsiveness to SCP-2 as compared to the parent plasma membrane and non-raft domains. This was not predicted based on model membrane lipid rafts/caveolae studies showing that SCP-2 is essentially ineffective in transferring sterol from cholesterol-rich, sphingomyelin-rich model membranes (Schroeder *et al.*, 1991a). Likewise, SCP-2 also does not enhance sterol transfer from biological membranes containing high levels of sphingomyelin as well as cholesterol (Kavecansky *et al.*, 1994).

Relatively little is known regarding the potential role of SCP-2 in regulating lipid raft/caveolae cholesterol distribution and dynamics in intact cells. One recent study utilized transfected cells overexpressing SCP-2 to show that SCP-2 expression significantly alters the lipid and protein constituents of caveolae/lipid rafts of living cells such that the initial rate, $t_{1/2}$, and size of the exchangeable spontaneous/diffusional sterol domain are reduced 2-fold, increased 1.8-fold, and decreased 2.1-fold, respectively, in lipid rafts/caveolae isolated without detergents from the SCP-2 overexpressing cells (Atshaves *et al.*, 2003). Furthermore, the non-detergent lipid rafts/caveolae from SCP-2 overexpressing cells are resistant to the direct action of additional SCP-2 added *in vitro*. Interestingly, these findings were consistent with earlier studies that SCP-2 overexpression in L-cells inhibits reverse cholesterol transport to HDL (Gallegos *et al.*, 2001b). The mechanism whereby this occurs indicates that SCP-2 plays a role in both protein-mediated and vesicular cholesterol transfer as follows (reviewed in Atshaves *et al.*, 2000): (i) HDL-mediated cholesterol efflux from intact L-cells is biexponential, with rapid (half-life of 1 min) and slow (half-life of 15 min) kinetic pools. The kinetics of these pools are consistent with protein-mediated and vesicular transfer of sterol, respectively. (ii) SCP-2 expression does not affect the relative sterol pool sizes, but accelerates the rate of cholesterol movement in the rapid, protein-mediated sterol efflux pool. (iii) In contrast, SCP-2 expression decreases the rate of movement of the slow vesicular pool. Since the majority (80–90%) of sterol transfer in the cultured cells exhibits vesicular kinetics transfer, the net effect of SCP-2 overexpression is to reduce cholesterol efflux (Atshaves *et al.*, 2000). Taken together, these data indicate that SCP-2 enhances molecular sterol transfer from lipid rafts/caveolae, but not non-rafts, *in vitro*. This direct response to SCP-2 is significantly reduced in lipid rafts/caveolae isolated from SCP-2-overexpressing cells. How SCP-2 inhibits vesicular cholesterol transfer in intact cells is not yet clear, but there are several possibilities, including the following: (i) altered lipid and protein distribution in lipid rafts/caveolae (Atshaves *et al.*, 2003); (ii) direct interaction of SCP-2 with caveolin to inhibit caveolin action in sterol transfer (Schroeder *et al.*, 2003); and (iii) SCP-2 binds and transfers ligands (phosphatidylinositol [PI], polyphosphoinositides, fatty acyl-CoAs) known to be involved in vesicle formation and budding (reviewed in Pfanner *et al.*, 1989, 1990; Frolov *et al.*, 1996a; Cockcroft, 1998; Seedorf *et al.*, 1998; Schroeder *et al.*, 2003) and stimulates phospholipase-mediated IP_3 production (Schroeder *et al.*, 2003).

VII. MULTIPHOTON IMAGING OF STEROL STRUCTURE AND DISTRIBUTION IN LIPID RAFTS/CAVEOLAE OF LIVING CELLS

Although plasma membrane subfractionation by both non-detergent and detergent methods suggests the presence of cholesterol-rich lipid rafts/caveolae in the plasma membrane, there are still important questions regarding lipid rafts/caveolae in living cells. For example, the question of whether such cholesterol-rich microdomains actually exist in the cell surface plasma membrane of living cells remains unanswered. While the existence of lipid rafts/caveolae in living cells has been demonstrated on the basis of protein and non-sterol lipid markers, whether the distribution of these markers coincides with that of cholesterol needs to be resolved. Another major question is the size of lipid rafts/caveolae in living cells. A variety of morphological, protein, and phospholipid markers have yielded widely divergent estimates. Depending on the specific marker and method used, the diameter of lipid rafts/caveolae is reported as 26–330 nm when GPI-anchored proteins are used as markers for lipid rafts (Sheets *et al.*, 1997; Pralle *et al.*, 2000), 500–2000 nm when MHC I molecules are used as markers for lipid rafts (Edidin, 2001), 60–80 nm when flask-shaped morphology electron microscopy is used as a marker for caveolae (Anderson, 1998; Smart *et al.*, 1999), 260–330 nm when ganglioside GM1 is used as a marker for caveolae (Sheets *et al.*, 1997), and 230 nm when phosphatidylethanolamine is used as a marker for non-raft domains (Sheets *et al.*, 1997; Atshaves *et al.*, 2003). However, none of these early studies in intact cells utilized cholesterol as a marker for lipid rafts/caveolae. This is especially surprising in view of the fact that cholesterol is highly enriched in both lipid rafts and caveolae. Part of the difficulty has been the inability to directly visualize cholesterol in the plasma membrane of living cells. Recent studies from our laboratories utilized dehydroergosterol, a naturally occurring fluorescent sterol, together with multiphoton microscopy to directly visualize sterol distribution in plasma membranes of living cells (Schroeder *et al.*, 1998; Frolov *et al.*, 2000). Dehydroergosterol is readily incorporated into plasma membranes (Hale and Schroeder, 1982; Atshaves *et al.*, 2003) and lipid rafts/caveolae (Atshaves *et al.*, 2003) of cultured cells. These advances, together with the use of membrane and lipid droplet markers, allow direct visualization of dehydroergosterol in plasma membranes and analysis of dehydroergosterol distribution therein (Atshaves *et al.*, 2003; Liu *et al.*, 2004).

In order to examine the distribution of dehydroergosterol in the plasma membrane of living cells, ECFP-Mem, a protein that targets the plasma membrane, was overexpressed in L-cells. pECFP-Mem is a mammalian overexpression vector that contains the enhanced cyan fluorescent protein (ECFP) variant of green fluorescent protein (GFP) coupled to a plasma membrane targeting vector "Mem" to form pECFP-Mem (Clontech, Palo Alto, CA). Mem encodes a protein containing the N-terminal 20 amino acids derived from neuromodulin (GAP-43; 1) (targeting sequence for the plasma membrane) that is fused to the N terminus of ECFP. The plasma membrane targeting is provided by the addition of a palmitoyl group to the GAP-43 protein during posttranslational processing (Liu *et al.*, 1994). Although predominantly located on the inner surface of the plasma membrane, other intracellular membranes are also targeted. The fluorescence emission from this protein in combination with Nile red staining was useful for demarcation of the plasma membrane, since Nile red at low concentrations provides only weak staining of the plasma membrane as compared to other intracellular features, while the ECFP-Mem brightly labels the plasma membrane and, to a much lesser degree, some other intracellular membranes. L-cells that were supplemented with dehydroergosterol as stated previously and stained with Nile red were imaged by multiphoton laser scanning microscopy. Multiphoton images of each fluorescent probe simultaneously visualized through three external detectors are shown in Fig. 4. Nile red localized very intensely in bright staining structures and less so throughout the cell in other lipid-containing structures (Fig. 4A). Nile red is known to localize primarily in lipid droplets (Atshaves *et al.*, 2000). The dehydroergosterol localizes most intensely at the cell surface and less so in intracellular membranes (Fig. 4B). The ECFP-Mem localizes most intensely at the plasma membrane and less so within the cell interior (Fig. 4C). Superposition of the three images reveals that dehydroergosterol and ECFP-Mem were highly colocalized at the cell surface plasma membrane (Fig. 4D).

To remove all intracellular contributions of dehydroergosterol and image only that occurring in the plasma membrane, a rank-statistic-based technique was used to segment the plasma membrane and remove the intracellular portion of each cell (Liu *et al.*, 1994). In brief, the technique consisted of initial manual selection of a square window of size $k \times k$ on a known plasma membrane area. Next, this window was used

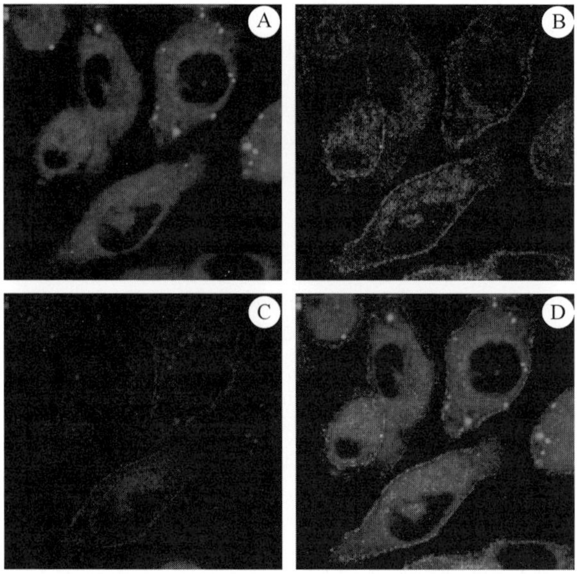

Figure 4. Multiphoton laser scanning microscopy of fluorescent sterol (DHE), a plasma membrane marker (ECFP-Mem), and a lipid marker (Nile red) in L-cell fibroblasts. The plasmid pECFP-Mem (Clontech) was stably expressed in L-cells (L $arpt^{-}tk^{-}$) by transfection using Superfect (Qiagen) according to the manufacturer's instructions. After G418 selection and PCR verification to select resistant clones, L-cells stably expressing ECFP-Mem were cultured for 2 days on two-well Lab-Tek chambered coverglasses (VWR, Sugarland, TX) with Higuchi medium containing 10% fetal bovine serum and supplemented with a total concentration of 20 μg/ml of dehydroergosterol (DHE) in the form of large unilamellar vesicles. These vesicles were prepared with 1-palmitoyl-2-oleoyl-sn-glycero-3-phosphocholine (POPC) and dehydroergosterol (65:35 POPC:DHE), wherein DHE was almost completely monomeric (McIntosh *et al.*, 2003). Cells were then incubated with 100–400 nM Nile red (Molecular Probes, Eugene, OR) for approximately 30 min. Multiphoton laser scanning microscopy (MPLSM) was performed using a Bio-Rad MRC1024 MP system attached Axiovert 135 (Zeiss Inc., New York, NY) microscope with a Zeiss 63 $\times$ Plan-Apochromat (1.4 N.A.) oil immersion objective. All three fluorescent probes (DHE, Nile red, ECFP-Mem) were simultaneously excited using multiphoton excitation at 920 nm with a femtosecond Coherent Mira 900 Ti:Sapphire laser pumped with a Spectra-Physics Millenia X. Fluorescence emission was detected through an external three-channel detector system provided by Dr. Warren Zipfel (Cornell University, Ithaca, NY). Emission of the three fluorescent probes was individually selected by the INDO dichroic filter set provided by Dr. Warren Zipfel. (A) Nile red emission detected at 525–650 nm. (B) Dehydroergosterol emission detected at 360–430 nm. (C) ECFP-Mem emission detected at 485–515 nm. (D) Merged image with Nile red (red), dehydroergosterol (green), and ECFP-Mem (blue). (See Color Insert.)

to search the image pixel by pixel with the pixel intensities of each two windows ranked from smallest to largest. After computation of the total and mean rank, the Miller's rank statistic was calculated to determine if the pixel density distributions of the two windows were different. Parameters were changed in order to optimize the results. For this analysis, two windows were selected on sections of plasma membranes on different cells with higher concentrations of (i) dehydroergosterol and (ii) ECFP-Mem. The results of the rank-statistic technique based upon the two selected plasma membrane regions were combined as shown (Fig. 5A). Some intracellular portions remained as evidenced by the amount of Nile red. Further definition of the plasma membrane occurred as a result of a binary mask created by subtraction of the green channel from the red channel and changing all resulting pixel intensities greater than 0 to 1 and those less than 0 to 0. This mask was applied to Fig. 5A and the result is shown in Fig. 5B. A particular segment was chosen (outlined in white in Fig. 5B) and a blowup was created (Fig. 5C).

Image thresholds were established by choosing the pixels containing dehydroergosterol emission intensities above the 85th percentile, and a blowup of the selected segment (Fig. 5D) and a spatial plot of the positional arrangement of these peak dehydroergosterol intensities (Fig. 6A) were created. A complete spatial randomness test was performed upon this dehydroergosterol peak data set by comparing the proportion of nearest neighbor distances to that of a simulated homogeneous Poisson process through Monte Carlo simulations. The graph showing the empirical distribution function from the data along with the Monte Carlo envelope is given in Fig. 6B.

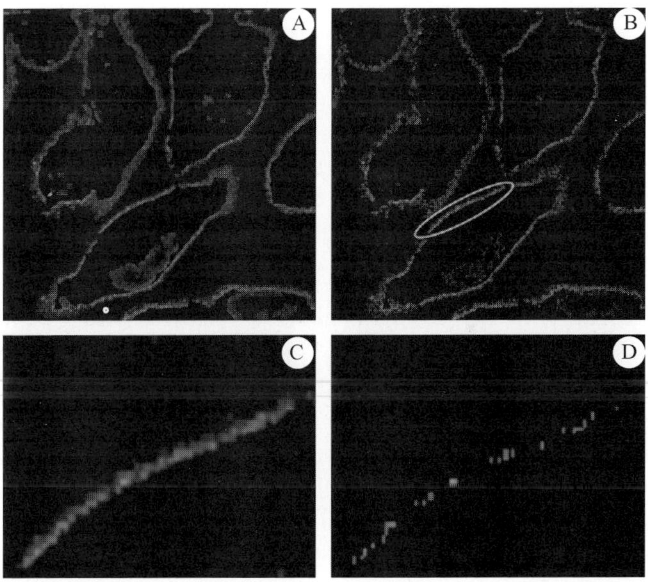

Figure 5. Statistical segmentation of the plasma membrane of L-cells expressing ECFP-Mem (blue), supplemented with dehydroergosterol (green), and subsequently stained with Nile red (red). (A) Initial segmentation result from the rank-statistic technique produced by a combination of two windows chosen from two different plasma membrane sections. (B) Final segmentation result after binary masking was applied to further remove remaining intracellular regions and enhancing definition of the plasma membrane. A typical segment (circled in white) of the plasma membrane was chosen for statistical image analysis. (C) A blowup of the selected segment (all three channels) showing the discontinuity of the dehydroergosterol along the plasma membrane for the full data set. (D) A blowup of the selected segment (all three channels) after 85th percentile thresholding of the dehydroergosterol intensities (peak DHE intensities) along the plasma membrane. (See Color Insert.)

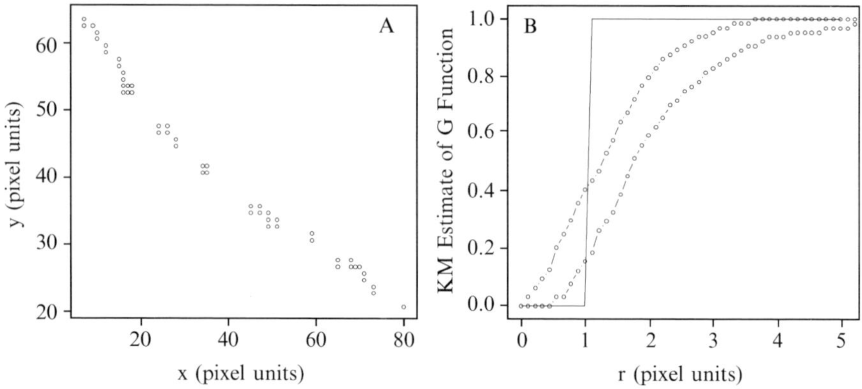

Figure 6. Statistical image analysis of plasma membrane segmented results. (A) Spatial plot of the location of peak spots of dehydroergosterol along the plasma membrane of the chosen segment. (B) Plot of the nearest neighbor empirical distribution function and Monte Carlo envelope (from 99 simulations) calculated for the peak dehydroergosterol data set (85th percentile thresholding). This graph illustrates the strong clustering properties of dehydroergosterol in the plasma membrane for nearest neighbor distances <3–4.

This graph illustrates that the nearest-neighbor distance proportions were larger than the homogeneous Poisson process envelope for distances less than 3–4 pixels. Since the smallest nearest distance between two pixels is 1, the curve below a distance of 1 was neglected. This statistical analysis indicates that the dehydroergosterol pixels with peak intensities above the 85th percentile exhibit strong statistical clustering patterns.

Multiphoton imaging of dehydroergosterol, extraction of plasma membrane dehydroergosterol independent of intracellular dehydroergosterol, and statistical analysis indicates that plasma membranes of L-cell fibroblasts contain cholesterol-rich domains. These dehydroergosterol-rich domains exhibit a strong tendency to distribute non-randomly and not interdisperse into either regular or random patterns. Although the limit of resolution of optical microscopy is about 200 nm, multiphoton imaging methods can resolve domains >200 nm in diameter. The size of the dehydroergosterol-rich domains in the plasma membranes of living L-cell fibroblasts is estimated at 600–900 nm (data not shown), well within the range (260–2000 nm) reported for larger sized lipid rafts determined using protein and other lipidic markers (Sheets *et al.*, 1997; Pralle *et al.*, 2000; Edidin, 2001).

VIII. CAVEOLIN INTERACTS WITH SELECT LIPIDS IN CAVEOLAE

Caveolin has been reported to bind and facilitate transport of fatty acids (Trigatti *et al.*, 1999) and cholesterol (Murata *et al.*, 1995; Monier *et al.*, 1996; Uittenbogaard *et al.*, 1998; Uittenbogaard and Smart, 2000) across plasma membrane. Photoaffinity labeling reveals that caveolin interacts with photoactivatable fatty acid in the plasma membrane (Trigatti *et al.*, 1991, 1999; Gerber *et al.*, 1993). Caveolin is the primary protein that interacts with a photoactivatable ganglioside GM1 in plasma membranes, and it is the only protein with which the photoactivatable ganglioside GM1 interacts in lipid rafts/caveolae isolated by a detergent-based method (Pitto *et al.*, 2000). Finally, caveolin also contributes to non-vesicular transport of molecular cholesterol and cholesteryl ester within the cell cytosol as a caveolin–cholesterol–heat shock protein chaperone complex (Uittenbogaard *et al.*, 1998) and a caveolin–cholesteryl-ester–annexin II complex (Uittenbogaard *et al.*, 2002), respectively. These data indicate that caveolin is highly associated with ganglioside GM1 in caveolae as well as binding fatty acids and cholesterol therein.

IX. PEPTIDE–PROTEIN INTERACTIONS WITH MODEL MEMBRANE LIPID RAFTS: SCP-2

It is well established that SCP-2 functions in the metabolism, transport, and/or secretion of cholesterol and other lipids (glycerides, straight- and branched-chain fatty acids, isoprenoids) (reviewed in Schroeder *et al.*, 1998; Gallegos *et al.*, 2001b; Stolowich *et al.*, 2002). Overexpression of SCP-2 in transfected cells enhances cholesterol uptake (Moncecchi *et al.*, 1996; Atshaves *et al.*, 1999), transport of cholesterol to and from the plasma membrane to the ER (Baum *et al.*, 1997; Murphy and Schroeder, 1997), intracellular cholesterol esterification and triacylglycerol formation (Moncecchi *et al.*, 1996; Atshaves *et al.*, 1999), mitochondrial cholesterol oxidation (Yamamoto *et al.*, 1991), and biliary cholesterol transport (Kawata *et al.*, 1991; Ito *et al.*, 1996; Puglielli *et al.*, 1996; Fuchs *et al.*, 1997, 1998, 2001; Muench *et al.*, 2000). Most studies concur that SCP-2 binds cholesterol in solution (Schroeder *et al.*, 1990a; Colles *et al.*, 1995; Stolowich *et al.*, 1999, 2002) and that this binding is essential for sterol transfer by SCP-2 (reviewed in Schroeder *et al.*, 1991a, 1996, 2001b; Woodford *et al.*, 1995; Stolowich *et al.*, 2002). Hence, SCP-2 apparently contains at least two important structural and functional domains, one to bind the ligand and the other to bind membranes. NMR structural analysis shows that the N-terminal region of SCP-2 contains two amphipathic α-helices (residues 9–22 and 25–30) and a short helix corresponding to amino acids 78–84 (Szyperski *et al.*, 1993). The N-terminal helices were postulated to function as the SCP-2 membrane-binding domain (Pastuszyn *et al.*, 1987; Szyperski *et al.*, 1993). Early studies showed that SCP-2-mediated intermembrane sterol transfer was optimal when membranes contained anionic phospholipids, suggesting that ionic interactions were involved in SCP-2 membrane binding (Butko *et al.*, 1990, 1992; Hapala *et al.*, 1990, 1994; Gadella and Wirtz, 1994). Further, membrane binding appeared to be a prerequisite for cholesterol transfer by SCP-2 (Woodford *et al.*, 1995). Support of a critical membrane domain localized to the N-terminal, helical region of SCP-2 came from a study of mutant SCP-2 molecules in which specific N-terminal disruptions and/or deletions concurrently abolished lipid transfer and reduced the α-helical content (Seedorf *et al.*, 1994). The N-terminal membrane-binding function of SCP-2 was further supported using circular dichroism (CD) and a direct binding assay with model membranes (Huang *et al.*, 1999a,b, 2002). In these studies, full-length SCP-2 and N-terminal synthetic peptides corresponding to residues 1–32 (both N-terminal α-helices), 1–24 (first N-terminal α-helix), 10–32 (deletion of the non-helical residues), and 1-E20–32 (substitution of leucine 20 with glutamic acid) were tested for binding to model membranes of different curvature and cholesterol and anionic phospholipid content. SCP-2 and peptides preferentially partitioned to membranes of high curvature and those rich in cholesterol and anionic phospholipids (Huang *et al.*, 1999a,b, 2002). Direct binding with the different membrane preparations and SCP-2 and peptides revealed that the N-terminal helices of SCP-2 comprised a membrane-binding domain. All of the SCP-2 peptides bound anionic vesicles with greater affinity than neutral zwitterionic lipids, indicating that electrostatic interactions are important in the membrane binding. The single helix-containing peptide (SCP-2 residues 1–24) had a weaker affinity for lipids than that of SCP-2 amino acids 1–32 containing both N-terminal helices; deletion of the second helix did not influence the first helix formation (Huang *et al.*, 1999a,b, 2002). In contrast, substitution of glutamic acid for leucine significantly disrupted the helix conformation and decreased the membrane interaction. Thus, the amphipathic character of the helix appears to be important to both structure and function. In addition to charged phospholipids, the presence of cholesterol was required for optimal interaction of SCP-2 and peptides with membranes, putatively through the formation of cholesterol-rich, lateral domains. This observation is in agreement with earlier data showing that SCP-2 stimulates cholesterol transfer more efficiently from cholesterol-rich membranes (reviewed in Schroeder *et al.*, 1991a). Last, SCP-2 and peptides show a preferential binding to membranes of high curvature. Taken together, the data show that SCP-2 and peptides optimally partition with anionic, cholesterol-rich, highly curved lipid vesicles (Huang *et al.*, 1999a,b, 2002), indicating a preferential interaction with caveolae/rafts. Co-localization of SCP-2 with caveolin-1 in hepatic cells (Schroeder *et al.*, 2003) and demonstration of a direct interaction of SCP-2–caveolin-1 by yeast two-hybrid assay (unpublished data) strongly support the conclusion of a caveolae–SCP-2 association (Zhou *et al.*, 2004).

The high sensitivity of lipid raft/caveolae sterol to SCP-2-mediated transfer may be due at least in part to factors such as caveolae curvature and distribution of anionic phospholipids (phosphatidylinositol, polyphosphoinositides, phosphatidylserine) (reviewed in Woodford *et al.*, 1995; Huang *et al.*, 1999a,b, 2002; Gallegos *et al.*, 2001b). Since direct binding of SCP-2 to anionic phospholipid-containing membranes is required for efficient sterol transfer (Woodford *et al.*, 1995; Huang *et al.*, 1999a,b, 2002), these data suggest that SCP-2 may target sterol transfer to lipid rafts/caveolae by direct interaction with membrane lipids. Alternately, SCP-2 may directly bind caveolin, as suggested by immunofluorescence studies showing that SCP-2 significantly colocalizes with caveolin-1 at the plasma membrane (Schroeder *et al.*, 2003). However, the limit of resolution of colocalization by confocal microscopy is about 200 nm, too large to directly demonstrate that SCP-2 binds caveolin-1.

X. PEPTIDE–PROTEIN INTERACTIONS WITH MODEL MEMBRANE LIPID RAFTS: CHOLERA TOXIN AND ROTAVIRUS NSP4, AN ENTEROTOXIN

In addition to intracellular host cell proteins, plasma membrane caveolae/rafts have been shown to be an important site of entry and exit, as well as function of a number of pathogens (viruses, protozoa, bacteria). Many toxins are endocytosed and transported to a specific site in the cell to exert their cytotoxic effect(s), thus contributing to the pathogenesis of the organism that produces them. These toxins employ cellular endocytotic mechanisms and membrane trafficking pathways for entry and transport. The most studied example is cholera toxin (CT), which is encoded by *Vibrio cholerae* and activates adenylate cyclase at basolateral membranes by catalyzing ADP-ribosylation of the regulatory GTPase, $G_{s\alpha}$ (Oh and Schnitzer, 2001). The increased intracellular cAMP promotes electrogenic chloride secretion as measured in Ussing chambers, the hallmark of secretory diarrhea (Welsh *et al.*, 1982; Morris and Frizzell, 1994). The bacterium does not invade the intestinal mucosa or aid in the delivery of CT into the host intestinal cell (Lencer, 2001). Rather, CT invades the intestinal epithelium as a fully folded holotoxin that subsequently moves retrograde through the biosynthetic pathway through the Golgi cisternae to the ER, where the enzymatic subunit is released and translocated to the basolateral membrane for disease induction (Lencer *et al.*, 1992, 1999; Orlandi, 1997; Wolf *et al.*, 2002).

Because caveolin-1 is found in the intestine (Field *et al.*, 1998) and CT has become a common marker for caveolae entry (Gustin and Goodman, 1981; McIntosh and Schnitzer, 1999; Matveev *et al.*, 2001; Nichols, 2002), although CT internalization is not exclusively caveolae mediated, it is reasonable to propose that caveolae have an important role in the entry of biological toxins to the gut. This proposition can be extended to other microbes that invade the gastrointestinal tract and other organs. There has been a recent explosion of studies on caveolae endocytosis. Reports of caveolae-bound molecules being delivered to the cytosol, to the endoplasmic reticulum (ER), across the cell, and to caveolin-1-positive endosomes or "caveosomes," as well as documentation that caveolae deliver molecules to components of clathrin-mediated endocytosis, illustrate some of these advances (Pelkmans *et al.*, 2001; Schnitzer, 2001; Pelkmans and Helenius, 2002; Stan, 2002). Vesicles rich in caveolin, GM_1, and the vesicle docking protein VAMP were shown to undergo induced caveolar fission in a cell free system with the addition of cytosol and GTP hydrolysis (Schnitzer *et al.*, 1996). The large GTPase dynamin appears to form a structural collar around the neck of invaginated caveolae, presumably to mediate caveolar fission to form free transport vesicles (Deidre *et al.*, 1998). Microinjection of dynamin-specific peptide antibodies induces an accumulation of caveolae at the plasma membrane (Henley *et al.*, 1998) and blocks caveolae-mediated endocytosis of CT (Gallusser and Kirchhausen, 1993; Henley *et al.*, 1998). In addition, a resident population of protein kinase C (PKCα) is needed for invagination of caveolae (Smart *et al.*, 1995). Tyrosine kinases also function in regulating caveolae-mediated endocytosis (Dangoria *et al.*, 1996; Tiruppathi *et al.*, 1997; Minshall *et al.*, 2000), whereby an antibody specific to phosphotyrosine at position 14 of caveolin-1 results in accumulation of caveolin-1-positive vesicles (Aoki *et al.*, 1999). Hence, there appears to be a correlation between caveolae internalization and the activation of signaling molecules localized to caveolae. Many signaling molecules directly bind caveolin (src tyrosine kinase, Gα subunits, H-Ras, nitrous oxide synthase, and EGF receptor tyrosine kinase) (Li *et al.*, 1995; Song *et al.*, 1996; Couett *et al.*, 1997; Okamoto *et al.*, 1998).

A number of experimental studies linked caveolae/rafts to intracellular signal transduction, cell adhesion, and regulation of polarized intracellular sorting (Stefanova *et al.*, 1991; Simons and Ikonen, 1997). Localization of G_α isoforms and G_β subunits in the plasma membrane to caveolae and non-caveolar rafts illustrates differential targeting of G proteins to the membrane microdomains (Oh and Schnitzer, 2001); G_q localizes to caveolae by interacting with caveolin, whereas G_i, G_s, and G_β concentrate within lipid rafts. However, in cells lacking caveolae, the three G proteins (G_i, G_s, G_q) complex with $G_{\beta\gamma}$ subunits and colocalize in rafts (Oh and Schnitzer, 2001). Similarly, the insulin receptor normally localized to caveolae is sorted to lipid rafts in cells that lack caveolae. Likewise, the kinase activity that is typically activated by caveolin binding is modulated by raft lipids (Vainio *et al.*, 2002).

In 1996, rotavirus nonstructural protein 4 (NSP4) was reported to be a viral enterotoxin that induces diarrhea by a calcium-mediated signaling pathway (Ball *et al.*, 1996). Like CT, the net result of the induced signaling pathway is promotion of chloride secretory currents as measured in Ussing chambers. NSP4 is an ER, transmembrane glycoprotein with an extended cytoplasmic domain (amino acids 86–175) that folds as a coiled coil (22% α-helical content by CD) (Taylor *et al.*, 1996). The tetrameric coiled-coil structure of NSP4 residues 95–137 has been confirmed by crystallographic studies (Bowman *et al.*, 2000). CD experiments also have been utilized to evaluate the interaction(s) of intact NSP4 and NSP4-specific synthetic peptides with model membranes of different composition and size (Huang *et al.*, 2001, 2004). CD analysis of purified, full-length NSP4 in physiological buffer shows that NSP4 contains 33% helical structure with 31% β-strand and 12% β-turn, consistent with the 22% helical content of residues 86–175. The enterotoxic peptide $NSP4_{114–135}$ likewise adopts a secondary structure with a high helical content in buffer calculated at 37% (Huang *et al.*, 2001). When placed in a hydrophobic solvent (50% trifluoroethanol), the α-helical content of $NSP4_{114–135}$ increases to 91%, indicating that the peptide may favor the more hydrophobic environment of membranes to promote structural changes. The isodichroic point at 205 nm suggests that the structural transformation is from random coil to α-helix, similar to that seen with the SCP-2 peptides (Huang *et al.*, 1999a,b; Feron *et al.*, 2001). These structural alterations were utilized to determine if NSP4, $NSP4_{114–135}$, or other NSP4 peptides bind lipid vesicles of different curvature and composition akin to the studies of SCP-2 and peptides described above.

CD analyses of NSP4 and $NSP4_{114–135}$ with large unilamellar vesicles (LUV) of 100–120 nm and small unilamellar vesicles (SUV) of 20–30 nm of different compositions indicate a specific interaction of defined membranes (Huang *et al.*, 2001, 2004). Full-length NSP4 preferentially partitions with anionic SUVs (high membrane curvature) containing 35% cholesterol as shown by significant conformational changes (43% helical) elicited upon binding. Similarly, $NSP4_{114–135}$ becomes more helical upon interacting with anionic SUVs with increasing cholesterol content. When compared to the structure in aqueous buffer, the helical content of $NSP4_{114-135}$ is 2-fold greater in the presence of anionic SUVs with 50 mol% cholesterol; however, cholesterol in the absence of negative phospholipids is insufficient for the peptide interaction (Huang *et al.*, 2001). An independent filtration binding assay confirmed these results, demonstrating that NSP4 and $NSP4_{114-135}$ interact exclusively with lipids and preferentially interact with vesicles that are highly curved, rich in anionic phospholipids and cholesterol, and thus resemble caveolae microdomains (Anderson *et al.*, 1992; Brown and Rose, 1992; Rothberg *et al.*, 1992; Anderson, 1998).

The structural basis of NSP4, SCP-2, and corresponding synthetic peptide interactions with highly curved membranes likely includes the more mobile packing and lower surface pressure of the phospholipids in the outer leaflet and the packing constraint in the inner leaflet (Machida and Ohnishi, 1980; Talbot *et al.*, 1997), resulting in an exposed hydrophobic core that may facilitate hydrophobic interactions. Highly curved membrane regions are prevalent in the intestinal microvillus (Lipka *et al.*, 1995; Boffelli *et al.*, 1997), the site of NSP4 enterotoxic activity (Chan *et al.*, 1988). Additional NSP4 peptides, including $NSP4_{120-147}$ and $NSP4_{131K}$, a mutant 114–135 peptide in which tyrosine (amino acid 131) is replaced with a positively charged lysine, similarly interact with anionic and cholesterol-rich SUVs when analyzed by CD and filtration binding assays (Huang *et al.*, 2004). These data correlate with the finding that NSP4 is associated with rafts (Triton-X resistant membrane fractions) in rotavirus-infected cells (Sapin *et al.*, 2002) and likely traffic to the cell surface in association with caveolae vesicles.

The membrane topology of caveolin-1 is unusual, with a central hydrophobic domain (amino acids 102–134) that is predicted to form a hairpin-like structure within the plasma membrane resulting in cytosolic N-(amino acids 1–101) and C-terminal (135–178) domains (Anderson, 1993; Okamoto *et al.*, 1998) such that caveolin-1 only incorporates into one membrane leaflet and the lipophilic moiety incompletely spans the bilayer. Robenek *et al.* (2004) evaluated whether caveolin-1 remains exclusively in the cytoplasmic leaflet during intracellular transport. In contrast to the plasma membrane, caveolin-1 was found on the exoplasmic leaflet of intracellular organelles, including those of the ER and Golgi. These data will aid in dissecting the intracellular transport of proteins associated with caveolae, such as SCP-2, CT, and NSP4.

XI. PROSPECTUS

In summary, more than 2000 publications over the past few years have focused on a host of biological functions compartmentalized to plasma membrane microdomains termed lipid rafts/caveolae. However, little more than a score of these reports address the lipids and lipid protein interactions in lipid rafts/caveolae. Recent data revealed the importance of both lipids and protein constituents of lipid rafts/caveolae in cholesterol transport (reviewed in Brown and London, 1998a, 2000; Smart and van der Westhuyzen, 1998; Anderson and Jacobson, 2002; Liu *et al.*, 2002; Wood *et al.*,

2002), interactions with intracellular cholesterol transporters (reviewed in Smart and van der Westhuyzen, 1998; Schroeder *et al.*, 2001b, 2003), and entry pathways for viral pathogens or toxins (Tian *et al.*, 1996; Huang *et al.*, 2001; Huang *et al.*, 2004; Swaggerty *et al.*, 2004). However, many important questions remain about the structure, fluidity, cholesterol dynamics, regulatory factors, and peptide–protein interactions in lipid rafts separate from caveolar lipid rafts. With one exception (Atshaves *et al.*, 2003), nothing is known regarding these properties in non-raft domains. There is a need for new methodology to simultaneously separate lipid rafts, caveolar lipid rafts, and non-rafts for biochemical and biophysical characterization. There is uncertainty regarding the relative proportion of lipid rafts, caveolae lipid rafts, and non-raft domains in membranes. Finally, there is an almost complete absence of information regarding the structure, size, and dynamics of cholesterol-rich domains in living cells. The recent advent of new multiphoton imaging microscopy techniques now provides the framework to begin to address these and other (e.g., size of cholesterol-rich rafts) questions in real time in living cells.

ACKNOWLEDGMENTS

This work was supported in part by the USPHS, National Institutes of Health grants GM31651 (FS), GM63236 (JMB), DK062812 (AMG), and the Center for Environmental and Rural Health ES09106 (AK).

REFERENCES

Abrami, L., Liu, S., Cosson, P., Leppla, S. H., and van der Groot, F. G. (2003). Anthrax toxin triggers endocytosis of its receptor via a lipid raft-mediated clathrin-independent process. *J. Cell Biol.* **160,** 321–328.

Ahmed, S. N., Brown, D. A., and London, E. (1997). On the origin of sphingolipid/cholesterol-rich detergent-insoluble cell membranes: Physiological concentrations of cholesterol and sphingolipid induce formation of a detergent-insoluble, liquid-ordered lipid phase in model membranes. *Biochemistry* **36,** 10944–10953.

Anderson, R. (1998). The caveolae membrane system. *Ann. Rev. Biochem.* **67,** 199–225.

Anderson, R. G. W. (1993). Caveolae: Where incoming and outgoing messengers meet. *Proc. Natl. Acad. Sci. USA* **90,** 10909–10913.

Anderson, R. G. W. and Jacobson, K. (2002). A role for lipid shells in targeting proteins to caveolae, rafts, and other lipid domains. *Science* **296,** 1821–1825.

Anderson, R. G. W., Kamen, B. A., Rothberg, K. G., and Lacey, S. W. (1992). Potocytosis: Sequestration and transport of small molecules by caveolae. *Science* **255,** 410–411.

Aoki, T., Nomura, R., and Fujimoto, T. (1999). Tyrosine phosphorylation of caveolin-1 in the endothelium. *Exp. Cell Res.* **253,** 629–636.

Atshaves, B. P., Gallegos, A., McIntosh, A. L., Kier, A. B., and Schroeder, F. (2003). Sterol carrier protein-2 selectively alters lipid composition and cholesterol dynamics of caveolae/lipid raft vs non-raft domains in L-cell fibroblast plasma membranes. *Biochemistry* **42,** 14583–14598.

Atshaves, B. P., Petrescu, A., Starodub, O., Roths, J., Kier, A. B., and Schroeder, F. (1999). Expression and intracellular processing of the 58 kDa sterol carrier protein 2/3-oxoacyl-CoA thiolase in transfected mouse L-cell fibroblasts. *J. Lipid Res.* **40,** 610–622.

Atshaves, B. P., Starodub, O., McIntosh, A. L., Roths, J. B., Kier, A. B., and Schroeder, F. (2000). Sterol carrier protein-2 alters HDL-mediated cholesterol efflux. *J. Biol. Chem.* **275,** 36852–36861.

Baenziger, J. U. and Fiete, D. (1979). Structural determinants of concanavalin A specificity for oligosaccharides. *J. Biol. Chem.* **254,** 2400–2407.

Ball, J. M., Tian, P., Zeng, C. Q. Z., Morris, A. P., and Estes, M. K. (1996). Age-dependent diarrhea induced by a rotaviral nonstructural protein. *Science* **272,** 101–104.

Baum, C. L., Reschly, E. J., Gayen, A. K., Groh, M. E., and Schadick, K. (1997). Sterol carrier protein-2 overexpression enhances sterol cycling and inhibits cholesterol ester synthesis and high density lipoprotein cholesterol secretion. *J. Biol. Chem.* **272,** 6490–6498.

Bavari, S., Bosio, C. M., Wiegand, E., Ruthel, G., Will, A. B., Geisbert, T. W., Hevey, M., Schmaljohn, C., and Aman, M. J. (2002). Lipid raft microdomains. A gateway for comportmentalized trafficking of Ebola and Marburg viruses. *J. Exp. Med.* **195,** 593–602.

Boffelli, D., Weber, F. E., Compassi, S., Werder, M., Schulthess, G., and Hauser, H. (1997). Reconstitution and further characterization of the cholesterol transport activity of the small-intestinal brush border membrane. *Biochemistry* **36,** 10784–10792.

Bowman, G. D., Nodelman, I. M., Levy, O., Lin, S. L., Tian, P., Zamb, T. J., Udem, S. A., Venkataraghavan, B., and Schutt, C. E. (2000). Crystal structure of the oligomerization domain of NSP4 from rotavirus reveals a core metal-binding site. *J. Mol. Biol.* **304,** 861–871.

Bretscher, M. S. and Munro, S. (1993). Cholesterol and the Golgi apparatus. *Science* **261,** 1280–1281.

Brown, D. A. and London, E. (1998a). Functions of lipid rafts in biological membranes. *Annu. Rev. Cell Dev. Biol.* **14,** 111–136.

Brown, D. A. and London, E. (1998b). Structure and origin of ordered lipid domains in biological membranes. *J. Membr. Biol.* **164,** 103–114.

Brown, D. A. and London, E. (2000). Structure and function of sphingolipid- and cholesterol-rich membrane rafts. *J. Biol. Chem.* **275,** 17221–17224.

Brown, D. A. and Rose, J. K. (1992). Sorting of GPI anchored proteins to glycolipid-enriched membrane subdomains during transport to the apical cell surface. *Cell* **68,** 533–544.

Brown, R. E. (1998). Sphingolipid organization in biomembranes: What physical studies of model membranes reveal. *J. Cell Sci.* **111,** 1–9.

Brunner, G., Ferber, E., and Resch, K. (1976). Fractionation of membrane vesicles I. *Differentiation* **5,** 161–164.

Butko, P., Hapala, I., Nemecz, G., and Schroeder, F. (1992). Sterol domains in phospholipid membranes: Dehydroergosterol polarization measures molecular sterol transfer. *J. Biochem. Biophys. Meth.* **24,** 15–37.

Butko, P., Hapala, I., Scallen, T. J., and Schroeder, F. (1990). Acidic phospholipids strikingly potentiate sterol carrier protein 2 mediated intermembrane sterol transfer. *Biochemistry* **29,** 4070–4077.

Chan, W. K., Au, K. S., and Estes, M. K. (1988). Topography of the simian rotavirus nonstructural glycoprotein (NS28) in the endoplasmic reticulum. *Virology* **164,** 435–442.

Cockcroft, S. (1998). Phosphatidylinositol transfer proteins: A requirement in signal transduction and vesicle traffick. *BioEssays* **20,** 423–432.

Colles, S. M., Woodford, J. K., Moncecchi, D., Myers-Payne, S. C., McLean, L. R., Billheimer, J. T., and Schroeder, F. (1995). Cholesterol interactions with recombinant human sterol carrier protein-2. *Lipids* **30,** 795–804.

Couett, J., Sargiacomo, M., and Lisanti, M. P. (1997). Interaction of a receptor tyrosine kinase, EGF-R, with caveolins. *J. Biol. Chem.* **272,** 30429–30438.

Dangoria, N. S., Breu, W. C., Anderson, H. A., Cishek, C. M., and Norkin, L. C. (1996). Extracellular simian virus 40 induces an ERK/MAP kinase-independent signaling pathway that activates primary response genes and promotes virus entry. *J. Gen. Virol.* **77,** 2173–2182.

Deidre, P. O., McIntosh, P., and Schnitzer, J. E. (1998). Dynamin at the neck of caveolae mediates their budding to form transport vesicles by GTP-driven fusion from the plasma membrane of endothelium. *J. Cell. Biol.* **141,** 101–114.

Eckert, G. P., Igbavboa, U., Muller, W., and Wood, W. G. (2003). Lipid rafts of purified mouse brain synaptosomes prepared with or without detergent reveal different lipid and protein domains. *Brain Res.* **962,** 144–150.

Edidin, M. (1990). Molecular associations and membrane domains. *Curr. Topics Memb. Transport* **36,** 81–96.

Edidin, M. (2001). Shrinking patches and slippery rafts: Scales of domains in the plasma membrane. *Trends Cell. Biol.* **11,** 492–496.

Empig, C. J. and Goldsmith, M. A. (2002). Association of the caveola vesicular system with cellular entry by filoviruses. *J. Virol.* **76,** 5266–5270.

Feron, O., Dessy, C., Desager, J.-P., and Balligand, J.-L. (2001). Hydroxy-methylglutaryl-coenzyme A reductase inhibition promotes endothelial nitric oxide synthase activation through a decrease in caveolin abundance. *Circulation* **103,** 113–118.

Field, F. J., Born, E., Murthy, S., and Mathur, S. N. (1998). Caveolin is present in intestinal cells: Role in cholesterol trafficking? *J. Lip. Res.* **39,** 1938–1950.

Fielding, C. J. and Fielding, P. E. (2000). Cholesterol and caveolae: Structural and functional relationships. *Biochim Biophys. Acta* **1529,** 210–222.

Fielding, C. J. and Fielding, P. E. (2001a). Caveolae and intracellular trafficking of cholesterol. *Adv. Drug Delivery Rev.* **49,** 251–264.

Fielding, C. J. and Fielding, P. E. (2001b). Cellular cholesterol efflux. *Biochim. Biophys. Acta* **1533,** 175–189.

Frolov, A., Cho, T. H., Billheimer, J. T., and Schroeder, F. (1996a). Sterol carrier protein-2, a new fatty acyl coenzyme A-binding protein. *J. Biol. Chem.* **271,** 31878–31884.

Frolov, A., Petrescu, A., Atshaves, B. P., So, P. T. C., Gratton, E., Serrero, G., and Schroeder, F. (2000). High density lipoprotein mediated cholesterol uptake and targeting to lipid droplets in intact L-cell fibroblasts. *J. Biol. Chem.* **275,** 12769–12780.

Frolov, A., Woodford, J. K., Murphy, E. J., Billheimer, J. T., and Schroeder, F. (1996b). Spontaneous and protein-mediated sterol transfer between intracellular membranes. *J. Biol. Chem.* **271,** 16075–16083.

Frolov, A. A., Woodford, J. K., Murphy, E. J., Billheimer, J. T., and Schroeder, F. (1996c). Fibroblast membrane sterol kinetic domains: Modulation by sterol carrier protein 2 and liver fatty acid binding protein. *J. Lipid Res.* **37,** 1862–1874.

Fuchs, M., Hafer, A., Muench, C., Kannenberg, F., Teichmann, S., Scheibner, J., Stange, E. F., and Seedorf, U. (2001). Disruption of the sterol carrier protein 2 gene in mice impairs biliary lipid and hepatic cholesterol metabolism. *J. Biol. Chem.* **276,** 48058–48065.

Fuchs, M., Lammert, F., Wang, D. Q. H., Paigen, B., Carey, M. C., and Cohen, D. E. (1997). Lith genes induce overexpression of sterol carrier protein 2 during cholesterol gallstone formation. *FASEB J.* **11,** A1060.

Fuchs, M., Lammert, F., Wang, D. Q. H., Paigen, B., Carey, M. C., and Cohen, D. E. (1998). Sterol carrier protein-2 participates in hypersecretion of biliary cholesterol during cholesterol gallstone formation in genetically gallstone susceptible mice. *Biochem. J.* **336,** 33–37.

Gadella, T. W. and Wirtz, K. W. (1994). Phospholipid binding and transfer by non-specific lipid transfer protein (SCP-2): A kinetic model. *Eur. J. Biochem.* **220,** 1019–1028.

Gallegos, A. M., Atshaves, B. P., Storey, S., McIntosh, A., Petrescu, A. D., and Schroeder, F. (2001a). Sterol carrier protein-2 expression alters plasma membrane lipid distribution and cholesterol dynamics. *Biochemistry* **40,** 6493–6506.

Gallegos, A. M., Atshaves, B. P., Storey, S. M., Starodub, O., Petrescu, A. D., Huang, H., McIntosh, A., Martin, G., Chao, H., Kier, A. B., and Schroeder, F. (2001b). Gene structure, intracellular localization, and functional roles of sterol carrier protein-2. *Prog. Lipid Res.* **40,** 498–563.

Gallegos, A. M., McIntosh, A. L., Atshaves, B. P., and Schroeder, F. (2004). Structure and cholesterol domain dynamics of caveolae. *Biochem. J.* **382,** 451–461.

Gallusser, A.. and Kirchhausen, T. (1993). The b1 and b2 subunits of the AP complexes are the clathrin coat assembly components. *EMBO J.* **12,** 5237–5244.

Garrigues, A., Escargueil, A. E., and Orlowski, S. (2002). The multidrug transporter, P-glycoprotein, actively mediates cholesterol redistribution in the cell membrane. *Proc. Natl. Acad. Sci. USA* **99,** 10347–10352.

Gerber, G. E., Mangroo, D., and Trigatti, B. L. (1993). Identification of high affinity membrane-bound fatty acid-binding proteins using a photoreactive fatty acid. *Mol. Cell. Biochem.* **123,** 39–44.

Glaser, M. (1993). Lipid domains in biological membranes. *Curr. Opin. Struct. Biol.* **3,** 475–481.

Gustin, M. C. and Goodman, B. P. (1981). Isolation of brush-border membrane from the rabbit descending colon epithelium. *J. Biol. Chem.* **256,** 10651–10656.

Hafer, A., Katzberg, N., Muench, C., Scheibner, J., Stange, E. F., Seedorf, U., and Fuchs, M. (2000). Studies with sterol carrier protein-2 (SCP-2) gene knockout mice identify liver fatty acid binding protein (FABP1) as intracellular cholesterol transporter contributing to biliary cholesterol hypersecretion and gallstone formation. *Gastroenterology* **118**(4, Part 1, Suppl. 2), 135.

Hale, J. E. and Schroeder, F. (1982). Asymmetric transbilayer distribution of sterol across plasma membranes determined by fluorescence quenching of dehydroergosterol. *Eur. J. Biochem.* **122,** 649–661.

Hapala, I., Butko, P., and Schroeder, F. (1990). Role of acidic phospholipids in intermembrane sterol transfer. *Chem. Phys. Lipids* **56,** 37–47.

Hapala, I., Kavecansky, J., Butko, P., Scallen, T. J., Joiner, C., and Schroeder, F. (1994). Regulation of membrane cholesterol domains by sterol carrier protein-2. *Biochemistry* **33,** 7682–7690.

Henley, J. R., Krueger, E. W., Oswald, B. J., and McNiven, M. A. (1998). Dynamin-mediated internalization of caveolae. *J. Cell Biol.* **141,** 85–99.

Hooper, N. M. (1999). Detergent-insoluble glycosphingolipid/cholesterol rich membrane domains, lipid rafts, and caveolae. *Mol. Memb. Biol.* **16,** 145–156.

Huang, H., Ball, J. A., Billheimer, J. T., and Schroeder, F. (1999a). Interaction of the N-terminus of sterol carrier protein-2 with membranes: Role of membrane curvature. *Biochem. J.* **344,** 593–603.

Huang, H., Ball, J. A., Billheimer, J. T., and Schroeder, F. (1999b). The sterol carrier protein-2 amino terminus: A membrane interaction domain. *Biochemistry* **38,** 13231–13243.

Huang, H., Gallegos, A., Zhou, M., Ball, J. M., and Schroeder, F. (2002). Role of sterol carrier protein-2 N-terminal membrane binding domain in sterol transfer. *Biochemistry* **41,** 12149–12162.

Huang, H., Schroeder, F., Estes, M. K., McPherson, T., and Ball, J. M. (2004). The interactions of rotavirus NSP4 C-terminal peptides with model membranes. *Biochem. J.* **380,** 723–733.

Huang, H., Schroeder, F., Zeng, C., Estes, M. K., Schoer, J., and Ball, J. A. (2001). Membrane interactions of a novel viral enterotoxin: Rotavirus nonstructural glycoprotein NSP4. *Biochemistry* **40,** 4169–4180.

Ito, T., Kawata, S., Imai, Y., Kakimoto, H., Trzaskos, J., and Matsuzawa, Y. (1996). Hepatic cholesterol metabolism in patients with cholesterol gallstones: Enhanced intracellular transport of cholesterol. *Gastroenterology* **110,** 1619–1627.

Kavecansky, J., Joiner, C. H., and Schroeder, F. (1994). Erythrocyte membrane lateral sterol domains: A dehydroergosterol fluorescence polarization study. *Biochemistry* **33,** 2880–2890.

Kawata, S., Imai, Y., Inada, M., Inui, M., Kakimoto, H., Fukuda, K., Maeda, Y., and Tarui, S. (1991). Modulation of cholesterol 7-a hydroxylase activity by nsLTP in human liver—possible altered regulation of its cytosolic level in patients with gallstones. *Clin. Chim. Acta* **197,** 201–208.

Kier, A. B., Sweet, W. D., Cowlen, M. S., and Schroeder, F. (1986). Regulation of transbilayer distribution of a fluorescent sterol in tumor cell plasma membranes. *Biochim. Biophys. Acta* **861,** 287 301.

Lavie, Y.. and Liscovitch, M. (2000). Changes in lipid and protein constituents of rafts and caveolae in multidrug resistant cancer cells and their functional consequences. *Glycoconjugate J.* **17,** 253–259.

Lencer, W. I. (2001). Microbes and microbial toxins: Paradigms for microbial-mucosal interactions V. Cholera: Invasion of the intestinal epithelial barrier by a stably folded protein toxin. *Am. J. Physiol.* **280,** G781–G786.

Lencer, W. I., Delp, C., Neutra, M. R., and Madara, J. L. (1992). Mechanism of cholera toxin action on a polarized human intestinal epithelial cell line: Role of vesicular traffic. *J. Cell Biol.* **117,** 1197–1209.

Lencer, W. I., Hirst, T. R., and Holmes, R. K. (1999). Membrane traffic and the cellular uptake of cholera toxin. *Biochim. Biophys. Acta* **1450,** 177–190.

Li, S., Okamoto, T., Chun, M., Sargiacomo, M., Casanova, J. E., Hansen, S. H. N. I., and Lisanti, M. P. (1995). Evidence for a regulated interaction between heterotrimeric G proteins and caveolin. *J. Biol. Chem.* 15693–15701.

Lin, S. L. and Tian, P. (2003). Detailed computational analysis of a comprehensive set of group A rotavirus NSP4 proteins. *Virus Genes* **26,** 271–282.

Lipka, G., Schulthess, G., Thurnhofer, H., Wacker, H., Wehrli, E., Zeman, K., Weber, F. E., and Hauser, H. (1995). Characterization of lipid exchange proteins isolated from small intestinal brush border membrane. *J. Biol. Chem.* **270,** 5917–5925.

Lisanti, M. P., Tang, Z., Scherer, P. E., Kubler, E., Koleske, A. J., and Sargiacomo, M. (1995). Caveolae, transmembrane signalling and cellular transformation. *Mol. Memb. Biol.* **12,** 121–124.

Liu, J. C., Zhang, W., Xu, H., Wu, D., Gruninger, T., McIntosh, A., and Schroeder, F. (2005). Structural analysis of sterol distribution in the plasma membrane of living cells. *Biochem.* **44,** 2864–2884.

Liu, P., Rudick, M., and Anderson, R. G. W. (2002). Multiple functions of caveolin-1. *J. Biol. Chem.* **277,** 41295–41298.

Liu, Y., Fisher, D. A., and Storm, D. R. (1994). Intracellular sorting of neuromodulin (GAP-43) mutants modified in the membrane targeting domain. *J. Neurosci.* **14,** 5807–5817.

London, E. and Brown, D. A. (2000). Insolubility of lipids in Trition X-100: Physical origin and relationship to sphingolipid/cholesterol membrane domains (rafts). *Biochim. Biophys. Acta* **1508,** 182–195.

Machida, K. and Ohnishi, S. I. (1980). Effect of bilayer membrane curvature on activity of phosphatidylcholine exchange protein. *Biochim. Biophys. Acta* **596,** 201–209.

Marjomaki, V., Pietiainen, V., Upla, P., Ivaska, J., Nissinen, L., Reunanen, H., Huttunen, P., Hyypia, T., and Heino, J. (2002). Internalization of Echovirus 1 in caveolae. *J. Virol.* **76,** 1856–1865.

Marsh, M. and Pelchen-Matthews, A. (2000). Endocytosis in viral replication. *Traffic* **1,** 525–532.

Marx, J. (2001). Caveolae: A once-elusive structure gets some respect. *Science* **294,** 1862–1865.

Masserini, M. and Ravasi, D. (2001). Review role of sphingolipids in the biogenesis of membrane domains. *Biochim. Biophys. Acta* **1532,** 149–161.

Matveev, S. V., Uittenbogaard, A., van der Westhuyzen, D. R., and Smart, E. J. (2001). Caveolin-1 negatively regulates SRB1 mediated selective uptake of high density lipoprotein-derived cholesteryl ester. *Eur. J. Biochem.* **268,** 5609–5616.

McIntosh, A., Gallegos, A., Atshaves, B. P., Storey, S., Kannoju, D., and Schroeder, F. (2003). Fluorescence and multiphoton imaging resolve unique structural forms of sterol in membranes of living cells. *J. Biol. Chem.* **278,** 6384–6403.

McIntosh, D. P. and Schnitzer, J. E. (1999). Caveolae require intact VAMP for targeted transport in vascular endothelium. *Am. J. Physiol.* **363,** H2222–H2232.

Minshall, R. D., Tiruppathi, C., Vogel, S. M., Niles, W. D., Gilchrist, A., Hamm, H. E., and Malik, A. B. (2000). Endothelial cell surface gp60 activates vesicle formation and trafficking via G(I)-coupled Src kinase signaling pathway. *J. Cell Biol.* **150,** 1057–1070.

Moncecchi, D. M., Murphy, E. J., Prows, D. R., and Schroeder, F. (1996). Sterol carrier protein-2 expression in mouse L-cell fibroblasts alters cholesterol uptake. *Biochim. Biophys. Acta* **1302,** 110–116.

Monier, S., Dietzen, D. J., Hastings, W. R., Lublin, D. M., and Kurzchalia, T. V. (1996). Oligomerization of VIP21-caveolin *in vitro* is stabilized by long chain fatty acylation or cholesterol. *FEBS Lett.* **388,** 143–149.

Morris, A. P. and Frizzell, R. A. (1994). Vesicle targeting and ion secretion in epithelial cells: Implications for cystic fibrosis. *Ann. Rev. Physiol.* **56,** 371–397.

Muench, C., Hafer, A., Katzberg, N., Scheibner, J., Stange, E. F., Seedorf, U., and Fuchs, M. (2000). Relevance of the sterol carrier protein-2 gene for bile acid synthesis and gallstone formation in genetically susceptible mice. *Gastroenterology* **118**(4, Part 1, Suppl. 2), 1167.

Murata, M., Peranen, J., Schreiner, R., Wieland, F., Kurzchalia, T. V., and Simons, K. (1995). VIP21/caveolin is a cholesterol-binding protein. *Proc. Natl. Acad. Sci USA* **92,** 10339–10343.

Murphy, E. J. and Schroeder, F. (1997). Sterol carrier protein-2 mediated cholesterol esterification in transfected L-cell fibroblasts. *Biochim. Biophys. Acta* **1345,** 283–292.

Nichols, B. J. (2002). A distinct class of endosome mediates clathrin-independent endocytosis to the Golgi complex. *Nature Cell Biol.* **4,** 374–378.

Norkin, L. C. (2001). Caveolae in the uptake and targeting of infectious agents and secreted toxins. *Adv. Drug Delivery Rev.* **49,** 301–315.

Oh, P. and Schnitzer, J. E. (1999). Immunoisolation of caveolae with high affinity antibody binding to the oligomeric caveolin cage. *J. Biol. Chem.* **274,** 23144–23154.

Oh, P. and Schnitzer, J. E. (2001). Segregation of heterotrimeric G proteins in cell surface microdomains. *Mol. Biol. Cell* **12,** 685–698.

Okamoto, T., Schlegel, A., Scherer, P. E., and Lisanti, M. P. (1998). Caveolins, a family of scaffolding proteins for organizing "preassembled signaling complexes" at the plasma membrane. *J. Biol. Chem.* **273,** 5419–5422.

Orlandi, P. A. (1997). Protein-disulfide isomerase-mediated reduction of the A subunit of cholera toxin in a human intestinal cell line. *J. Biol. Chem.* **272,** 4591–4599.

Pastuszyn, A., Noland, B. J., Bazan, F., Fletterick, R. J., and Scallen, T. J. (1987). Primary sequence and structural analysis of sterol protein-2 from rat liver: Homology with immunoglobins. *J. Biol. Chem.* **262,** 13219–13227.

Pelkmans, L. and Helenius, A. (2002). Endocytosis via caveolae. *Traffic* **3,** 311–320.

Pelkmans, L., Kartenbeck, J., and Helenius, A. (2001). Caveolar endocytosis of simian virus 40 reveals a new two-step vesicular-transport pathway to the ER. *Nature Cell Biol.* **3,** 473–483.

Pfanner, N., Glick, B. S., Arden, S. R., and Rothman, J. E. (1990). Fatty acylation promotes fusion of transport vesicles with Golgi cisternae. *J. Cell Biol.* **110,** 955–961.

Pfanner, N., Orci, L., Glick, B. S., Amherdt, M., Arden, S. R., Malhotra, V., and Rothman, J. E. (1989). Fatty acyl CoA is required for budding of transport vesicles from Golgi cisternae. *Cell* **59,** 95–102.

Pike, L. J., Han, X., Chung, K.-N., and Gross, R. W. (2002). Lipid rafts are enriched in arachidonic acid and plasmenylethanolamine and their composition is independent of caveolin-1 expression: A quantitative electrospray ionization/mass spectrometric analysis. *Biochemistry* **41,** 2075–2088.

Pitto, M., Brunner, J., Ferraretto, A., Ravasi, D., Palestini, P., and Masserini, M. (2000). Use of a photoactivatable GM1 ganglioside analogue to assess lipid distribution in caveolae bilayer. *Glycoconjugate J.* **17,** 215–222.

Pralle, A., Keller, P., Florin, E.-L., Simons, K., and Horber, J. K. H. (2000). Sphingolipid-cholesterol rafts diffuse as small entities in the plasma membrane of mammalian cells. *J. Cell Biol.* **148,** 997–1007.

Puglielli, L., Rigotti, A., Amigo, L., Nunez, L., Greco, A. V., Santos, M. J., and Nervi, F. (1996). Modulation on introhepatic cholesterol trafficking: Evidence by *in vivo* antisense treatment for the involvement of sterol carrier protein-2 in newly synthesized cholesterol transfer into bile. *Biochem. J.* **317,** 681–687.

Resch, K., Loracher, A., Mahler, B., Stoeck, M., and Rode, H. N. (1978). Functional mosaicism of the lymphocyte plasma membrane. Characterization of membrane subfractions obtained by affinity chromatography on concanavalin A-sepharose. *Biochim. Biophys. Acta* **511,** 176–193.

Riddell, D. R., Christie, G., Hussain, I., and Dingwall, C. (2002). Compartmentalization of beta-secretase (Asp2) into low-buoyant density, noncaveolar rafts. *Curr. Biol.* **11,** 1288–1293.

Robenek, M. J., Schlattmann, K., Zimmer, K.-P., Plenz, G., Troyer, D., and Robenek, H. (2004). Cholesterol transporter caveolin-1 transits the lipid bilayer during intracellular cycling. *FASEB J.* **17,** 1940–1942.

Rothberg, K. G., Heuser, J. E., Donzell, W. C., Ying, Y. S., Glenney, J. R., Jr., and Anderson, R. G. W. (1992). Caveolin, a protein component of caveolae membrane coats. *Cell* **68,** 673–682.

Sandvig, K. and van Deurs, B. (1999). Endocytosis and intracellular transport of ricin: Recent discoveries. *FEBS Lett.* **452,** 67–70.

Sapin, C., Colard, O., Delmas, O., Tessier, C., Breton, M., Enouf, V., Chwetzoff, S., Ouanich, J., Cohen, J., Wolf, C., and Trugnan, G. (2002). Rafts promote assembly and atypical targeting of a nonenveloped virus, rotavirus, in Caco-2 cells. *J. Virol.* **76,** 4591–4602.

Scheiffele, P., Roth, M. G., and Simons, K. (1997). Interaction of influenza virus haemagglutinin with sphingolipid-cholesterol membrane domains via its transmembrane domain. *EMBO J.* **16,** 5501–5508.

Schnitzer, J. E. (2001). Caveolae: From basic trafficking mechanisms to targeting transcytosis for tissue-specific drug and gene delivery *in vivo*. *Adv. Drug Delivery Rev.* **49,** 265–280.

Schnitzer, J. E., Oh, P., and McIntosh, D. P. (1996). Role of GTP hydrolysis in fission of caveolae directly from plasma membranes. *Science* **274,** 239–242.

Schroeder, F. (1984). Fluorescent sterols: Probe molecules of membrane structure and function. *Prog. Lipid Res.* **23,** 97–113.

Schroeder, F., Butko, P., Nemecz, G., and Scallen, T. J. (1990a). Interaction of fluorescent delta 5,7,9(11),22-ergostatetraen-3b-ol with sterol carrier protein-2. *J. Biol. Chem.* **265,** 151–157.

Schroeder, F., Fontaine, R. N., and Kinden, D. A. (1982). LM fibroblast plasma membrane subfractionation by affinity chromatography on ConA-sepharose. *Biochim. Biophys. Acta* **690,** 231–242.

Schroeder, F., Frolov, A., Schoer, J., Gallegos, A., Atshaves, B. P., Stolowich, N. J., Scott, A. I., and Kier, A. B. (1998). Intracellular cholesterol binding proteins, cholesterol transport and membrane domains. In: *Intracellular Cholesterol Trafficking* (Chang, T. Y. and Freeman, D. A., Eds.), pp. 213–234. Kluwer Academic Publishers, Boston.

Schroeder, F., Frolov, A., Starodub, O., Russell, W., Atshaves, B. P., Petrescu, A. D., Huang, H., Gallegos, A., McIntosh, A., Tahotna, D., Russell, D., Billheimer, J. T., Baum, C. L., and Kier, A. B. (2000). Pro-sterol carrier protein-2 role of the N-terminal presequence in structure, function, and peroxisomal targeting. *J. Biol. Chem.* **275,** 25547–25555.

Schroeder, F., Frolov, A. A., Murphy, E. J., Atshaves, B. P., Jefferson, J. R., Pu, L., Wood, W. G., Foxworth, W. B., and Kier, A. B. (1996). Recent advances in membrane cholesterol domain dynamics and intracellular cholesterol trafficking. *Proc. Soc. Exp. Biol. Med.* **213,** 150–177.

Schroeder, F., Gallegos, A. M., Atshaves, B. P., McIntosh, A., Petrescu, A. D., Huang, H., Chao, H., Yang, H., Frolov, A., and Kier, A. B. (2001a). Recent advances in membrane microdomains: Rafts, caveolae and intracellular cholesterol trafficking. *Exp. Biol. Med.* **226,** 873–890.

Schroeder, F., Gallegos, A. M., Atshaves, B. P., Storey, S. M., McIntosh, A., Petrescu, A. D., Huang, H., Starodub, O., Chao, H., Yang, H., Frolov, A., and Kier, A. B. (2001b). Recent advances in membrane cholesterol microdomains: Rafts, caveolae, and intracellular cholesterol trafficking. *Exp. Biol. Med.* **226,** 873–890.

Schroeder, F., Jefferson, J. R., Kier, A. B., Knittell, J., Scallen, T. J., Wood, W. G., and Hapala, I. (1991a). Membrane cholesterol dynamics: Cholesterol domains and kinetic pools. *Proc. Soc. Exp. Biol. Med.* **196,** 235–252.

Schroeder, F., Kier, A. B., and Sweet, W. D. (1990b). Role of polyunsaturated fatty acids and lipid peroxidation in LM fibroblast plasma membrane structure. *Arch. Biochem. Biophys.* **276,** 55–64.

Schroeder, F. and Nemecz, G. (1989). Interaction of sphingomyelins and phosphatidylcholines with fluorescent dehydroergosterol. *Biochemistry* **28,** 5992–6000.

Schroeder, F. and Nemecz, G. (1990). Transmembrane cholesterol distribution. In: *Advances in Cholesterol Research* (Esfahami, M. and Swaney, J., Eds.), pp. 47–87. Telford Press, Caldwell, NJ.

Schroeder, F., Nemecz, G., Wood, W. G., Joiner, C., Morrot, G., Ayraut-Jarrier, M., and Devaux, P. F. (1991b). Transmembrane distribution of sterol in the human eurythrocyte. *Biochim. Biophys. Acta* **1066,** 183–192.

Schroeder, F. and Sweet, W. D. (1988). The role of membrane lipid and structure asymmetry on transport systems. In: *Advances in Biotechnology of Membrane Ion Transport* (Jorgensen, P. L. and Verna, R, Eds.), pp. 183–195. Serono Symposia, New York, NY.

Schroeder, F., Zhou, M., Swaggerty, C. L., Atshaves, B. P., Petrescu, A. D., Storey, S., Martin, G. G., Huang, H., Helmkamp, G. M., and Ball, J. M. (2003). Sterol carrier protein-2 functions in phosphatidylinositol transfer and signaling. *Biochemistry* **42,** 3189–3202.

Schroeder, R., London, E., and Brown, D. (1994). Interactions between saturated acyl chains confer detergent resistance on lipids and glycosylphosphatidylinositol (GPI)-anchored proteins: GPI-anchored proteins in liposomes and cells show similar behavior. *Proc. Natl. Acad. Sci. USA* **91,** 12130–12134.

Seedorf, U., Raabe, M., Ellinghaus, P., Kannenberg, F., Fobker, M., Engel, T., Denis, S., Wouters, F., Wirtz, K. W. A., Wanders, R. J. A., Maeda, N., and Assmann, G. (1998). Defective peroxisomal catabolism of branched fatty acyl coenzyme A in mice lacking the sterol carrier protein-2/sterol carrier protein-x gene function. *Genes Dev.* **12,** 1189–1201.

Seedorf, U., Scheek, S., Engel, T., Steif, C., Hinz, H. J., and Assmann, G. (1994). Structure-activity studies of human sterol carrier protein 2. *J. Biol. Chem.* **269,** 2613–2618.

Sheets, E. D., Lee, G. M., Simons, R., and Jacobson, K. (1997). Transient confinement of a glycosylphosphatidylinositol-anchored protein in the plasma membrane. *Biochemistry* **36,** 12449–12458.

Shin, J.-S. and Abraham, S. N. (2002). Caveolae as portals of entry for microbes. *Microbes Infect.* **3,** 755–761.

Sieczkarski, S. B. and Whittaker, G. R. (2002). Dissecting virus entry via endocytosis. *J. Gen. Virol.* **83,** 1535–1545.

Simons, K. and Ikonen, E. (1997). Functional rafts in cell membranes. *Nature* **387,** 569–572.

Singer, S. J. and Nicolson, G. L. (1972). The fluid mosaic model of the structure of cell membranes. *Science* **175,** 720–731.

Smart, E. J., Graf, G. A., McNiven, M. A., Sessa, W. C., Engelman, J. A., Scherer, P. E., Okamoto, T., and Lisanti, M. P. (1999). Caveolins, liquid-ordered domains, and signal transduction. *Mol. Cell. Biol.* **19,** 7289–7304.

Smart, E. J. and van der Westhuyzen, D. R. (1998). Scavenger receptors, caveolae, caveolin, and cholesterol trafficking. In: *Intracellular Cholesterol Trafficking* (Chang, T. Y. and Freeman, D. A., Eds.), pp. 253–272. Kluwer Academic Publishers, Boston.

Smart, E. J., Ying, Y., Mineo, C., and Anderson, R. G. W. (1995). A detergent-free method for purifying caveolae membrane from tissue culture cells. *Proc. Natl. Acad. Sci. USA* **92,** 10404–10408.

Song, K. S., Li, S., Okamoto, T., Quilliam, L. A., Sargiacomo, M., and Lisanti, M. P. (1996). Copurification and direct interaction of ras with caveolin, an integral membrane protein of caveolae microdomains. *J. Biol. Chem.* **271,** 9690–9697.

Stan, R.-V. (2002). Structure and function of caveolae. *Microsc. Res. Tech.* **57,** 350–364.

Stan, R.-V., Roberts, W. G., Predescu, D., Ihida, K., Saucan, L., Ghitescu, L., and Palade, G. E. (1997). Immunoisolation and partial characterization of endothelial plasmalemmal vesicles (caveolae). *Mol. Biol. Cell* **8,** 595–605.

Stefanova, I., Horesjsi, V., Ansotegui, I. J., Knapp, W., and Stockinger, H. (1991). GPI-anchored cell surface molecules complexed to protein tyrosine kinases. *Science* **254,** 1016–1019.

Stolowich, N. J., Frolov, A., Petrescu, A. D., Scott, A. I., Billheimer, J. T., and Schroeder, F. (1999). Holo-sterol carrier protein-2: ^{13}C-NMR investigation of cholesterol and fatty acid binding sites. *J. Biol. Chem.* **274,** 35425–35433.

Stolowich, N. J., Petrescu, A. D., Huang, H., Martin, G., Scott, A. I., and Schroeder, F. (2002). Sterol carrier protein-2: Structure reveals function. *Cell. Mol. Life Sci.* **59,** 193–212.

Sviridov, D. (1999). Intracellular cholesterol trafficking. *Histol. Histophathol.* **14,** 305–319.

Swaggerty, C. L., Huang, H., Lim, W. S., Schroeder, F., and Ball, J. A. (2004). Comparison of SIVmac239 (352–382) and SIVsmmPBjJ41(360–390) enterotoxic synthetic peptides. *Virology* **320,** 243–257.

Sweet, W. D. and Schroeder, F. (1988). Lipid domains and enzyme activity. In: *Advances in Membrane Fluidity: Lipid Domains and the Relationship to Membrane Function* (Aloia, R. C., Cirtain, C. C., and Gordon, L. M., Eds.), pp. 17–42. Alan R. Liss, Inc., New York, NY.

Sweet, W. D., Wood, W. G., and Schroeder, F. (1987). Charged anesthetics selectively alter plasma membrane order. *Biochemistry* **26,** 2828–2835.

Szyperski, T., Scheek, S., Johansson, J., Assmann, G., Seedorf, U., and Wuthrich, K. (1993). NMR determination of the secondary structure and the three-dimensional polypeptide backbone fold of the human sterol carrier protein 2. *FEBS Lett.* **335,** 18–26.

Talbot, W. A., Zheng, L. X., and Lentz, B. R. (1997). Acyl chain unsaturation and vesicle curvature alter outer leaflet packing and promote poly(ethylene glycol)-mediated membrane fusion. *Biochemistry* **36,** 5827–5836.

Taylor, J. A., O'Brien, J. A., and Yeager, M. (1996). The cytoplasmic tail of NSP4, the endoplasmic reticulum-localized non-structural glycoprotein of rotavirus, contains distinct virus binding and coiled coil domains. *EMBO J.* 4469–4476.

Thompson, T. E., Huang, C., and Litman, B. J. (1974). Bilayers and biomembranes: Compositional asymmetries induced by surface curvature. In: *The Cell Surface in Development* (Moscona, A. A, Ed.), pp. 1–16. John Wiley & Sons, Inc., New York, NY.

Tian, P., Ball, J. M., Zeng, D., and Estes, M. (1996). The rotavirus nonstructural glycoprotein NSP4 possesses membrane destabilization activity. *J. Virol.* **70,** 6973–6981.

Tiruppathi, C., Song, W., Bergenfeldt, M., Sass, P., and Malik, A. B. (1997). Gp60 activation mediates albumin transcytosis in endothelial cells by tyrosine kinase dependent pathway. *J. Biol. Chem.* **272,** 25968–25975.

Trigatti, B., Anderson, R. G. W., and Gerber, G. E. (1999). Identification of caveolin-1 as a fatty acid binding protein. *Biochem. Biophys. Res. Commun.* **255,** 34–39.

Trigatti, B. L., Mangroo, D., and Gerber, G. E. (1991). Photoaffinity labeling and fatty acid permeation in 3T3-L1 adipocytes. *J. Biol. Chem.* **266,** 22621–22625.

Uittenbogaard, A., Everson, W. V., Matveev, S. V., and Smart, E. J. (2002). Cholesteryl ester is transported from caveolae to internal membranes as part of a caveolin-annexin II lipid-protein complex. *J. Biol. Chem.* **277,** 4925–4931.

Uittenbogaard, A. and Smart, E. J. (2000). Palmitoylation of caveoin-1 is required for choleserol binding, chaperone complex formation, and rapid transport of cholesterol to caveolae. *J. Biol. Chem.* **275,** 25595–25599.

Uittenbogaard, A., Ying, Y. S., and Smart, E. J. (1998). Characterization of a cytosolic heat-shock protein-caveolin chaperone complex. *J. Biol. Chem.* **273,** 6525–6532.

Vainio, S., Heino, S., Mansson, J. E., Fredman, P., Kuismanen, E., Vaarala, O., and Ikonen, E. (2002). Dynamic association of human insulin receptor with lipid rafts in cells lacking caveolae. *EMBO Rep.* **3,** 95–100.

Waugh, M. G., Lawson, D., and Hsuan, J. J. (1999). Epidermal growth factor receptor activation is localized within low-buoyant density, non-caveolar membrane domains. *Biochem. J.* **337,** 591–597.

Welsh, M. J., Smith, P. L., and Frizzell, R. A. (1982). Chloride secretion by canine tracheal epithelium II. the cellular electrical potential profile. *J. Membr. Biol.* **70,** 227–238.

Wolf, A. A., Fujinaga, Y., and Lencer, W. I. (2002). Uncoupling of the cholera toxin-GM1 ganglioside receptor complex from endocytosis, retrograde Golgi trafficking, and downstream signal transduction by depletion of membrane cholesterol. *J. Biol. Chem.* **277,** 16249–16256.

Wood, W. G., Eckert, G. P., Igbavboa, U., and Mueller, W. E. (2003). Amyloid beta-protein interactions with membranes and cholesterol: Causes or casualties of Alzheimer's disease. *Biochim. Biophys. Acta* **1610,** 281–290.

Wood, W. G., Schroeder, F., Avdulov, N. A., Chochina, S. V., and Igbavboa, U. (1999). Recent advances in brain cholesterol dynamics: Transport, domains, and Alzheimer's disease. *Lipids* **34,** 225–234.

Wood, W. G., Schroeder, F., Hogy, L., Rao, A. M., and Nemecz, G. (1990). Asymmetric distribution of a fluorescent sterol in synaptic plasma membranes: Effects of chronic ethanol consumption. *Biochim. Biophys. Acta* **1025,** 243–246.

Wood, W. G., Schroeder, F., Igbavboa, U., Avdulov, N. A., and Chochina, S. V. (2002). Brain membrane cholesterol domains, aging and amyloid beta-peptides. *Neurobiol. Aging* **23,** 685–694.

Wood, W. G., Schroeder, F., Rao, A. M., Igbavboa, U., and Avdulov, N. A. (1996). Membranes and ethanol: lipid domains and lipid protein interactions. In: *Pharmacological Effects of Ethanol on the Nervous System* (Dietrich, R. and Erwin, V. G., Eds.), pp. 13–27. CRC Press, Boca Raton, FL.

Woodford, J. K., Colles, S. M., Myers-Payne, S., Billheimer, J. T., and Schroeder, F. (1995). Sterol carrier protein-2 stimulates intermembrane sterol transfer by direct membrane interaction. *Chem. Phys. Lipids* **76,** 73–84.

Wustner, D., Herrmann, A., Hao, M., and Maxfield, F. R. (2002). Rapid nonvesicular transport of sterol between the plasma membrane domains of polarized hepatic cells. *J. Biol. Chem.* **277,** 30325–30336.

Yamamoto, R., Kallen, C. B., Babalola, G. O., Rennert, H., Billheimer, J. T., and Strauss, J. F. I. (1991). Cloning and expression of a cDNA encoding human sterol carrier protein 2. *Proc. Natl. Acad. Sci. USA* **88,** 463–467.

Yokoyama, S. (2000). Release of cellular cholesterol: Molecular mechanism for cholesterol homeostasis in cells and in the body. *Biochim. Biophys. Acta* **1529,** 231–244.

Zhou, M., Parr, R. D., Petrescu, A. D., Payne, H. R., Atshaves, B. P., Kier, A. B., Ball, J. M., and Schroeder, F. (2004). Sterol carrier protein-2 directly interacts with caveolin-1 *in vitro* and *in vivo*. *Biochem.* **43,** 7288–7306.

Chapter 2

Caveolae and the Regulation of Cellular Cholesterol Homeostasis

William V. Everson and Eric J. Smart

University of Kentucky, Sanders Brown Center on Aging, Lexington, Kentucky 40536-0230

Advances in Molecular and Cell Biology, Vol. 36

ISBN: 0-444-51500-3
DOI: 10.1016/S0079-6603(05)36002-8

I. INTRODUCTION

Caveolae and lipid rafts are specialized domains of membranes found in the majority of cells. These specialized domains play a central role in signal transduction, protein–protein interactions, and dynamic trafficking in cells (Smart *et al.*, 1999; Parton and Richards, 2003). Caveolae were initially described by their EM morphology as invaginations in the plasma membrane with a unique flask-shaped structure associated with transcytosis (Bruns and Palade, 1968a,b). Identification and cloning of a caveola coat-associated protein, caveolin-1 (Glenney and Soppet, 1992; Rothberg *et al.*, 1992), initiated a new era of investigation of these structures that expanded over the next several years to the identification of caveolae and/or lipid rafts in most cell types. Caveolae and lipid rafts are defined as membrane domains enriched in cholesterol and sphingolipid. These were formerly identified as liquid-ordered domains (Schroeder *et al.*, 1996, 1998), but their distribution in cells was thought to be quite limited, in comparison to current understanding that lipid rafts and/or caveolae are present in most cells. In this chapter, lipid rafts are referred to as regions of membranes enriched in cholesterol and sphingolipids, whether these domains are localized in the plasma membrane or within intracellular membranes. The term caveolae is used to refer to a specialized subset of lipid rafts identified by the presence of the marker protein caveolin-1. This chapter examines the organization of lipid rafts and caveolae and their associated trafficking pathways, and the role cholesterol plays in modifying the distribution, stability, and functions of these domains.

II. ORGANIZATION

Lipid raft domains are heterogeneous in different cell types and even within a single cell type with respect to their stability, size, percentage area of the cell surface, and functional roles (Smart *et al.*, 1999; Fernandez *et al.*, 2002) (Fig. 1). Additionally, studies using polarized cells have shown that functionally distinct lipid raft domains are present in apical and lateral or basolateral membranes (Parton *et al.*, 1994; Oliferenko *et al.*, 1999; Smart *et al.*, 1999; Parton and Richards, 2003) and that these different surface domains are linked to distinct vesicle trafficking pathways (Fig. 1). Some domains appear to be static (Mundy *et al.*, 2002), whereas other domains are capable of disappearing from the cell surface and reforming in a very dynamic fashion (Parton, 1994; Parton *et al.*, 1994; Smart *et al.*, 1994).

III. ENDOTHELIAL TRANSCYTOSIS

The earliest function associated with caveolae was transcytosis (Bruns and Palade, 1968a,b). As endothelial cells in culture reach confluence, caveolin becomes highly enriched and cholesterol accumulates at lateral membranes (Corvera *et al.*, 2000). Caveolae have been identified as the site from which endosomes bud to form the extensive vesiculo-vacuolar organelle (VVO) in the venular endothelium, which

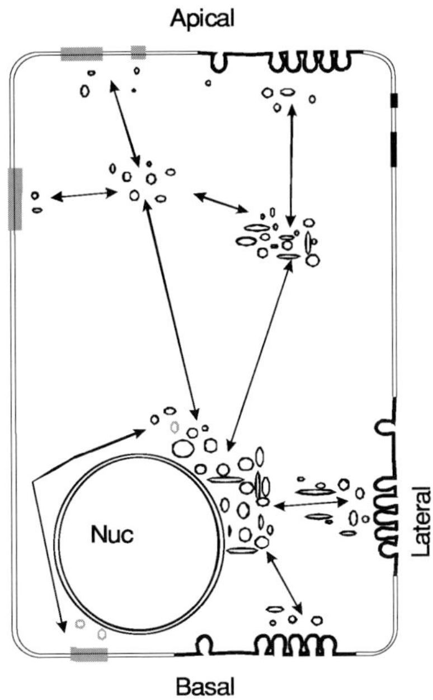

Figure 1. Diversity of caveolae and lipid rafts trafficking pathways in cells. Caveolae and lipid rafts have been shown to exist in small domains, large domains, and extensive aggregates within the plasma membrane, and they are distributed in apical, lateral, and basolateral membranes. This figure shows the diversity of lipid rafts (heavy grey line) and caveolae (heavy black line) in different regions of the cell and some of the vesicle-trafficking pathways that have been associated with both lipid rafts and caveolae. Within the cell, the number of intermediate trafficking steps in caveolae and lipid raft endocytosis and exocytosis has not been examined in a comprehensive fashion.

constitutes the major organelle regulating endothelial transcytosis (Fig. 2). The VVO is localized to a region of the cytosol adjacent to the lateral membrane at cell–cell junctions and occupies as much as 18% of the cytosol volume (Dvorak *et al.*, 1996; Dvorak and Feng, 2001; Feng *et al.*, 2002). This network has been shown to form a convoluted, connected passageway through which leukocytes may cross the endothelial barrier during extravasation from the blood (Dvorak *et al.*, 1996; Dvorak and Feng, 2001; Feng *et al.*, 2002). The formation of this network and its dual role in transcytosis and extravasation suggest that caveolae are essential to one of the major functions of the endothelium as a barrier between the blood and the interstitial space.

IV. PROTEINS ASSOCIATED WITH CAVEOLAE AND LIPID RAFTS CONFER STABILITY

Many proteins that interact with lipid rafts increase the stability of lipid rafts. One of the most common modifications found in proteins associated with lipid rafts is acylation (Shaul *et al.*, 1996; Uittenbogaard and Smart, 2000; McCabe and

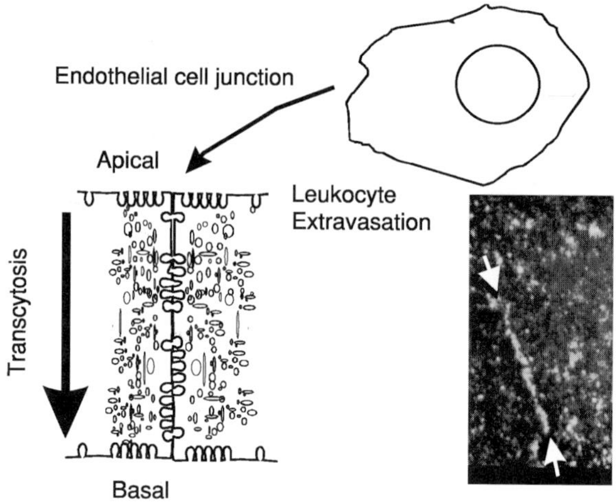

Figure 2. Association of endothelial transcytosis and leukocyte extravasation with an extensive tubular vesicular network. On the left is a representation of the region where two endothelial cells are in contact. In this region, an extensive tubular-vesicular network develops that is enriched in caveolin-1, annexin II, cholesterol, and several raft-associated proteins, as described in the text. This structure, which has been termed the vesicular vacuolar organelle (VVO), can regulate transcytosis and has been associated with leukocyte extravasation. The inset on the right shows a region of human umbilical vein endothelial cells cultured to confluence, immunostained with antibody against caveolin-1, and examined by laser scanning confocal microscopy. The arrowheads show a region of the lateral border where caveolin has developed into a bright line of vesicles, corresponding to the arrangement of caveolin-immunopositive vesicles and tubular vesicular structures that comprise the VVO.

Berthiaume, 2001; Uittenbogaard *et al.*, 2002). Caveolin-1 is acylated at three sites and contains a 21-amino-acid hydrophobic domains that is thought to insert into the membrane (Smart *et al.*, 1999; Fernandez *et al.*, 2002). Many proteins found in caveolae also contain protein interaction domains (Smart *et al.*, 1999; Fernandez *et al.*, 2002) or the presence of domains that allow formation of stable protein complexes that form just beneath the membrane layer (Fig. 3). Actin has been found to be associated with caveolae and lipid rafts in subcellular fractions (Chang *et al.*, 1994; Lisanti *et al.*, 1994; Smart *et al.*, 1999), and in studies employing electron microscopy, it has been shown to form a plexus that may confer added stability or allow formation of larger aggregates of lipid rafts (Kobayashi and Robinson, 1991; Kobayashi *et al.*, 1998; Robinson *et al.*, 1999; Mundy *et al.*, 2002; Parton and Richards, 2003).

V. ANNEXIN II IS A PROMINENT PROTEIN IN CAVEOLAE/LIPID RAFTS AND LIPID RAFT VESICLES

Annexin II was initially identified as a protein associated with trafficking to the plasma membrane (PM) in chromaffin cells, a model cell system used to investigate exocytosis (Creutz *et al.*, 1983, 1987). Annexin II, along with p11, an 11-kDa protein in

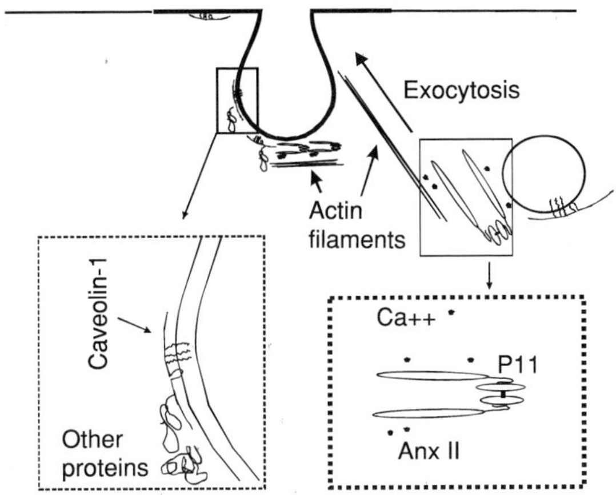

Figure 3. Proteins associated with caveolae confer stability. Left insert: Caveolin-1 interacts with caveolae by insertion of a hydrophobic domain into the membrane and insertion of three acyl chains bound to cysteines. In addition, caveolin can form multimers and interact through several defined domains with other caveolae-associated proteins, which can generate signaling platforms. Right insert: Annexin II interacts with p11/S100A10 to form a heterotetramer, which gives it bifunctional capability, allowing one of the monomeric subunits of annexin II to interact with actin or phospholipid membranes in a calcium-dependent fashion and the other monomeric subunit to interact with membranes in a cholesterol- or calcium-dependent fashion. Annexin II can therefore bridge between vesicles and actin filaments in microfilament-mediated trafficking steps as shown. Annexin II also can form larger aggregates at the cell surface and bring stability to caveolae and clusters of caveolae.

the s100 family now classified as S100A10, was shown shortly after its discovery to form bridges between vesicles and actin filaments and between vesicles and the fusion domain in the plasma membrane (Gerke and Moss, 2002). One report showed that annexin II can be released from chromaffin vesicles by addition of methyl-beta-cyclodextrin, suggesting that these vesicle membranes also contain lipid raft domains (Ayala-Sanmartin *et al.*, 2001). This suggests that annexin II may play a key in the regulation of lipid raft membrane transport to the cell surface. In addition, annexin II may play a role in caveolae stability by providing a linkage between lipid rafts and the actin cytoskeletal meshwork (Fig. 3). Annexin II was the most abundant protein found in several tissues in which actin-binding proteins were purified based on calcium-dependent association with actin filaments aggregated by addition of excess calcium (Shadle *et al.*, 1985). Intriguingly, annexin II has been identified in most studies using subcellular fractionation (Chang *et al.*, 1994; Lisanti *et al.*, 1994), including cells that lack caveolin-1 and thus only contain lipid rafts (Harder and Gerke, 1994). Reports have shown that annexin II interacts with membranes in either a calcium- or cholesterol-dependent fashion (Ayala-Sanmartin, 2001; Ayala-Sanmartin *et al.*, 2001; Zeuschner *et al.*, 2001) and that annexin II/p11, caveolin-1, and cholesterol accumulate together at lateral membranes as endothelial cells become confluent (Corvera *et al.*, 2000). The mechanistic basis on which annexin II shifts between a calcium-dependent interaction with phospholipids to a cholesterol-dependent, calcium-independent interaction with lipid rafts has not yet been defined mechanistically.

VI. FOCAL ADHESIONS ARE SPECIALIZED CAVEOLAE

Focal adhesions are cholesterol and sphingolipid rich, and contain caveolin (Ushio-Fukai *et al.*, 2001). Focal adhesions represent sites at which stress fibers composed of actin attach indirectly to the inner leaflet, through the proteins vinculin, paxillin, and talin. Cells attach to the extracellular matrix through heterodimers of specific integrins (Giannone *et al.*, 2003; Mostafavi-Pour *et al.*, 2003). In smooth muscle, angiotensin II stabilizes focal adhesions, through activation of src accompanied by phosphorylation of caveolin-1 and paxillin in the focal adhesions (Ushio-Fukai *et al.*, 2001). Depletion of cholesterol disrupts the localization and phosphorylation of both of these proteins in focal adhesions.

Caveolin-1 can form oligomers (Fernandez *et al.*, 2002) and is able to interact with a number of proteins to form a network of interacting proteins that confers stability to caveolae (Fig. 3). Annexin II forms heterotetramers with p11/S100A10, which is stabilized by formation of a cystine bond between the p11 subunits and can also form a triskelion architecture on the cell surface in large aggregates (Gerke and Moss, 2002). In its heterotetrameric form, one 36-kDa subunit can interact in a cholesterol-dependent fashion with the membrane, while the other 36-kDa subunit interacts in a calcium-dependent fashion with actin filaments. This bifunctional interaction between actin filaments and membranes is important in the transport of vesicles along actin filaments to the cell surface during exocytosis and in conferring stability to aggregates of caveolae through interaction with the actin plexus that forms beneath the membrane (Fig. 3). This interaction is quite evident in the pathophysiology of infection of cells by enteropathogenic bacteria. As these bacteria, which do not invade cells, attach to the cell surface, lipid rafts coalesce along with annexin II, and a large actin plexus that confers stability to these structures forms (Zeuschner *et al.*, 2001).

VII. CHOLESTEROL AND CAVEOLIN INTERNALIZATION

Alteration in cholesterol profoundly affects caveolin and caveolae. Over a decade ago, experiments with bacterial cholesterol oxidase uncovered the association of caveolin with a rapid, dynamic trafficking pathway initiated from caveolae (Smart *et al.*, 1994; Fig. 4). In cultured human fibroblast cells, addition of bacterial cholesterol oxidase induced a rapid translocation of caveolin-1 out of caveolae to vesicles in the endoplasmic reticulum Golgi intermediate compartment (ER-GIC), a region between the Golgi and the ER (Smart *et al.*, 1994). Removal of cholesterol oxidase led to a loss of the modified cholesterol and restoration of normal cholesterol amounts in caveolae over a period of about 90 min. This was accompanied by a return of caveolin to

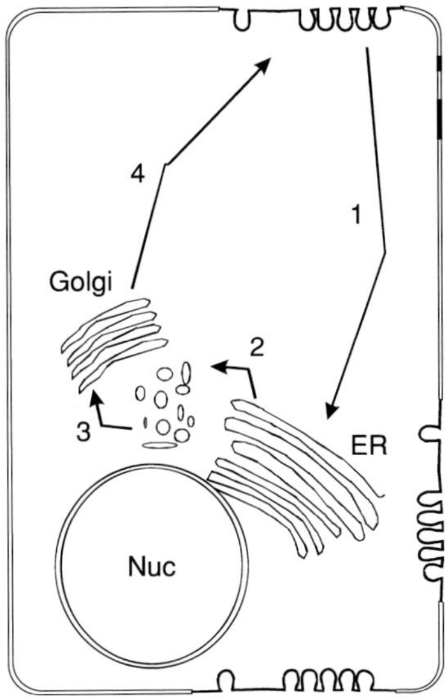

Figure 4. Cholesterol and caveolin trafficking in fibroblasts. Caveolin cycles through four trafficking steps in response to alterations in caveolae cholesterol amounts. The four steps are (1) movement of caveolae to the ER for internalization of caveolin-1 and cholesterol, (2) transfer of caveolin/cholesterol between the ER and the ER-GIC, (3) movement of caveolin/cholesterol from the ER-GIC to the Golgi, and (4) recycling of caveolin/cholesterol from the Golgi to cell surface caveolae.

caveolae. By a combination of temperature shifting and inhibitors, the sequential sites to which caveolin translocate were revealed. Initially, caveolin-1 rapidly translocated to the rough ER, then slowly, over 15–20 min, translocated to the ER-GIC, to the Golgi, and then returned to surface caveolae. These studies were among the first to reveal a caveolin-trafficking cycle between cell surface caveolae and the ER/Golgi membranes, for transport of both caveolin and cholesterol.

VIII. CAVEOLIN CHAPERONE COMPLEXES, CHOLESTEROL TRAFFICKING, AND CAVEOLAE ORGANIZATION

Caveolae trafficking was further expanded by the discovery of non-vesicle trafficking pathways that transport cholesterol between caveolae and intracellular sites (Fig. 5). These pathways link cholesterol transfer between lipoproteins and caveolae, to specific receptors and to two different chaperone complexes that ferry cholesterol between the cell surface and intracellular organelles. The first chaperone complex that was discovered ferries newly synthesized cholesterol from the ER through the cytosol to the plasma membrane at caveolae (Uittenbogaard *et al.*, 1998). This chaperone complex was isolated from the cytosol, distinct from any caveola membrane, plasma membrane, intracellular membrane, or vesicle fraction. The efflux chaperone complex was shown to be composed of cholesterol, caveolin-1, and three additional proteins,

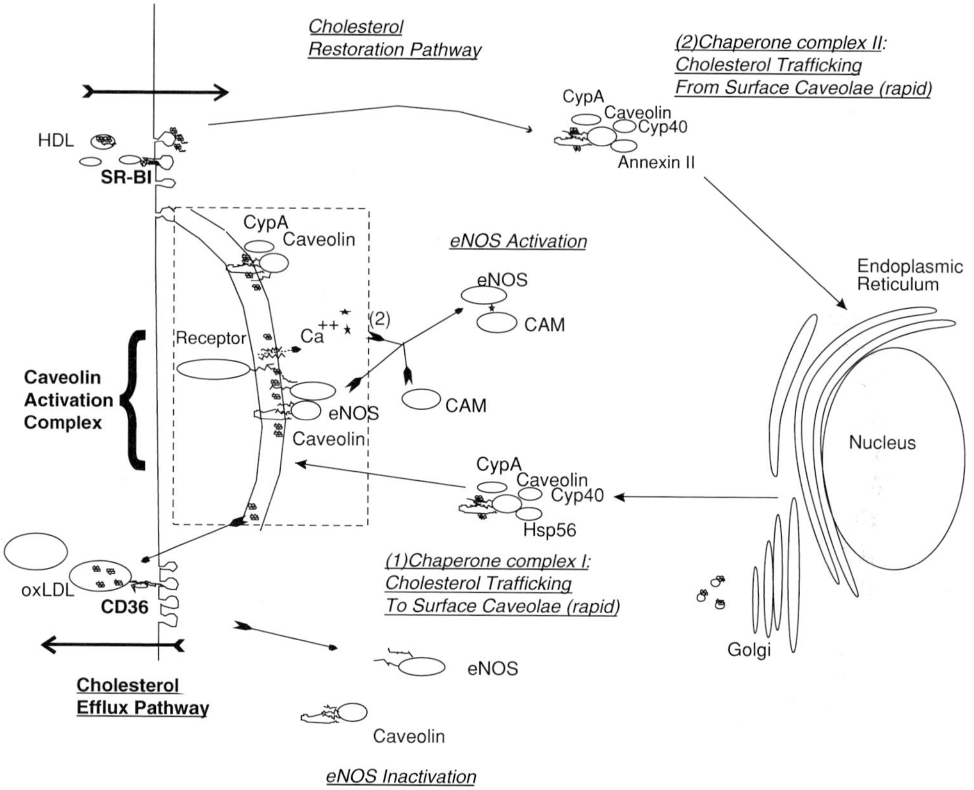

Figure 5. Caveolin chaperone complexes, scavenger receptors, caveolae-cholesterol homeostasis, and eNOS activation. In the trafficking steps defined in Figure 4, step 1 and step 4 have been demonstrated to be regulated by two different chaperone complexes. Chaperone complex I is responsible for transport of newly synthesized cholesterol from the ER to surface caveolae. This complex consists of the four proteins shown, CypA, caveolin-1, Cyp40, and Hsp56. Chaperone complex II transports newly delivered cholesteryl ester from cell surface caveolae to the ER. Chaperone complex II consists of CypA, caveolin-1, Cyp40, and annexin II. The amount of cholesterol in caveolae is critical for the maintenance of a caveolin activation complex that links acetylcholine receptor to the enzyme endothelial nitric oxide synthase (eNOS). When cholesterol homeostasis is maintained, acetylcholine binding to its receptor triggers changes that lead to a localized calcium influx, binding of calmodulin (CAM) to eNOS, and translocation of active eNOS into the cell. When cholesterol becomes depleted, both caveolin and eNOS internalize, while the acetylcholine receptor moves out of caveolae and becomes dispersed throughout the plasma membrane. In this condition, acetylcholine binding to its receptor does not trigger activation of eNOS. The cholesterol efflux pathway links CD36 to oxLDL, whereas the cholesterol restoration pathway links SR-BI to HDL, as described in the text. The amount of cholesterol in caveolae necessary to keep the caveolin activation complex linked in caveolae is determined by the balance between the efflux of cholesterol and the influx of cholesterol.

cyclophilin A (cypA), cyclophilin 40 (cyp40), and heat shock protein 56 (hsp56). The activity of this complex could be inhibited by compounds that interact with cyclophilins (cyclosporin) or bind to and inhibit the function of hsp56 (rapamycin). Either of these inhibitors prevented the formation of the chaperone complex and blocked cholesterol movement from the Golgi to PM caveolae. The delivery of cholesterol via this complex allows cholesterol to efflux to oxidized low density lipoprotein (LDL) in a CD36-dependent fashion, causing loss of caveolae cholesterol (Kincer *et al.*, 2002). This complex and the interaction of scavenger receptors with cholesterol efflux and cholesterol homeostasis in caveolae have been shown to have an important role in the regulation of caveolae function by an effect on caveolae cholesterol homeostasis, described in an earlier review article (Everson and Smart, 2001). A second chaperone complex has been identified; it consists of four proteins, caveolin-1, cyclophilin A, cyclophilin 40, and annexin II. High-density lipoprotein (HDL) can deliver cholesteryl ester to caveolae via Scavenger receptor, class B, type I (SR-BI). Cholesteryl ester is then transported via this complex into the cell, where it can be converted to cholesterol and transported back to the cell surface to caveolae. These two complexes regulate cholesterol uptake and delivery from caveolae in fibroblasts and other caveolin-1-containing cells, but not in liver, which lacks caveolin-1.

IX. CAVEOLIN ACYLATION AND CHAPERONE COMPLEX-MEDIATED CHOLESTEROL TRAFFICKING

Interestingly, different sites and modifications of caveolin-1 are important in regulation of the two chaperone complexes that control caveolae cholesterol homeostasis (Everson and Smart, 2001). Caveolin is acylated at three cysteines, and site-directed mutagenesis studies have revealed that two of the sites, Cys 143 and Cys 156, are required for formation and activity of the chaperone complex that delivers cholesterol from the Golgi to caveolae (Uittenbogaard and Smart, 2000). The third acylation, at Cys 133, is essential for the formation and activity of the chaperone complex that internalizes cholesterol ester into cells (Uittenbogaard *et al.*, 2002). Although the role of acylation sites in the chaperone complexes is clear, the role of these acylation sites in vesicle trafficking is not fully understood. The mechanisms and roles of other site-specific modifications that regulate these complex trafficking pathways and that may relate to routing to specific intracellular domains within the ER or the Golgi are fruitful areas for further exploration.

X. CHOLESTEROL TRANSPORT IN SPECIALIZED CELLS

Some cells have a major role in the overall regulation of systemic cholesterol levels. One of the critical tissues in cholesterol homeostasis in the whole animal is the intestine. A third chaperone complex that is involved in the net uptake and transport of cholesterol in the intestine has been identified (Smart *et al.*, 2004). Annexin II and caveolin-1 form a chaperone complex that binds cholesterol in the intestine. Ablation of either annexin II or caveolin-1 in the zebrafish leads to the loss of the intestinal uptake of cholesterol. In the intestine of the zebrafish, caveolin and annexin II do not resolve at their normal molecular weights, but resolve as a complex at 55–58 kDa on sodium dodecyl sulfate (SDS) gels. This band, after digestion and analysis, yielded several peptides corresponding to regions of each individual protein, and its size was consistent with that of a heterodimeric complex that is stable to heat, SDS, and reducing agents. A band that corresponds to this heterodimer was also found in the mouse intestine, but not in the aorta, though in the mouse, both annexin II and caveolin-1 are present in monomeric forms as well. The drug ezetimibe has been shown to affect cholesterol levels in the circulation by disrupting intestinal cholesterol transport. Surprisingly, ezetimibe disrupts this novel annexin II–caveolin-1 chaperone complex.

XI. CAVEOLIN VESICLES CONNECT THE SURFACE TO A NETWORK OF INTRACELLULAR TUBULES AND VESICLES

VIP21 was discovered as a protein associated with vesicles trafficking to the apical membrane in epithelial cells (A431 and MDCK cells). Cells treated with antibody to induce alkaline phosphatase clustering in caveolae caused VIP-21/caveolin-1 to internalize via tubular vesicles to a region adjacent to the nucleus (Parton *et al.*, 1994). The extent of loss from the surface and translocation of caveolin-1 into cells was accelerated by addition of okadaic acid, an inhibitor of protein phosphatases, or by addition of hypertonic media. A combination of both okadaic acid with hypertonic media led to almost complete internalization of caveolin-1. The vesicles were found in large clusters near the nucleus but distinct from the Golgi and ER. Internalized caveolin-1 was found to be associated with numerous types of intracellular structures, including tubules, vesicles, and aggregates of tubular-vesicular structures in a peri-nuclear region of the cytosol.

Many additional studies have shown that lipid rafts or caveolae can rapidly redistribute between the surface and the interior of the cell in similar tubular vesicular structures. For example, a study by Kobayashi *et al.* (Kobayashi *et al.*, 1998) showed that alkaline phosphatase, a glycosylphosphatidylinositol (GPI) anchored protein and thus a marker of lipid rafts, and superoxide were activated and co-localized within a complex network of tubular vesicular structures that developed in response to activation of neutrophils by phorbol ester. This also revealed that lipid rafts move rapidly from intracellular vesicles to the surface and from the surface to intracellular tubules and to lipid droplets in response to different stimuli.

XII. ANNEXIN II AND LIPID RAFT ENDOCYTOSIS AND EXOCYTOSIS

Annexin II in conjunction with p11 has been associated with endocytosis and exocytosis for over two decades (Gerke and Moss, 2002). Annexin II was shown to regulate the non-ATP-dependent fusion of vesicles to target membranes during endocytosis (Mayorga *et al.*, 1994). Annexin II and p11 have also been shown to associate with the CD44 receptor, which has been localized to lipid rafts distinct from

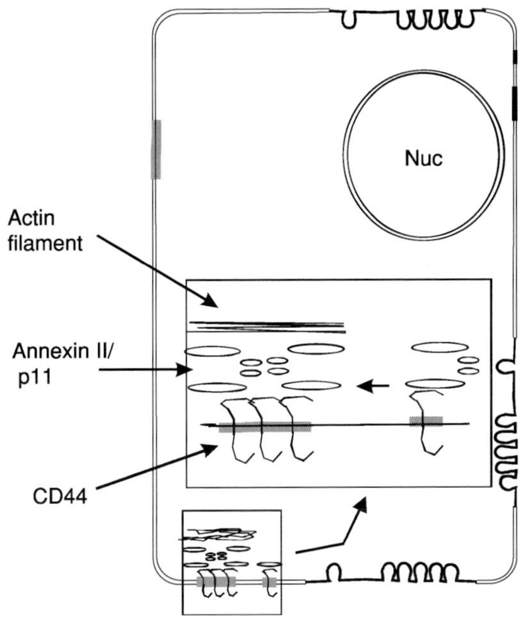

Figure 6. Annexin II and p11 interact to stabilize CD44 in lipid rafts. CD44 is a transmembrane receptor found in lipid rafts. This receptor has been shown to be stabilized by interaction of annexin II/p11 with lipid rafts and further stabilized as annexin II/p11 forms interactions with an actin filament plexus that forms as CD44 clusters. The inset shows the arrangement of CD44 in isolated rafts and the development of a more stable structure as these rafts cluster, and annexin II/p11 forms aggregates on the membrane and provides a bifunctional linkage to the underlying cytoskeleton.

caveolae in the basolateral membrane (Oliferenko *et al.*, 1999) (Fig. 6). Antibody clustering of CD44 induces clustering of annexin II and formation of an actin plexus, mediated by annexin II (Oliferenko *et al.*, 1999). Disruption of cholesterol caused both CD44 and annexin II to dissociate and disperse uniformly throughout the membrane. Disturbance of the normal distribution of annexin II and p11 by a mutant that causes it to aggregate was accompanied by redistribution to the aggregate of both CD44 and actin. Thus, annexin II and p11 are implicated in the localization of CD44 to lipid rafts.

As noted above, very early studies in the field showed that annexin II is associated with caveolae, and much of the work on the role of annexin II in endocytosis and exocytosis in the decade preceding the discovery of caveolin-1 takes on new meaning when considering its prominent association with caveolae and lipid rafts. Many aspects of its mechanistic association with signal transduction and trafficking are becoming clear as its role as an endocytic trafficking protein involved in the regulation of lipid raft signaling and trafficking is considered. Several other annexins co-purify in lipid rafts by subcellular fractionation, and most of the mammalian annexins were found to associate with chromaffin granules in a calcium-dependent fashion (Creutz *et al.*, 1983, 1987). These other annexins may act synergistically or independently to contribute to the stability of lipid rafts, to serve roles in the organization of signaling molecules at the membrane, to aid in signal propagation, or to regulate lipid raft vesicle trafficking.

XIII. PM CHOLESTEROL CONTENT ALTERS CAVEOLA VESICLE INTERNALIZATION

Cholera toxin, which binds to ganglioside M1 (GM1), a membrane component of caveolae/lipid rafts (Dupree *et al.*, 1993), has been used extensively to examine the endocytosis of caveolae or lipid raft components. For example, Orlandin and Fishman examined the role of cholesterol in caveolae/lipid raft-mediated uptake of cholera toxin (Orlandi and Fishman, 1998). They showed that in both cells that contain caveolae and cells lacking caveolin-1 but containing lipid rafts, cholera toxin was taken up via caveolin-1-positive and GM1-positive lipid raft vesicles, respectively, and transported to the ER. This trafficking pathway could be inhibited by removal of cholesterol from surface caveolae or lipid rafts by the addition of filipin. A study using green fluorescent protein (GFP) constructs of several different glycosyl phosphatidylinositol (GPI)-anchored proteins revealed that these lipid raft-associated proteins cycle back and forth between the PM and the Golgi (Nichols *et al.*, 2001) in a cholesterol-dependent fashion. By the use of photobleaching, several different lipid raft localized receptors were shown to rapidly cycle between the PM and the Golgi. However, the intracellular distribution differed, with some trafficking through endosomes containing transferrin and others trafficking to the Golgi in endosomes that are independent of the clathrin-associated endocytic pathways.

XIV. MICROFILAMENTS AND MICROTUBULES REGULATE DISTINCT STEPS IN THE CAVEOLAE TRAFFICKING

Microtubules and microfilaments serve as the "railways" for communication and vesicle transport in distinct segments of the overall trafficking cycles. Nocodazole, an inhibitor of microtubule formation, caused a block in Golgi to PM caveolin-vesicle trafficking (Conrad *et al.*, 1995). When cholesterol oxidase was added in the presence of nocodazole, caveolin internalized to the ER but did not move in the segment between the ER and the Golgi. When nocodazole was added after cholesterol oxidase had caused caveolin to accumulate in the Golgi, it did not disrupt movement from the Golgi to the PM. Thus, the segment between the ER and the Golgi was shown to depend on microtubules. Studies with caveolin fused to GFP have shown that both actin filaments and microtubules are important in the organization and distribution of caveolin among the surface, intracellular vesicles, and intracellular sites (Mundy *et al.*, 2002).

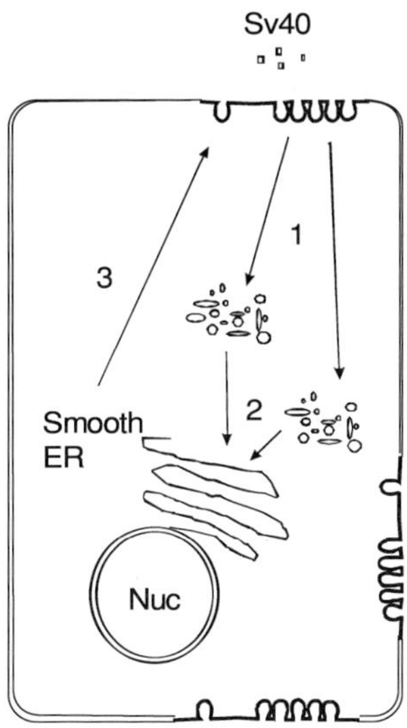

Figure 7. SV40 and trafficking to the caveosome and smooth ER. This figure shows the vesicle-mediated pathway by which SV40 is taken up from caveolae to the caveosome and smooth ER, as discussed in the text. The steps in SV40 trafficking are (1) uptake via caveolae vesicles to the caveosome, (2) budding of vesicles that transport SV40 to the smooth ER, and (3) return of SV40 and release at the surface for invasion of surrounding cells.

XV. UPTAKE OF SV40 AND TRAFFICKING TO THE CAVEOSOME AND SMOOTH ER

An elegant study by Pelkmans *et al.* examined the uptake of SV40 (Pelkmans *et al.*, 2001). In this study, a vesicle-mediated pathway between the cell surface and the ER was defined (Fig. 7). SV40 is a virus that requires endocytosis within the mammalian cell to replicate and returns to the surface for release and infection of surrounding cells. SV40-GFP was taken up into clusters of vesicles that formed stable structures, termed caveosomes, and then was slowly transported to the smooth ER. These clusters of vesicles were not limited to a perinuclear region but were distributed throughout the cytosol. Intriguingly, these caveosomes remained stable for several hours, and from them, nocodazole-sensitive vesicles formed, which transited to the smooth ER where SV40 accumulated.

XVI. AUTOCRINE MOTILITY FACTOR TRAFFICKS VIA CAVEOLA VESICLES TO TWO REGIONS OF THE ER

Upon binding of autocrine motility factor (AMF) to its receptor, which is localized in caveolae (Benlimame *et al.*, 1998), the receptor internalizes via a vesicle pathway to the smooth ER. Within the ER, AMF-R localizes to a region contiguous with membranes of the mitochondrial reticulum (Wang *et al.*, 2000). This region of these two organelles is critical in the coordinate response of the ER and the mitochondria to calcium fluxes. A follow-up study on AMF-R endocytosis compared to cholera toxin (CTX) endocytosis revealed the existence of two vesicle pathways to the ER that are distinct from the caveosome pathway. CTX connects to the ER via retrograde trafficking in caveola vesicles to the Golgi, whereas AMF-R trafficks via caveola vesicles to an endosome that is distinct from the caveosome in two respects: it is not inhibited by nocodazole, and the net transport to the smooth ER occurs within 5 min (Wang *et al.*, 2000).

XVII. ORIENTATION OF CHOLESTEROL AND PROTEINS IN THE LIPID RAFT BILAYER

An important aspect of caveolae structure and cholesterol trafficking is the orientation of lipid components and proteins in the membrane. Efficient cholesterol uptake and efflux require several steps that are not well understood. Uptake of cholesteryl ester from HDL requires the following: (1) binding of a lipoprotein to a specific receptor, (2) transfer of cholesteryl ester from the lipoprotein into the outer leaflet, (3) transfer from the outer leaflet to the inner leaflet, (4) transfer of cholesteryl ester from the inner leaflet to the chaperone complex, (5) transport of the chaperone complex to the ER, (6) delivery of cholesteryl ester to the ER cytosol membrane, and (7) transfer of cholesteryl ester from the cytosol membrane of the ER to the lumenal membrane. One of the most intriguing findings in the initial studies on caveolin-1-mediated cholesterol trafficking was the observation, based on immunogold localization in EM studies, that caveolin-1 flipped across the membrane to the interior of the ER and the inner membrane of the Golgi (Smart *et al.*, 1994; Conrad *et al.*, 1995). This event requires that cholesterol and caveolin associate with the cytosolic face of the membrane at the cell surface, but upon reaching the Golgi or ER membrane, both caveolin-1 and cholesterol were found localized to the intralumenal membrane, indicating they had both flipped across the membrane. Upon removal of cholesterol oxidase, caveolin expression and cholesterol content of caveolae returned to normal. During transport to cell surface caveolae, reorientation to the proper leaflet requires that caveolin-1 flips back across the membrane to the inner cytosol leaflet. For this to occur, another series of transfers across the membrane is required. It appears that as caveolin and cholesterol are transported to the surface, both caveolin and cholesterol flip across the membrane, which requires the presence of a mechanism to allow caveolin-1 to once again flip across the membrane, as it is found localized almost entirely associated with the inner leaflet. A recent report has shown that caveolin-1 is tightly associated with cholesterol during cholesterol transfer from the outer to the inner leaflet of the plasma membrane, suggesting that caveolin might be responsible for this event (Robenek *et al.*, 2003). These aspects of caveolin-1 and cholesterol trafficking within the membrane are intriguing and have yet to be fully explored.

XVIII. CHOLESTEROL TRAFFICKING AND CAVEOLAE PHYSIOLOGY: eNOS SIGNALING IN VASODILATION

The maintenance of caveolae cholesterol homeostasis by the activity of two chaperone complexes described above has been shown to be critical in the physiological regulation of vasodilation and hypertension (Kincer *et al.*, 2002). In cell culture models, the interaction between HDL and LDL and the role of SR-BI and CD36 have been clearly established. Oxidized LDL interacts with CD36 to cause cholesterol efflux from caveolae, and HDL binding to SR-BI allows uptake of cholesteryl ester and repletion of caveolae cholesterol. The coupling of endothelial nitric oxide synthase (eNOS) to the acetylcholine receptor in caveolae allows acetylcholine to induce eNOS activity leading to vasodilation, but when cholesterol is depleted, eNOS uncouples from the caveolae activation complex and acetylcholine does not activate eNOS or cause vasodilation (Everson and Smart, 2001) (Fig. 5). The relevance of this set of pathways to eNOS activation *in vivo* has recently been demonstrated by studies using apoE and CD36-deficient mice fed either a chow diet or a high-fat diet (Kincer *et al.*, 2002). ApoE null mice fed a high-fat diet became hypercholesteremic and did not respond to acetylcholine with a decrease in blood pressure, in contrast to apoE null mice fed a normal diet, which did not become hypercholesteremic. Caveolae isolated from *in vivo* vessels did not contain eNOS and were depleted of cholesterol, indicating that hypercholesteremia had caused cholesterol efflux from caveolae accompanied by disruption of the caveolae activation complex to eNOS. Age-matched apoE/CD36 null mice fed a chow or high-fat diet responded to acetylcholine with a decrease in blood pressure, indicating that loss of CD36 prevented depletion of caveolae cholesterol. Further analysis demonstrated that the plasma low-density lipoprotein fraction was responsible for the depletion of cholesterol and loss of eNOS-mediated, acetylcholine-induced decrease in blood pressure. In mice with both apoE and CD36 ablated, the endothelial cells lacking CD36 did not transfer cholesterol out of caveolae; thus even under conditions of hypercholesteremia, there was no loss of coupling of the caveolin receptor complex to eNOS, vasodilation in response to acetylcholine was maintained, and the mice did not become hypertensive (Kincer *et al.*, 2002).

XIX. "LIQUID ORDERED DOMAIN" FLUIDITY AND eNOS SIGNALING

In addition to proteins that control caveolae organization and signaling by regulating the content of cholesterol, signaling can be activated or inhibited by changing the stability of the lipid components alone. There are hydrophobic interactions that confer stability to the membrane component of lipid rafts, e.g., to these "liquid ordered domains." By use of fluorescent tagged lipids that intercalate into liquid ordered domains, the stability of these domains has been measured and shown to respond to compounds such as ethanol and estrogen (Wood and Schroeder, 1988; Wood *et al.*, 1989; Wang *et al.*, 2004) that interact directly with these lipid structures. Changes in the stability are referred to as increases or decreases in membrane fluidity, reflecting the ability of molecules to move more or less freely within these domains (Wood and Schroeder, 1988; Wood *et al.*, 1989). Changes in the fluidity can alter the activity of proteins that depend on protein–lipid interactions, such as the activity of transmembrane channels, or the coupling or inhibition of activities that are dependent on protein–protein interactions, such as the caveolin activation complex that controls acetylcholine-mediated vasodilation.

Ethanol has a quite dramatic effect on the membrane fluidity of liquid ordered domains (lipid rafts) at physiological concentrations (Wood and Schroeder, 1988; Wood *et al.*, 1989). Ethanol activates eNOS at physiological concentrations in cultured endothelial cells and in the peripheral circulation *in vivo* (Davda *et al.*, 1993; Greenberg *et al.*, 1993), suggesting that a portion of its effect may be mediated by the direct biophysical effect it has on membrane fluidity. Estrogen induces a rapid, non-genomic activation of eNOS in cells in culture and in blood vessels, both *in situ* and *in vivo* (Van Buren *et al.*, 1992; Caulin-Glaser *et al.*, 1997; Lang *et al.*, 1997; Gong *et al.*, 2003). We have recently shown that estrogen is one of the active components in HDL that is transferred into caveolae in response to HDL binding to SR-BI (Gong *et al.*, 2003). In these studies, estrogen was shown to transfer in a saturable fashion, consistent with movement from HDL and rapid chemical partitioning into caveolae. Estrogen alters the stability of synthetic lipid rafts at physiological concentrations (Wang *et al.*, 2004), suggesting that some of the rapid, non-genomic effects of estrogen may be mediated by its direct effects on the lipid components as it partitions into them.

XX. CONCLUSION

The diversity of trafficking pathways that initiate at caveolae and lipid rafts, as well as the variety of vesicle and non-vesicle pathways that provide transport and communications between these cell surface domains and multiple intracellular sites, indicates that there is no single prototypical pathway or cell type that represents all caveolae/lipid rafts. It is becoming apparent that the structure and organization of lipid rafts in cells are as diverse and varied as the many functions that have been associated with these structures. The cell requires a multiplicity of communication pathways in order to integrate its many responses to the wide variety of hormones, nutrients, and circulating factors that activate signal propagation and trafficking within the cell. Caveolae and lipid rafts have emerged over the last decade as the dominant structures in which proteins that regulate the response of cells to external factors are localized and organized. Several important questions remain to be answered, particularly with respect to the regulation and interactions between the chaperone complex-mediated transport of cholesterol and the many vesicle-mediated trafficking pathways. The variety of linkages between the cell surface and distinct intracellular organelles, or even regions of organelles, that have emerged define a highly diversified network of communication pathways between the cell surface and the interior that allow the cell to integrate responses to extracellular signals and nutrients. Further studies with tissue-specific changes in expression of genes and mutants will help clarify the functional roles of lipid rafts of different types and localization in individual cells, tissues, and organs.

REFERENCES

Ayala-Sanmartin, J. (2001). Cholesterol enhances phospholipid binding and aggregation of annexins by their core domain. *Biochem. Biophys. Res. Commun.* **283**(1), 72–79.

Ayala-Sanmartin, J., Henry, J. P., and Pradel, L. A. (2001). Cholesterol regulates membrane binding and aggregation by annexin 2 at submicromolar Ca(2+) concentration. *Biochim. Biophys.* **1510**(1–2), 18–28.

Benlimame, N., Le, P. U., and Nabi, I. R. (1998). Localization of autocrine motility factor receptor to caveolae and clathrin–independent internalization of its ligand to smooth endoplasmic reticulum. *Mol. Biol. Cell* **9**(7), 1773–1786.

Bruns, R. R. and Palade, G. E. (1968a). Studies on blood capillaries. I. General organization of blood capillaries in muscle. *J. Cell Biol.* **37**(2), 244–276.

Bruns, R. R. and Palade, G. E. (1968b). Studies on blood capillaries. II. Transport of ferritin molecules across the wall of muscle capillaries. *J. Cell Biol.* **37**(2), 277–299.

Caulin-Glaser, T., Garcia-Cardena, G., Sarrel, P., Sessa, W. C., and Bender, J. R. (1997). 17 beta-estradiol regulation of human endothelial cell basal nitric oxide release, independent of cytosolic Ca^{2+} mobilization. *Circ. Res.* **81**(5), 885–892.

Chang, W. J., Ying, Y. S., Rothberg, K. G., Hooper, N. M., Turner, A. J., Gambliel, H. A., De Gunzburg, J., Mumby, S. M., Gilman, A. G., and Anderson, R. G. (1994). Purification and characterization of smooth muscle cell caveolae. *J. Cell Biol.* **126**(1), 127–138.

Conrad, P. A., Smart, E. J., Ying, Y., Anderson, R. W. G., and Bloom, G. S. (1995). Caveolin cycles between plasma membrane caveolae and the Golgi complex by microtubule-dependent and microtubule-independent steps. *J. Cell Biol.* **131**(6 Pt 1), 1421–1433.

Corvera, S., DiBonaventura, C., and Shpetner, H. S. (2000). Cell confluence-dependent remodeling of endothelial membranes mediated by cholesterol. *J. Biol. Chem.* **275**(40), 31414–31421.

Creutz, C. E., Dowling, L. G., Sando, J. J., Villar-Palasi, C., Whipple, J. H., and Zaks, W. J. (1983). Characterization of the chromobindins. Soluble proteins that bind to the chromaffin granule membrane in the presence of Ca^{2+}. *J. Biol. Chem.* **258**(23), 14664–14674.

Creutz, C. E., Zaks, W. J., Hamman, H. C., Crane, S., Martin, W. H., Gould, K. L., Oddie, K. M., and Parsons, S. J. (1987). Identification of chromaffin granule-binding proteins. Relationship of the chromobindins to calelectrin, synhibin, and the tyrosine kinase substrates p35 and p36. *J. Biol. Chem.* **262**(4), 1860–1868.

Davda, R. K., Chandler, L. J., Crews, F. T., and Guzman, N. J. (1993). Ethanol enhances the endothelial nitric oxide synthase response to agonists. *Hypertension* **21**(6 Pt 2), 939–943.

Dupree, P., Parton, R. G., Raposo, G., Kurzchalia, T. V., and Simons, K. (1993). Caveolae and sorting in the trans-Golgi network of epithelial cells. *EMBO J.* **12**(4), 1597–1605.

Dvorak, A. M. and Feng, D. (2001). The vesiculo-vacuolar organelle (VVO). A new endothelial cell permeability organelle. *J. Histochem. Cytochem.* **49**(4), 419–432.

Dvorak, A. M., Kohn, S., Morgan, E. S., Fox, P., Nagy, J. A., and Dvorak, H. F. (1996). The vesiculo-vacuolar organelle (VVO), a distinct endothelial cell structure that provides a transcellular pathway for macromolecular extravasation. *J. Leukoc. Biol.* **59**(1), 100–115.

Everson, W. V. and Smart, E. J. (2001). Influence of caveolin, cholesterol, and lipoproteins on nitric oxide synthase: Implications for vascular disease. *Trends Cardiovasc. Med.* **11**(6), 246–250.

Feng, D., Nagy, J. A., Dvorak, H. F., and Dvorak, A. M. (2002). Ultrastructural studies define soluble macromolecular, particulate, and cellular transendothelial cell pathways in venules, lymphatic vessels, and tumor–associated microvessels in man and animals. *Microsc. Res. Tech.* **57**(5), 289–326.

Fernandez, I., Ying, Y., Albanesi, J., and Anderson, R. G. (2002). Mechanism of caveolin filament assembly. *Proc. Natl. Acad. Sci.* **99**(17), 11193–11198.

Gerke, V. and Moss, S. E. (2002). Annexins: From structure to function. *Physiol. Rev.* **82**(2), 331–371.
Giannone, G., Jiang, G., Sutton, D. H., Critchley, D. R., and Sheetz, M. P. (2003). Talin1 is critical for force-dependent reinforcement of initial integrin–cytoskeleton bonds but not tyrosine kinase activation. *J. Cell Biol.* **163**(2), 409–419.
Glenney, J. R., Jr. and Soppet, D. (1992). Sequence and expression of caveolin, a protein component of caveolae plasma membrane domains phosphorylated on tyrosine in Rous sarcoma virus-transformed fibroblasts. *Proc. Natl. Acad. Sci.* **89**(21), 10517–10521.
Gong, M. C., Wilson, M. E., Kelly, T., Su, W., Dressman, J., Kincer, J. F., Matveev, S., Guo, L., Guerin, T. M., Li, X-A., Zhu, W., Uittenbogaard, A., and Smart, E. J. (2003). HDL-associated estradiol stimulates endothelial NO synthase and vasodilation in an SR-BI-dependent manner. *J. Clin. Invest.* **111**(10), 1579–1587.
Greenberg, S. S., Xie, J., Wang, Y., Kolls, J., Shellito, J., Nelson, S., and Summer, W. R. (1993). Ethanol relaxes pulmonary artery by release of prostaglandin and nitric oxide. *Alcohol* **10**(1), 21–29.
Harder, T. and Gerke, V. (1994). The annexin II2p11(2) complex is the major protein component of the triton X-100-insoluble low-density fraction prepared from MDCK cells in the presence of Ca^{2+}. *Biochim. Biophys. Acta* **1223**(3), 375–382.
Kincer, J. F., Uittenbogaard, A., Dressman, J., Guerin, T. M., Febbraio, M., Guo, L., and Smart, E. J. (2002). Hypercholesterolemia promotes a CD36-dependent and endothelial nitric-oxide synthase-mediated vascular dysfunction. *J. Biol. Chem.* **277**(26), 23525–23533.
Kobayashi, T. and Robinson, J. M. (1991). A novel intracellular compartment with unusual secretory properties in human neutrophils. *J. Cell Biol.* **113**(4), 743–756.
Kobayashi, T., Robinson, J. M., and Seguchi, H. (1998). Identification of intracellular sites of superoxide production in stimulated neutrophils. *J. Cell Sci.* **111**(Pt 1), 81–91.
Lang, U., Baker, R. S., and Clark, K. E. (1997). Estrogen-induced increases in coronary blood flow are antagonized by inhibitors of nitric oxide synthesis. *Eur. J. Obstet. Gynecol. Reprod. Biol.* **74**(2), 229–235.
Lisanti, M. P., Scherer, P. E., Vidugiriene, J., Tang, Z., Hermanowski-Vosatka, A., Tu, Y. H., Cook, R. F., and Sargiacomo, M. (1994). Characterization of caveolin-rich membrane domains isolated from an endothelial-rich source: Implications for human disease. *J. Cell Biol.* **126**(1), 111–126.
Mayorga, L. S., Beron, W., Sarrouf, M. N., Colombo, M. I., Creutz, C., and Stahl, P. D. (1994). Calcium-dependent fusion among endosomes. *J. Biol. Chem.* **269**(49), 30927–30934.
McCabe, J. B. and Berthiaume, L. G. (2001). N-terminal protein acylation confers localization to cholesterol, sphingolipid–enriched membranes but not to lipid rafts/caveolae. *Mol. Biol. Cell* **12**(11), 3601–3617.
Mostafavi–Pour, Z., Askari, J. A., Parkinson, S. J., Parker, P. J., Ng, T. T., and Humphries, M. J. (2003). Integrin-specific signaling pathways controlling focal adhesion formation and cell migration. *J. Cell Biol.* **161**(1), 155–167.
Mundy, D. I., Machleidt, T., Ying, Y. S., Anderson, R. G., and Bloom, G. S. (2002). Dual control of caveolar membrane traffic by microtubules and the actin cytoskeleton. *J. Cell Sci.* **115**(Pt 22), 4327–4339.
Nichols, B. J., Kenworthy, A. K., Polishchuk, R. S., Lodge, R., Roberts, T. H., Hirschberg, K., Phair, R. D., and Lippincott-Schwartz, J. (2001). Rapid cycling of lipid raft markers between the cell surface and Golgi complex. *J. Cell Biol.* **153**(3), 529–541.
Oliferenko, S., Paiha, K., Harder, T., Gerke, V., Schwarzier, C., Schwarz, H., Beug, H., Gunthert, U., and Huber, L. A. (1999). Analysis of CD44-containing lipid rafts: Recruitment of annexin II and stabilization by the actin cytoskeleton. *J. Cell Biol.* **146**(4), 843–854.
Orlandi, P. A. and Fishman, P. H. (1998). Filipin-dependent inhibition of cholera toxin: Evidence for toxin internalization and activation through caveolae-like domains. *J. Cell Biol.* **141**(4), 905–915.
Parton, R. G. (1994). Ultrastructural localization of gangliosides; GM1 is concentrated in caveolae. *J. Histochem. Cytochem.* **42**(2), 155–166.
Parton, R. G., Joggerst, B., and Simons, K. (1994). Regulated internalization of caveolae. *J. Cell Biol.* **127** (5), 1199–1215.
Parton, R. G. and Richards, A. A. (2003). Lipid rafts and caveolae as portals for endocytosis: New insights and common mechanisms. **4**(11), 724–738.

Pelkmans, L., Kartenbeck, J., and Helenius, A. (2001). Caveolar endocytosis of simian virus 40 reveals a new two-step vesicular-transport pathway to the ER. *Nat. Cell Biol.* **3**(5), 473–483.

Robenek, M. J., Schlattmann, K., Zimmer, K. P., Plenz, G., Troyer, D., and Robenek, H. (2003). Cholesterol transporter caveolin-1 transits the lipid bilayer during intracellular cycling. *FASEB J.* **17**(13), 1940–1942.

Robinson, J. M., Kobayashi, T., Seguchi, H., and Takizawa, T. (1999). Evaluation of neutrophil structure and function by electron microscopy: Cytochemical studies. *J. Immunol. Meth.* **232**(1–2), 169–178.

Rothberg, K. G., Heuser, J. E., Donzell, W. C., Ying, Y., Glenney, J. R., and Anderson, R. G. W. (1992). Caveolin, a protein component of caveolae membrane coats. *Cell* **68**(4), 673–682.

Schroeder, F., Frolov, A. A., Murphy, E. J., Atshaves, B. P., Jefferson, J. R., Pu, L., Wood, W. G., Foxworth, W. B., and Kier, A. B. (1996). Recent advances in membrane cholesterol domain dynamics and intracellular cholesterol trafficking. *Proc. Soc. Exp. Biol. Med.* **213**(2), 150–177.

Schroeder, R. J., Ahmed, S. N., Zhu, Y., London, E., and Brown, D. A. (1998). Cholesterol and sphingolipid enhance the Triton X–100 insolubility of glycosylphosphatidylinositol-anchored proteins by promoting the formation of detergent–insoluble ordered membrane domains. *J. Biol. Chem.* **273**(2), 1150–1157.

Shadle, P. J., Gerke, V., and Weber, K. (1985). Three $Ca2^{+}$-binding proteins from porcine liver and intestine differ immunologically and physicochemically and are distinct in Ca^{2+} affinities. *J. Biol. Chem.* **260**(30), 16354–16360.

Shaul, P. W., Smart, E. J., Robinson, L. J., German, Z., Yuhanna, I. S., Ying, Y., Anderson, R. G., and Michel, T. (1996). Acylation targets emdothelial nitric-oxide synthase to plasmalemmal caveolae. *J. Biol. Chem.* **271**(11), 6518–6522.

Smart, E. J., De Rose, R. A., and Farber, S. A. (2004). Annexin 2-caveolin 1 complex is a target of ezetimibe and regulates intestinal cholesterol transport. **101**(10), 3450–3455.

Smart, E. J., Graf, G. A., McNiven, M. A., Sessa, W. C., Engelman, J. A., Scherer, P. E., Okamoto, T., and Lisanti, M. P. (1999). Caveolins, liquid-ordered domains, and signal transduction. *Mol. Cell. Biol.* **19** (11), 7289–7304.

Smart, E. J., Ying, Y. S., Conrad, P. A., and Anderson, R. G. (1994). Caveolin moves from caveolae to the Golgi apparatus in response to cholesterol oxidation. *J. Cell Biol.* **127**(5), 1185–1197.

Uittenbogaard, A., Everson, W. V., Matveev, S. V., and Smart, E. J. (2002). Cholesteryl ester is transported from caveolae to internal membranes as part of a caveolin-annexin II lipid-protein complex. **277**(7), 4925–4931.

Uittenbogaard, A. and Smart, E. J. (2000). Palmitoylation of caveolin-1 is required for cholesterol binding, chaperone complex formation, and rapid transport of cholesterol to caveolae. **275**(33), 25595–25599.

Uittenbogaard, A., Ying, Y., and Smart, E. J. (1998). Characterization of a cytosolic heat-shock protein-caveolin chaperone complex. Involvement in cholesterol trafficking. *J. Biol. Chem.* **273**(11), 6525–6532.

Ushio–Fukai, M., Hilenski, L., Santanam, N., Becker, P. L., Ma, Y., Griendling, K. K., and Alexander, R. W. (2001). Cholesterol depletion inhibits epidermal growth factor receptor transactivation by angiotensin II in vascular smooth muscle cells: Role of cholesterol-rich microdomains and focal adhesions in angiotensin II signaling. *J. Biol. Chem.* **276**(51), 48269–48275.

Van Buren, G. A., Yang, D. S., and Clark, K. E. (1992). Estrogen–induced uterine vasodilatation is antagonized by L–nitroarginine methyl ester, an inhibitor of nitric oxide synthesis. *Am. J. Obstet. Gynecol.* **167**(3), 828–833.

Wang, H. J., Guay, G., Pogan, L., Sauve, R., and Nabi, I. R. (2000). Calcium regulates the association between mitochondria and a smooth subdomain of the endoplasmic reticulum. **150**(6), 1489–1498.

Wang, J., Megha, and London, E. (2004). Relationship between sterol/steroid structure and participation in ordered lipid domains (lipid rafts), implications for lipid raft structure and function. *Biochemistry* **43** (4), 1010–1018.

Wood, W. G., Gorka, C., and Schroeder, F. (1989). Acute and chronic effects of ethanol on transbilayer membrane domains. **52**(6), 1925–1930.

Wood, W. G. and Schroeder, F. (1988). Membrane effects of ethanol: Bulk lipid versus lipid domains. *Life Sci.* **43**(6), 467–475.

Zeuschner, D., Stoorvogel, W., and Gerke, V. (2001). Association of annexin 2 with recycling endosomes requires either calcium- or cholesterol-stabilized membrane domains. *Eur. J. Cell Biol.* **80**(8), 499–507.

Section II

Caveolae Organization and the Regulation of Endocytosis

Chapter 3

The Caveolae Internalization Machinery

L. S. Orlichenko and M. A. McNiven*

Department of Biochemistry and Molecular Biology, and Center for Basic Research in Digestive Diseases, Mayo Clinic College of Medicine, Rochester, Minnesota 55905

I. INTRODUCTION: CAVEOLAE AS ENDOCYTIC ORGANELLES

Caveolae are generally regarded as immobile plasma membrane micro-domains (Thomsen *et al.*, 2002) associated with the structural proteins caveolin-1 and caveolin-2. Under certain conditions these flask-shaped organelles have the capacity to release from the plasma membrane as endocytic vesicles subsequent to the sequestration of a variety of agents and ligands. Some substrates internalized by caveolae include infectious agents such as SV40 and EV1 viruses (Anderson *et al.*, 1996; Marjomaki *et al.*, 2002), toxins such as cholera toxin (Parton *et al.*, 1994), and some receptors including autocrine motility factor receptor (Benlimame *et al.*, 1998) and the albumin receptor (Tiruppathi *et al.*, 1997). Understanding this internalization process is key to defining the entry of many important infectious or trophic factors into mammalian cells. As caveolae have also been associated with cell signaling events through interactions with a variety of signaling molecules including receptor tyrosine kinases (RTKs) (Sargiacomo *et al.*, 1993), these structures are implicated in cell growth cascades and neoplasia (Galbiati *et al.*, 1998). Whether caveolae scission and internalization alters these signaling cascades is unclear. This chapter does not attempt to describe all aspects of caveolae function, but focuses on the structural and enzymatic

Advances in Molecular and Cell Biology, Vol. 36

ISBN: 0-444-51500-3
DOI: 10.106/S0079-6603(05)36003-X

mechanisms by which these important structures form and are released from the plasma membrane into the cell interior.

II. CAVEOLIN RECRUITMENT TO THE PLASMA MEMBRANE

Caveolin represents the major structural protein component of caveolae. Three caveolin isoforms have been identified in mammals: caveolin-1 (Cav-1), caveolin-2 (Cav-2), and caveolin-3 (Cav-3). Cav-1 and Cav-2 are represented by long (α) and short (β) variants resulting from alternative initiation of translation (Scherer *et al.*, 1995). All caveolin isoforms contain a scaffolding domain sufficient for membrane targeting (Schlegel *et al.*, 1999). The scaffolding sequence is a part of the longer oligomerization domain needed for protein homo- and hetero-oligomerization and caveolae formation (Monier *et al.*, 1995). Expression of a caveolin mutant with a truncation of the N-terminal part of the oligomerization domain (deletion of residues 1–53 in human caveolin) (Cav-3 DGV) results in disappearance of caveolae in Madin-Darby canine kidney (MDCK) cells (Lahtinen *et al.*, 2003).

The N terminus of the caveolins contains several tyrosine residues. Tyrosine 14 in Cav-1 and tyrosine 19 in Cav-2 were found to be major phosphorylation sites for Src family tyrosine kinases (Li *et al.*, 1996; Lee *et al.*, 2002b). Caveolin-1 can also undergo phosphorylation at serine 80 that results in its intracellular retention and a subsequent co-sequestration of caveolin-2. As expected, the intracellular sequestration of both caveolar proteins away from the cell surface prevents the formation of caveolae (Schlegel *et al.*, 2001). A study has demonstrated that a critical concentration of the caveolins at the plasma membrane is required for caveolar assembly. It was found that more than 30% of total cellular caveolins residing within a floating fraction of sucrose gradient correlated with caveolae assembly in Caco-2 cells (Breuza *et al.*, 2002). This suggested that the enrichment of caveolin molecules into lipid rafts at the plasma membrane created a microenvironment that favored the association of these proteins into a coat and thus into the formation of invaginated caveolae.

Although Cav-1 is required for caveolae formation, some data have suggested that expression of caveolin-2 might affect caveolae biogenesis in MDCK, Chinese hampster ovary (CHO), and HepG2 cells (Fujimoto *et al.*, 2000; Lahtinen *et al.*, 2003). The ability of Cav-2 to modulate caveolae assembly, shape, and size was found to rely on Cav-2 phosphorylation status (Sowa *et al.*, 2003). Mutations of the primary phosphorylation sites of caveolin-2 on serine 23 and 36 reduced the number of plasmalemma-attached caveolae without affecting its interaction with caveolin-1 or its biochemical properties (Sowa *et al.*, 2003). Lee *et al.* showed that Src-induced phosphorylation of caveolin-2 on tyrosine 19 might contribute to caveolae disassembly through differential interaction of c-Src, Ras-GTPase activating protein (Ras-GAP), and Nck with caveolin-2 (Lee *et al.*, 2002b).

Caveolae are cholesterol- and glycosphingolipid (GSL)-rich smooth invaginations of the plasma membrane that partition into raft fractions. Chang *et al.* found that the presence of cholesterol is required for caveolae invagination (Chang *et al.*, 1992).

Cholesterol-binding drugs are among the inhibitors of caveolae formation (Schnitzer *et al.*, 1994; Anderson *et al.*, 1996; Kiss and Geuze, 1997). Several studies have found that specific local lipid composition is necessary for caveolae assembly and plasma membrane invagination (Meyer *et al.*, 1998) as well as for activation of caveolae internalization (Sharma *et al.*, 2004). Recently, Sharma *et al.* have demonstrated that elevation in cellular cholesterol level, or treatment of the cells with exogenous natural or synthetic glycosphingolipids, selectively stimulates caveolar endocytosis (Sharma *et al.*, 2004).

In addition to GSL-stimulated internalization, caveolae internalization can be induced through experimental manipulation such as antibody cross-linking of proteins in caveolae membrane domains, inhibition of tyrosine phosphatases, or addition of hypertonic medium (Parton *et al.*, 1994; Aoki *et al.*, 1999; Kang *et al.*, 2000; Richterova *et al.*, 2001).

III. CAVEOLAR SCISSION

A. Structural Components

1. The Large GTPase Dynamin as a Central Player in Caveolar Scission

Dynamin is a 96-kDa GTPase shown to be essential for many cellular processes, including budding of endocytic (Herskovits *et al.*, 1993) and secretory (Cao *et al.*, 2000) transport vesicles, podosome formation (McNiven *et al.*, 2004), actin rearrangements (Lee and De Camilli, 2002; Schafer *et al.*, 2002), and cytokinesis (Thompson *et al.*, 2002). Its importance in vesicle budding and endocytosis was first discovered through the identification of the temperature-sensitive mutant *shibire,* characterized by a paralytic phenotype in the fruit fly *Drosophila melanogaster* (van der Bliek and Meyerowitz, 1991). The fly exhibited a rapid paralysis at the non-permissive temperature that was reversible upon shifting to the permissive temperature. The biological defect was identified as a block in synaptic vesicle scission and recycling from the presynaptic plasma membrane. The enzyme currently has five identifiable domains (Fig. 1). At the N-terminal end is a highly conserved GTP-binding motif (G1–G3, residues 1–299) needed for GTP binding and hydrolysis. This nucleotide hydrolysis cycle leads to a substantial conformational change contributing to membrane constriction and vesicle fission (Carr and Hinshaw, 1997). A lysine 44 to alanine substitution in the first GTP-binding motif yields a dominant-negative mutant dynamin protein (K44A), which displays impaired GTPase activity as a result of a greatly reduced GTP-binding affinity (Damke *et al.*, 1994). Mutations of the third GTP-binding motif (substitution lysine 206 to aspartate) (K206E) or removal of three GTP-binding motifs (amino acids 1–272) (ΔN272) drastically reduces vesicle fission from cellular membranes (Herskovits *et al.*, 1993). The GTPase domain is followed by a middle domain (residues 300–520) required for self-assembly and oligomerization (Zhang and Hinshaw, 2001). Self-oligomerization supports dynamin's ability to wrap around the neck of budding vesicles (Carr and Hinshaw, 1997; Okamoto *et al.*, 1999b).

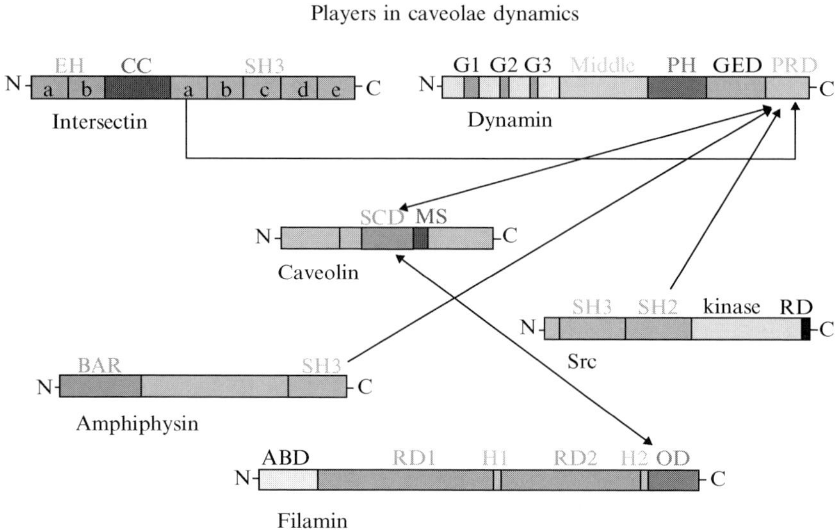

Figure 1. Multiple structural proteins interact with the caveolins to mediate caveolae formation and fusion. Arrows point to different caveolin-associated proteins and corresponding domains for which direct interactions have been documented. EH, Eps15 homology domain; CC, coiled-coil domain; SH2, SH3, Src homology 2, 3 domains; G1-3, GTPase domains 1-3; Middle, middle domain; PH, pleckstrin homology domain; GED, GTPase effector domain; PRD, proline-rich domain; SCD, scaffolding domain; MS, membrane-spanning domain; Kinase, Src kinase catalytic domain; RD, Src kinase regulatory domain; BAR, Bin/Amphiphysin/Rvs domain; ABD, actin-binding domain; RD1-2, rod domain 1, 2; HD1-2, hinge domain 1, 2; OD, oligomerization domain. (See Color Insert.)

A pleckstrin homology domain (PH, residues 521–622) is implicated in phosphatidylinositol 4,5-biphosphate (PI(4,5)P2) binding, which is believed to target dynamin to the plasma membrane and also greatly increase GTPase activity (Barylko *et al.*, 1998; Lee *et al.*, 1999). Indeed, expression of a dynamin PH domain mutant (K535M), which does not exhibit PI(4,5)P2 binding or membrane-stimulated GTPase activity, leads to an effective block of transferrin receptor internalization (Lee *et al.*, 1999). Dynamin also contains a coiled-coil GTPase effector domain (GED; residues 623–745). Interactions between the GTPase domain, the middle domain, and the GED can also stimulate the GTPase catalytic activity of dynamin (Muhlberg *et al.*, 1997). Finally, a C-terminal proline-arginine-rich domain (PRD) (residues 746–864) containing several SH3-binding sites facilitates recruitment of several accessory proteins necessary to stimulate GTPase activity, to target dynamin to specific membrane domains (Hill *et al.*, 2001; Szaszak *et al.*, 2002), or to support dynamin function in membrane constriction (Hill *et al.*, 2001).

Three dynamin isoforms have been identified in vertebrates. Dynamin 1 (Dyn1) is predominantly expressed in neuronal tissue (Scaife and Margolis, 1990) and is concentrated in the presynapse (Powell and Robinson, 1995). Dynamin 2 (Dyn2) is ubiquitously expressed (Cook *et al.*, 1994), while dynamin 3 (Dyn3) is expressed in lung, brain, and testis (Cook *et al.*, 1996; Kamitani *et al.*, 2002). Lower organisms such as *Drosophila melanogaster* and *Caenorhabditis elegans* contain a single dynamin isoform that exhibits significant alternative splicing and is assumed to perform the multiple functions of the distinct dynamin forms found in mammalian cells (Clark *et al.*, 1997; Staples and Ramaswami, 1999). The C-terminal PRD domain is the most divergent among the dynamin family members (Smirnova *et al.*, 1999), which suggests that different isoforms of dynamin may interact with various binding partners for recruitment to distinct cytoplasmic organelles. Additional diversity from each dynamin member is also provided by alternative splicing (Cao *et al.*, 1998).

How does dynamin mediate caveolae scission? Early studies of this process were accomplished using a reconstituted *in vitro* model of caveolae fission from lung endothelial plasma membranes by Schnitzer and colleagues (Schnitzer *et al.*, 1996). This work provided important insights into the role of GTP hydrolysis during this process. In permeabilized endothelial cells, the non-hydrolysable GTP analog GTPγS prevented caveolae budding and endocytosis of cholera toxin B to endosomes, whereas GTP stimulated internalization. Shortly after, two independent groups showed that dynamin function is required for caveolae-dependent endocytosis *in vivo* (Henley *et al.*, 1998; Oh *et al.*, 1998). In cultured bovine lung microvascular endothelial cells (BLMEC) (Oh *et al.*, 1998), expression of a rat dominant-negative K44E dynamin mutant lacking GTPase activity (Dyn1K44E) prevented internalization of cholera toxin B via caveolae. Further, a role of Dyn2 in caveolae fission was also demonstrated *in vitro*, using a purified rat lung plasma membrane caveolae preparation. In this assay, scission was markedly inhibited through the addition of cytosol derived from HeLa cells that overexpressed human dominant-negative Dyn2baK44A. Similar effects of inhibition of caveolae-dependent endocytosis of cholera toxin B were achieved in mouse hepatocytes via the injection of Dyn2 antibodies (Henley *et al.*, 1998). Consistent with this block was a marked accumulation of large numbers of toxin-positive caveolae at the plasma membrane, many forming long continuous chains of linked caveolae (Fig. 2c).

Since these seminal studies, the participation of the large GTPase dynamin in caveolae-dependent endocytosis has been demonstrated for a variety of ligands and receptors (Table I), including albumin in rat lung microvessel endothelial (RLMVEC) cells (John *et al.*, 2003; Shajahan *et al.*, 2004), G-protein-coupled receptors (GPCR) including M2 muscarinic acetylcholine receptor (M2 mAchR) in rat ventricular myocytes and COS-7 cells (Dessy *et al.*, 2000), nonmammalian gonadotropin-releasing hormone receptor (GnRH-R) in COS-7 and human embryonic kidney (HEK) 293 cells (Pawson *et al.*, 2003), and autocrine motility factor receptor (AMF-R) in transformed NIH-3T3 fibroblasts (Le *et al.*, 2002). Recent studies suggested that the GTP hydrolysis by dynamin is an obligatory requirement even for the processes of constitutive caveolae-dependent internalization. For example, constitutive endocytosis via caveolae of the transforming growth factor-β receptor (TGF-β RII) in mink lung epithelial Mv1Lu cells (Di Guglielmo *et al.*, 2003) and membrane type 1 matrix metalloproteinase (MT1-MMP) in transfected human fibrosarcoma bbHT1080 cells (Remacle *et al.*, 2003) have all been shown to be inhibited by mutant dynamin expression.

As described above, inhibiting the GTPase activity of dynamin has profound inhibitory effects on the caveolae-based internalization of a variety of toxins and

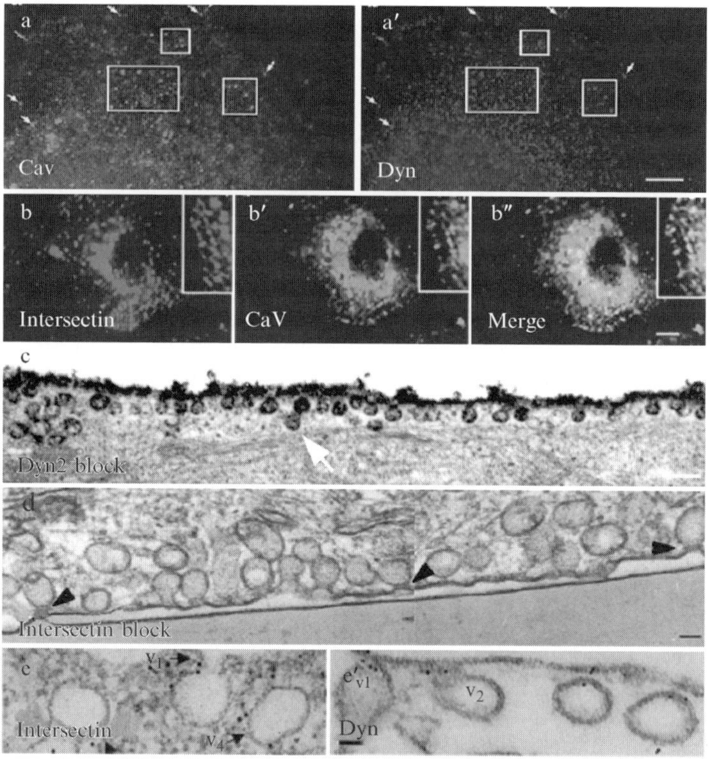

Figure 2. Dynamin and intersectin in caveolae membrane dynamics. (a, a′) Dynamin localizes to caveolae in cultured hepatocytes. Fluorescence micrographs representing laser scanning confocal microscopy of cultured hepatocytes that were double-labeled with a monoclonal anti-caveolin antibody (a) and the polyclonal anti-Pan61 antibody (a′) to label endogenous dynamin. A significant number of vesicular structures are labeled with both antibodies (arrows and outlined areas), indicating a colocalization of dynamin and caveolin. Scale bars, 8.0 μm. (b–b″) Subcellular distribution of intersectin by double immunofluorescence in cultured endothelial cells (ECs). Intersectin displays a punctate staining pattern both at the plasma membrane and throughout the cytoplasm (b). Caveolin immunostaining in cultured ECs (b′). The merged image (b″) reflects extensive colocalization for both intersectin and caveolin (panel b″ and inset b″). Scale bars, (b–b″) 10 μm; (insets) 5 μm. (c) Accumulation of surface-attached caveolae in cultured hepatocytes injected with anti-dynamin antibodies. Electron micrographs of dynamin antibody-injected cells that were fixed and stained with ruthenium red. Dark vesicles reveal both surface (small arrows) and deeper (large arrows) membrane invaginations that are continuous with the plasma membrane. Scale bars, 0.15 μm. (d) Caveolae proliferation in ECs overexpressing wt intersectin. Electron micrographs show large numbers of caveolae forming along the cell border. Note the caveolar profiles displaying staining dense rings (arrowheads). Scale bars, 50 nm. (e, e′) Immuno-EM localization of dynamin (e) and intersectin (e′) to the necks of formed caveolae in cultured ECs. Immunogold labeling shows gold particles (6 nm) that are preferentially associated with the neck region of caveolae open to the EC surface (e, v1). Frequently, more than two gold particles labeled a caveolar profile or its neck. Scale bar, (e, e′) 50 nm. (a, a′), and (c) reprinted with permission from Henley *et al.*, 1998. (b–b″) and (d–e′) reprinted with permission from Predescu *et al.*, 2003. (See Color Insert.)

Table I
Dynamin-Based Studies on Caveolar-Mediated Ligand Internalization. Scission / Internalization

Dynamin isoforms/ mutants	Receptors/ ligands	Cell line/ tissue	Inhibition/ % vs wt[a]	References
RDyn1K44E	Cholera toxin	BLMEC	Yes/71%	Oh *et al.*, 1998
HDyn2baK44A	*In vitro* fission	Rat lung/Hela extract	Yes/80%	Oh *et al.*, 1998
HDyn1aaK44A	AMF-R	NIH-3T3 (ras, abl)	Yes	Le *et al.*, 2002
	Cholera toxin	NIH-3T3	Yes/80%	Le and Nabi, 2003
RDyn2aaK44A	Albumin	RLMVEC	Yes/60%	Shajahan *et al.*, 2004
	Cholera toxin		Yes/80%	
	SV40	CV-1	Yes/85%	Pelkmans *et al.*, 2002
HDyn1K44A	M2 mAchR	Rat ventricular myocytes	Yes/70%	Dessy *et al.*, 2000
		COS-7	Yes/70%	Dessy *et al.*, 2000
RDyn1K44A	GnRH-R	COS-7	Yes/70%	Pawson *et al.*, 2003
		HEK 293	ND?	
RDyn1K44A	B2R	HEK 293	No/22%	Lamb *et al.*, 2001
DynK44A	TGF-β RII	Mv1Lu	Yes	Di Guglielmo *et al.*, 2003
DynK44A	TGF-β	293	No	Zwaagstra *et al.*, 2001
RDyn2aaK44A	MTI-MMP	HT1080	Yes	Remacle *et al.*, 2003
RDyn2aaY231F/ Y597F	Albumin, Cholera toxin	RLMVEC	Yes/45%	Shajahan *et al.*, 2004

[a]If not specified by authors, % of internalized ligand / receptor calculated from corresponding columns in references. R, rat; H, human.

receptor systems. Additional interesting insights have been obtained by the expression of dynamin mutants with alterations in phosphorylation sites. Caveolae are enriched in non-receptor tyrosine kinases, including Src (Sargiacomo *et al.*, 1993). The importance of the Src kinase activity in caveolae-dependent internalization was demonstrated in endothelial RLMVEC cells (Shajahan *et al.*, 2004). Treatment of the cells with Src kinase inhibitor PP2 abolished the phosphorylation and blocked albumin uptake via caveolae, while overexpression of dominant-negative K295M/Y527F Src inhibited 45% of albumin uptake compared to cells expressing wild-type Src (Shajahan *et al.*, 2004). The study also highlighted the importance of c-Src-based phosphorylation of dynamin in caveolae internalization (Shajahan *et al.*, 2004). c-Src has been shown to bind to Dyn1 (Foster-Barber and Bishop, 1998), resulting in phosphorylation at several specific tyrosine residues, including Y231 in the GTPase domain and Y597 within the PH domain (Ahn *et al.*, 1999; Shajahan *et al.*, 2004). Shajahan *et al.*

have demonstrated that expression of rat Y231F/Y597F Dyn2aa mutants in RLMVEC cells impaired albumin and cholera toxin uptake and reduced transendothelial albumin transport.

As intermolecular interactions between the GTPase and adjacent GED, PH, and middle domains have been reported to regulate the GTPase activity of the enzyme (Muhlberg *et al.*, 1997), it is tempting to speculate that phosphorylation of Tyr231 might regulate the GTPase activity of dynamin by controlling these interactions. Phosphorylation of Tyr597 might also alter the ability of dynamin to interact with PI(4,5)P2 and/or effector proteins, with or without modulation of dynamin GTPase activity. Shajahan *et al.* found that Src-induced phosphorylation of dynamin promoted dynamin translocation to the plasma membrane and its association with caveolin-1 in endothelial RLMVEC cells (Shajahan *et al.*, 2004). Consistent with this finding was the discovery that overexpression of non-Src phosphorylatable dynamin Y231F/Y597F reduced the binding of dynamin to caveolin. These results suggested that phosphorylation of dynamin by Src might not alter dynamin GTPase activity, but rather the efficient recruitment of dynamin to the budding caveolae and subsequent release of caveolae from the plasma membrane as endocytic vesicles.

2. Dynamin-Interacting Partners: Intersectin and Amphiphysin as Regulators of Caveolae Fission

Intersectin is a multiple Eps15 homology (EH) and SH3-domain-containing protein that is known to be an essential component of the endocytic machinery in neuronal and non-neuronal cells (Yamabhai *et al.*, 1998). It has two isoforms, "long" and "short," each of which has two different splice variants (intersectin-1 and intersectin-2) (Guipponi *et al.*, 1998). The long intersectin-1 isoform is neuronal specific, while the long intersectin-2 appears to be more widely expressed (Pucharcos *et al.*, 2001). The short isoforms show the highest expression in the nervous system and contain two EH domains, a coiled-coil (CC) domain, and five SH3 domains (O'Bryan *et al.*, 2001) (Fig. 1). EH domains promote association of intersectins with endocytic accessory proteins. The CC domain stimulates both homotypic and heterotypic interactions with other CC-containing proteins. As these proteins have multiple SH3 domains, it has been demonstrated that dynamin binds to intersectin, which together mediate the formation of clathrin-coated pits (Okamoto *et al.*, 1999a).

It was recently found that short intersectin-1 also associates with caveolae (Fig. 2b) and mediates caveolae scission. In this study, immunoelectron microscopy (immuno-EM) images showed that intersectin-1 associated preferentially with the caveolar neck in rat lung endothelial cells (Predescu *et al.*, 2003) (Fig. 2e′), both before and after scission from the plasma membrane. In support of these morphological observations, a cell-free system depleted of intersectin failed to support caveolae fission from the plasma membrane. Interestingly, overexpression of wild-type short intersectin-1 also impaired caveolae internalization and enhanced accumulation of caveolar vesicles at the membrane

(Fig. 2d). Further, this study suggested that the intersectin–dynamin complexes are important in regulating the fission and internalization of budding vesicles (Predescu *et al.*, 2003), as both immunoprecipitation and blot overlay indicated that intersectin recruited dynamin to the neck of the vesicles by the SH3A domain. Endothelial cells overexpressing the SH3 domain of intersectin showed large caveolae clusters accumulated under the plasma membrane, unable to move through the cell cortex into the cytosol. In addition to recruitment, intersectin might also regulate dynamin GTPase activity and self-assembly, as it was shown that dynamin remained functionally impaired and caveolae fission did not occur as long as dynamin was bound to intersectin (Predescu *et al.*, 2003). These findings indicated an important regulatory role of intersectin in caveolae release from the plasma membrane via its interaction with dynamin.

Another dynamin-interacting partner shown to be important in caveolae dynamics is amphiphysin. Amphiphysin is a Bin/amphiphysin/Rvs (BAR) domain and SH3 domain-containing protein that functions as an adaptor between the dynamins and the plasma membrane (David *et al.*, 1996) and has a putative role in membrane deformation at endocytic sites (Farsad *et al.*, 2001). *In vitro* studies demonstrated that the N-terminal BAR domain of amphiphysin binds and invaginates lipid membranes into narrow tubules (Farsad *et al.*, 2001) and a variable central region binds clathrin and adapter proteins, while the C-terminal SH3 domain binds to dynamin (Ramjaun *et al.*, 1997). All of these properties reflect a role for amphiphysin in endocytosis. Two amphiphysin isoforms have been identified. Amphiphysin-1 is ubiquitously expressed, and amphiphysin-2 is predominantly found in neurons (N-amphiphysin-2) (Wigge and McMahon, 1998). Muscle expresses a distinct spliced variant of amphiphysin, M-amphiphysin-2 (Butler *et al.*, 1997) (Fig. 1). Interestingly, the binding sites for clathrin and AP-2 sites are not present in M-amphiphysin-2. Instead, muscle amphiphysin contains a unique exon 10 that has a high basic amino acid content and has an overall resemblance to PI(4,5)P_2-binding sequences (Lee *et al.*, 2002a).

A role of amphiphysin function in caveolae dynamics has been gathered from studying tubular plasma membrane invaginations in striated muscles called transverse tubules or T-tubules. T-tubules develop from plasma membrane curvatures that resemble strings of caveolae (Ishikawa, 1968) (Fig. 3a). Moreover, a critical role for caveolin-3 in T-tubule biogenesis has been indicated previously (Parton *et al.*, 1997). A recent study showed that both *in vivo* and *in vitro* M-amphiphysin-2 has an important physiological function in membrane deformation (Lee *et al.*, 2002a). When expressed in cultured myoblasts, M-amphiphysin-2 induced deep, tubulovesicular plasma membrane invaginations (Fig. 3b). Double immunogold labeling of differentiated myotubes with antibodies to M-amphiphysin-2 and Cav-3 revealed intensive co-labeling of the tubular network (Fig. 3c). The expression of the mutant M-amphiphysin-2 without an SH3 domain in CHO cells caused extended tubule elongation unlike the expression of full-length M-amphiphysin-2. This suggests a mechanism of how amphiphysin could recruit dynamin 2 via the SH3 domain during T-tubule formation in muscle cells.

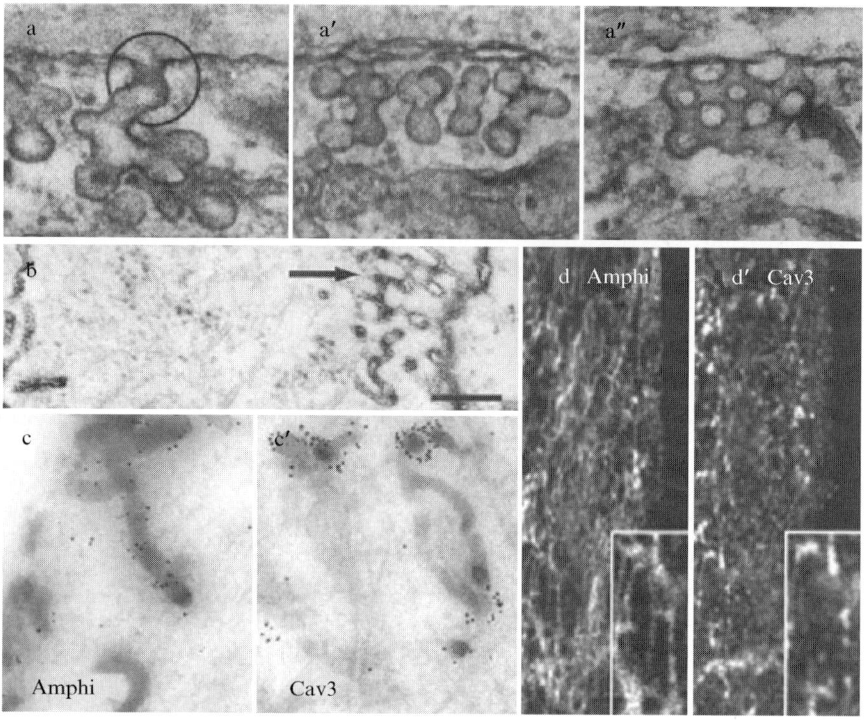

Figure 3. Amphiphysin and caveolar fusion in muscle. (a–a″) EM of caveolar-like structures displaying various fusion intermediates during T-tubule formation in cultured chick myoblasts. Fused caveolae-like reticulum maintains a continuous connection with the cell surface (circle). 76,000×. (b) EMs of differentiated C2C12 myotubes after incubation with ruthenium red demonstrates the presence of deep, tubulovesicular plasma membrane invaginations (arrow). Scale bar, 200 nm. (c, c′) Localization of amphiphysin 2 and caveolin-3 in differentiated C2C12 myotubes. Ultrathin frozen sections of myotubes show colocalization of amphiphysin 2 (c) and caveolin-3 (c′) as revealed by single immunogold labeling. Amphiphysin 2 and caveolin-3 are concentrated on the tubular and vesicular portion, respectively, of the HRP-labeled network (d, d′). Immunofluorescence confocal microscopy of differentiated C2C12 myotubes demonstrating localization of endogenous amphiphysin 2 on tubular elements and partial overlap amphiphysin 2 (d) with caveolin-3 (d′). The insets of (d, d′) show that puncta of caveolin-3 immunoreactivity are often aligned with amphiphysin 2-positive tubules. (a–a″) reprinted with permission from Ishikawa, 1968. (b–d′) reprinted with permisson from Lee *et al.*, 2002.

3. Caveolin

The role of the caveolin proteins in caveolae fission was demonstrated in a study carried out in knockout animals. Caveolin-deficient mice showed the impaired endocytosis of albumin (Zhao *et al.*, 2002). This phenotype could be reversed by expression of caveolin-1 cDNA (Zhao *et al.*, 2002). This observation was supported by studies in which expression of a caveolin mutant with a truncation of the N-terminal part of the oligomerization domain Cav-3 DGV effectively inhibited EV1 and SV40 virus infection via caveolae in human osteosarcoma (SAOS) and African green monkey (CV1) cells (Roy *et al.*, 1999; Marjomaki *et al.*, 2002).

Caveolin protein participation in vesicle release has been thought to occur through regulation of Src and dynamin activities (Minshall *et al.*, 2003; Shajahan *et al.*, 2004). Kim and Bertics demonstrated that dynamin and caveolin form a complex and co-immunoprecipitate in murine B82L fibroblasts (Kim and Bertics, 2002). In support of this study, recent observations have found that the PRD of dynamin interacts directly with residues 76–165 of caveolin-1 (Yao *et al.*, 2005). Interestingly, co-immunoprecipitation of dynamin and caveolin-1 was not affected by phosphorylation status, and caveolin-1 remained associated with dynamin up to 60 min after epidermal growth factor (EGF) addition in B82L fibroblasts (Kim and Bertics, 2002). These results are complemented by a study that found that Src-induced phosphorylation of dynamin promoted dynamin association with caveolin-1 in endothelial RLMVEC cells and that overexpression of non-Src phosphorylatable dynamin Y231F/Y597F reduced the binding of dynamin to caveolin (Shajahan *et al.*, 2004). The apparent differences could be attributed to the use of endothelial versus fibroblastic cell lines and can be reconciled by the alternative signaling and recruitment of different caveolin-and/or dynamin-binding partners in response to EGF treatment in these cells, which might affect caveolin/dynamin-binding affinity for each other. Nevertheless, tyrosine kinase activation is necessary for induction of caveolae-dependent internalization of various ligands and receptors (Minshall *et al.*, 2003). Thus, it is plausible to imply that tyrosine phosphorylation of caveolin and dynamin might be important for caveolae fission from the plasma membrane. In combination, all of these findings suggest that a caveolin–dynamin–intersectin complex, in close association with an actin scaffold, could mediate ligand-stimulated caveolar scission.

4. Actin

The importance of the actin cytoskeleton in caveolae fission was first indicated in a report showing that the internalization of caveolae could be blocked by treatment with the actin-depolymerizing drug cytochalasin D (CytD) (Parton *et al.*, 1994). The phosphatase inhibitor okadaic acid (OA) was shown to specifically affect caveolar endocytosis rather than general plasma membrane internalization. Parton *et al.* monitored the effect of CytD on caveolar internalization induced by okadaic acid in A431 cells. When cells were treated with CytD, the okadaic acid-induced internalization of alkaline phosphatase was blocked. This suggested that CytD prevented caveolae endocytosis of alkaline phosphatase, and the protein remained on the cell surface. CytD also blocked the okadaic acid-induced internalization of cholera toxin B without affecting caveolar morphology, when observed by electron microscope. Consistent with these results, treatment of MDCK cells with okadaic acid caused redistribution of caveolin-1 from the cell borders to the cytosol, implying caveolae internalization.

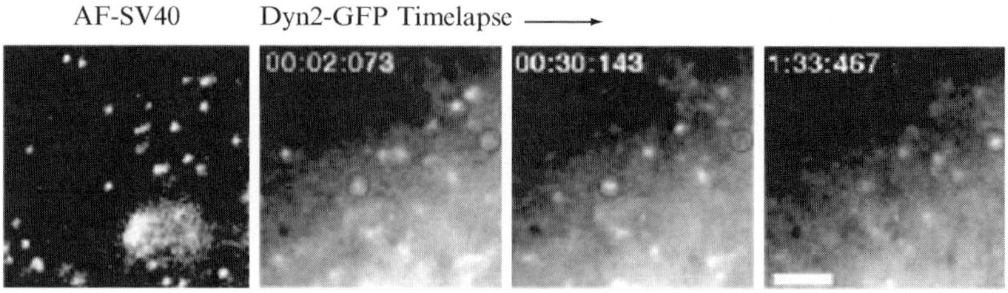

Figure 4. Dynamin is actively recruited to forming caveolae in SV40-infected cells. Fluorescence images of living cells expressing Dyn2-GFP show that Dyn2 is recruited to caveolae-bound AF-SV40 viral particles. Dyn2 recruitment (red circles) appears as "blinking" at the cell surface as actin and caveolin are recruited, then disappear as caveolar scission occurs. Scale bars, 2.5 μm. Images reprinted with permission from Helenius *et al.*, 2002. (See Color Insert.)

The change in caveolin-1 distribution upon OA treatment was completely blocked by CytD (Parton *et al.*, 1994).

More recently, a graphic study by Pelkmans *et al.* using evanescent microscopy demonstrated a sequential temporal recruitment of dynamin and actin to budding caveolae containing SV40 virus (Fig. 4) (Pelkmans *et al.*, 2002). Actin transiently associated with SV40-containing caveolae within a time frame similar to that of dynamin. The recruitment of dynamin and actin to the virus-containing caveolae was dependent upon local tyrosine kinase activation, as this process was inhibited by genistein, a tyrosine kinase inhibitor. It was suggested that actin was likely to function in later stages of caveolae internalization, possibly during the fission process (Engqvist-Goldstein and Drubin, 2003) (Fig. 4). These images provided provocative information linking the association of dynamin, actin, and caveolin in the process of caveolae scission. How such events are sequenced and mediated is currently undefined, although dynamin has been shown to interact with a variety of different actin-binding proteins, such as syndapin (Qualmann *et al.*, 1999), ABP1 (Kessels *et al.*, 2001), and cortactin. As both dynamin and cortactin bind to c-Src and are Src substrates (Foster-Barber and Bishop, 1998; McNiven *et al.*, 2000), this complex may have an important role in caveolar scission. These proteins have been found to participate in clathrin-based endocytosis (Cao *et al.*, 2003), although no data has implicated these actin-binding proteins in caveolae function. Recently, Predescu *et al.* showed that caveolae internalization might be regulated by another component of the actin regulatory machinery, Wiskott-Aldrich syndrome protein (WASP), by direct interaction of WASP with the SH3A domain of intersectin (Predescu *et al.*, 2003).

An additional actin-binding/cross-linking protein, filamin, has been shown to bind to Cav-1 directly via the C terminus (Stahlhut and van Deurs, 2000) (Fig. 1). In support of these biochemical findings, immuno-EM of NIH 3T3 cells treated with Rho-stimulating *Escherichia coli* cytotoxic necrotizing factor 1 (CNF-1) demonstrated a close association of filamin on forming caveolae. How this association might aid in caveolae formation and scission needs to be determined.

B. Signaling Components

1. Src Kinase

The vast majority of data demonstrate that caveolae internalization in different cell types is regulated by tyrosine phosphorylation. Early studies on caveolar endocytosis showed that OA, a general phosphatase inhibitor, could stimulate the internalization of cross-linked alkaline phosphatase (AP) (Parton *et al.*, 1994). An inhibitor specific for tyrosine kinases, genistein, reversed this effect. Later, several studies demonstrated that Src family tyrosine kinases might regulate caveolar internalization (Li *et al.*, 2001). Expression of a catalytically inactive Src mutant prevented caveolar vesicle fission and albumin endocytosis in endothelial cells and in human skin fibroblasts (Minshall *et al.*, 2000; Shajahan *et al.*, 2004; Sharma *et al.*, 2004). These results suggested an important role for Src family kinases in initiating the processes of caveolae internalization.

2. PKCα

Caveolae fission appears to depend on the activity of protein kinase Cα (PKCα) at least in some cell types (Smart *et al.*, 1995). A recent study has observed that chelerytherine chloride and Go 6976, general and specific inhibitors of PKC, and expression of a dominant-negative PKCα (DN-PKCα) inhibited albumin internalization and led to a reduction in surface caveolae in human skin fibroblast (HSF) cells (Sharma *et al.*, 2004). Consistent with these results, EV1 virus entry via caveolae and infection of SAOS cells was also inhibited by DN-PKCα (Upla *et al.*, 2004). These effects were specific to the PKCα isoform. Expression of a dominant-negative PKCε used as a negative control had no effect on EV1 infection, while safingol, a selective inhibitor for the α isoforms of PKC, reduced the number of EV1-infected cells to background levels (Upla *et al.*, 2004). As it was demonstrated that filamin is a target for phosphorylation by PKCα (Tigges *et al.*, 2003), the activation of PKC might alter caveolae internalization through reorganization of actin filaments.

In contrast to the Upla and Sharma findings, Parton *et al.* observed that the general PKC activator PMA did not significantly change the surface level of cross-linked alkaline phosphatase in A431 cells, suggesting that caveolae internalization in these cells is not mediated by PKCα (Parton *et al.*, 1994). These data might indicate that stimulation of PKCα alone is not sufficient to induce caveolae internalization and that the process requires activation of additional proteins.

3. Phospholipases

A p122/RhoGAP is a phospholipase Cδ (PLCδ)-interacting protein (Sekimata *et al.*, 1999) that possesses GTPase activating protein (GAP) activity for Rho and enhances the enzymatic activity of PLCδ1. It is localized to caveolae mainly via its GAP domain (Yamaga *et al.*, 2004). Interestingly, transient expression of p122/RhoGAP regulates caveolae internalization in fibroblastic baby hampster kidney (BHK) cells, suggesting that p122/RhoGAP might affect caveolae-dependent endocytosis through reorganization of actin filaments (Yamaga *et al.*, 2004).

The phospholipase A_2 (PLA_2) enzyme catalyzes the hydrolysis of glycerophospholipids important in second messenger generation and membrane vesiculation. Sphingomyelin (SM) inhibits PLA_2 due to the binding of the enzyme to SM. Cholesterol activates the enzyme by displacing SM from its association with PLA_2. Therefore, PLA_2 activity differs in various lipid microdomains that might regulate the vesiculation of rafts, including caveolae. A potential role for PLA_2 in caveolar vesicle budding was found by Staneva *et al.*, who showed that phospholipase alone can promote raft budding and fission *in vitro* (Staneva *et al.*, 2004). However, its putative role in caveolar fission *in vivo* needs to be determined.

IV. PERSPECTIVES

Only recently has it been demonstrated that caveolae are not permanent fixtures of the plasma membrane, but are dynamic vesicle carriers that may fuse, bud, detach, and translocate to provide cells with valuable communication with the outside milieu. We now know that caveolae may exist as both static and dynamic organelles that release from the plasma membrane or that may fuse into tubular continuums that bridge dorsal and ventral cell surfaces. The field has identified a limited number of different structural and enzymatic components that contribute to caveolar dynamics. It is currently unclear how the different caveolin gene products co-assemble into functional flask-shaped containers and how these accessory proteins might contribute to the assembly-scission process. Most intriguing is the fact that the caveolin proteins may be highly localized and sequestered to specific cellular domains in cells but are not assembled into caveolae. Other cellular regions may have many caveolae and relatively modest levels of caveolin protein. What are the signals for caveolar assembly and scission, and how are these processes compartmentalized? How are caveolae transported into the cell interior, and how are interactions with specific molecular motor enzymes initiated? There is much to do toward understanding caveolar dynamics.

REFERENCES

Ahn, S., Maudsley, S., Luttrell, L. M., Lefkowitz, R. J., and Daaka, Y. (1999). Src-mediated tyrosine phosphorylation of dynamin is required for beta2-adrenergic receptor internalization and mitogen-activated protein kinase signaling. *J. Biol. Chem.* **274,** 1185–1188.

Anderson, H. A., Chen, Y., and Norkin, L. C. (1996). Bound simian virus 40 translocates to caveolin-enriched membrane domains, and its entry is inhibited by drugs that selectively disrupt caveolae. *Mol. Biol. Cell* **7,** 1825–1834.

Aoki, T., Nomura, R., and Fujimoto, T. (1999). Tyrosine phosphorylation of caveolin-1 in the endothelium. *Exp. Cell Res.* **253,** 629–636.

Barylko, B., Binns, D., Lin, K. M., Atkinson, M. A., Jameson, D. M., Yin, H. L., and Albanesi, J. P. (1998). Synergistic activation of dynamin GTPase by Grb2 and phosphoinositides. *J. Biol. Chem.* **273,** 3791–3797.

Benlimame, N., Le, P. U., and Nabi, I. R. (1998). Localization of autocrine motility factor receptor to caveolae and clathrin-independent internalization of its ligand to smooth endoplasmic reticulum. *Mol. Biol. Cell* **9,** 1773–1786.

Breuza, L., Corby, S., Arsanto, J. P., Delgrossi, M. H., Scheiffele, P., and A., Le Bivic. (2002). The scaffolding domain of caveolin 2 is responsible for its Golgi localization in Caco-2 cells. *J. Cell Sci.* **115,** 4457–4467.

Butler, M. H., David, C., Ochoa, G. C., Freyberg, Z., Daniell, L., Grabs, D., Cremona, O., and DeCamilli, P. (1997). Amphiphysin II (SH3P9; BIN1), a member of the amphiphysin/Rvs family, is concentrated in the cortical cytomatrix of axon initial segments and nodes of ranvier in brain and around T tubules in skeletal muscle. *J. Cell Biol.* **137,** 1355–1367.

Cao, H., Garcia, F., and McNiven, M. A. (1998). Differential distribution of dynamin isoforms in mammalian cells. *Mol. Biol. Cell* **9,** 2595–2609.

Cao, H., Orth, J. D., Chen, J., Weller, S. G., Heuser, J. E., and McNiven, M. A. (2003). Cortactin is a component of clathrin-coated pits and participates in receptor-mediated endocytosis. *Mol. Cell. Biol.* **23,** 2162–2170.

Cao, H., Thompson, H. M., Krueger, E. W., and McNiven, M. A. (2000). Disruption of Golgi structure and function in mammalian cells expressing a mutant dynamin. *J. Cell Sci.* **113**(Pt 11), 1993–2002.

Carr, J. F. and Hinshaw, J. E. (1997). Dynamin assembles into spirals under physiological salt conditions upon the addition of GDP and gamma-phosphate analogues. *J. Biol. Chem.* **272,** 28030–28035.

Chang, W. J., Rothberg, K. G., Kamen, B. A., and Anderson, R. G. (1992). Lowering the cholesterol content of MA104 cells inhibits receptor-mediated transport of folate. *J. Cell Biol.* **118,** 63–69.

Clark, S. G., Shurland, D. L., Meyerowitz, E. M., Bargmann, C. I., and van der Bliek, A. M. (1997). A dynamin GTPase mutation causes a rapid and reversible temperature-inducible locomotion defect in C. elegans. *Proc. Natl. Acad. Sci. USA* **94,** 10438–10443.

Cook, T., Mesa, K., and Urrutia, R. (1996). Three dynamin-encoding genes are differentially expressed in developing rat brain. *J. Neurochem.* **67,** 927–931.

Cook, T. A., Urrutia, R., and McNiven, M. A. (1994). Identification of dynamin 2, an isoform ubiquitously expressed in rat tissues. *Proc. Natl. Acad. Sci. USA* **91,** 644–648.

Damke, H., Baba, T., Warnock, D. E., and Schmid, S. L. (1994). Induction of mutant dynamin specifically blocks endocytic coated vesicle formation. *J. Cell Biol.* **127,** 915–934.

David, C., McPherson, P. S., Mundigl, O., and de Camilli, P. (1996). A role of amphiphysin in synaptic vesicle endocytosis suggested by its binding to dynamin in nerve terminals. *Proc. Natl. Acad. Sci. USA* **93,** 331–335.

Dessy, C., Kelly, R. A., Balligand, J. L., and Feron, O. (2000). Dynamin mediates caveolar sequestration of muscarinic cholinergic receptors and alteration in NO signaling. *EMBO J.* **19,** 4272–4280.

Di Guglielmo, G. M., Le Roy, C., Goodfellow, A. F., and Wrana, J. L. (2003). Distinct endocytic pathways regulate TGF-beta receptor signalling and turnover. *Nat. Cell Biol.* **5,** 410–421.

Engqvist-Goldstein, A. E. and Drubin, D. G. (2003). Actin assembly and endocytosis: From yeast to mammals. *Annu. Rev. Cell Dev. Biol.* **19,** 287–332.

Farsad, K., Ringstad, N., Takei, K., Floyd, S. R., Rose, K., and De Camilli, P. (2001). Generation of high curvature membranes mediated by direct endophilin bilayer interactions. *J. Cell Biol.* **155,** 193–200.

Foster-Barber, A. and Bishop, J. M. (1998). Src interacts with dynamin and synapsin in neuronal cells. *Proc. Natl. Acad. Sci. USA* **95,** 4673–4677.

Fujimoto, T., Kogo, H., Nomura, R., and Une, T. (2000). Isoforms of caveolin-1 and caveolar structure. *J. Cell Sci.* **113**(Pt 19), 3509–3517.

Galbiati, F., Volonte, D., Engelman, J. A., Watanabe, G., Burk, R., Pestell, R. G., and Lisanti, M. P. (1998). Targeted downregulation of caveolin-1 is sufficient to drive cell transformation and hyperactivate the p42/44 MAP kinase cascade. *EMBO J.* **17,** 6633–6648.

Guipponi, M., Scott, H. S., Chen, H., Schebesta, A., Rossier, C., and Antonarakis, S. E. (1998). Two isoforms of a human intersectin (ITSN) protein are produced by brain-specific alternative splicing in a stop codon. *Genomics* **53,** 369–376.

Henley, J. R., Krueger, E. W., Oswald, B. J., and McNiven, M. A. (1998). Dynamin-mediated internalization of caveolae. *J. Cell Biol.* **141,** 85–99.

Herskovits, J. S., Burgess, C. C., Obar, R. A., and Vallee, R. B. (1993). Effects of mutant rat dynamin on endocytosis. *J. Cell Biol.* **122,** 565–578.

Hill, E., van Der Kaay, J., Downes, C. P., and Smythe, E. (2001). The role of dynamin and its binding partners in coated pit invagination and scission. *J. Cell Biol.* **152,** 309–323.

Ishikawa, H. (1968). Formation of elaborate networks of T-system tubules in cultured skeletal muscle with special reference to the T-system formation. *J. Cell Biol.* **38,** 51–66.

John, T. A., Vogel, S. M., Tiruppathi, C., Malik, A. B., and Minshall, R. D. (2003). Quantitative analysis of albumin uptake and transport in the rat microvessel endothelial monolayer. *Am. J. Physiol. Lung Cell Mol. Physiol.* **284,** L187–L196.

Kamitani, A., Yamada, H., Kinuta, M., Watanabe, M., Li, S. A., Matsukawa, T., McNiven, M., Kumon, H., and Takei, K. (2002). Distribution of dynamins in testis and their possible relation to spermatogenesis. *Biochem. Biophys. Res. Commun.* **294,** 261–267.

Kang, Y. S., Ko, Y. G., and Seo, J. S. (2000). Caveolin internalization by heat shock or hyperosmotic shock. *Exp. Cell Res.* **255,** 221–228.

Kessels, M. M., Engqvist-Goldstein, A. E., Drubin, D. G., and Qualmann, B. (2001). Mammalian Abp1, a signal-responsive F-actin-binding protein, links the actin cytoskeleton to endocytosis via the GTPase dynamin. *J. Cell Biol.* **153,** 351–366.

Kim, Y. N. and Bertics, P. J. (2002). The endocytosis-linked protein dynamin associates with caveolin-1 and is tyrosine phosphorylated in response to the activation of a noninternalizing epidermal growth factor receptor mutant. *Endocrinology* **143,** 1726–1731.

Kiss, A. L. and Geuze, H. J. (1997). Caveolae can be alternative endocytotic structures in elicited macrophages. *Eur. J. Cell Biol.* **73,** 19–27.

Lahtinen, U., Honsho, M., Parton, R. G., Simons, K., and Verkade, P. (2003). Involvement of caveolin-2 in caveolar biogenesis in MDCK cells. *FEBS Lett.* **538,** 85–88.

Le, P. U., Guay, G., Altschuler, Y., and Nabi, I. R. (2002). Caveolin-1 is a negative regulator of caveolae-mediated endocytosis to the endoplasmic reticulum. *J. Biol. Chem.* **277,** 3371–3379.

Lee, A., Frank, D. W., Marks, M. S., and Lemmon, M. A. (1999). Dominant-negative inhibition of receptor-mediated endocytosis by a dynamin-1 mutant with a defective pleckstrin homology domain. *Curr. Biol.* **9,** 261–264.

Lee, E. and De Camilli, P. (2002). Dynamin at actin tails. *Proc. Natl. Acad. Sci. USA* **99,** 161–166.

Lee, E., Marcucci, M., Daniell, L., Pypaert, M., Weisz, O. A., Ochoa, G. C., Farsad, K., Wenk, M. R., and De Camilli, P. (2002a). Amphiphysin 2 (Bin1) and T-tubule biogenesis in muscle. *Science* **297,** 1193–1196.

Lee, H., Park, D. S., Wang, X. B., Scherer, P. E., Schwartz, P. E., and Lisanti, M. P. (2002b). Src-induced phosphorylation of caveolin-2 on tyrosine(19). Phospho-caveolin-2 (Tyr(P)19) is localized near focal adhesions, remains associated with lipid rafts/caveolae, but no longer forms a high molecular mass hetero-oligomer with caveolin-1. *J. Biol. Chem.* **277,** 34556–34567.

Li, H., Brodsky, S., Basco, M., Romanov, V., De Angelis, D. A., and Goligorsky, M. S. (2001). Nitric oxide attenuates signal transduction: Possible role in dissociating caveolin-1 scaffold. *Circ. Res.* **88,** 229–236.

Li, S., Seitz, R., and Lisanti, M. P. (1996). Phosphorylation of caveolin by src tyrosine kinases. The alpha-isoform of caveolin is selectively phosphorylated by v-Src *in vivo*. *J. Biol. Chem.* **271,** 3863–3868.

Marjomaki, V., Pietiainen, V., Matilainen, H., Upla, P., Ivaska, J., Nissinen, L., Reunanen, H., Huttunen, P., Hyypia, T., and Heino, J. (2002). Internalization of echovirus 1 in caveolae. *J. Virol.* **76,** 1856–1865.

McNiven, M. A., Baldassarre, M., and Buccione, R. (2004). The role of dynamin in the assembly and function of podosomes and invadopodia. *Front. Biosci.* **9,** 1944–1953.

McNiven, M. A., Kim, L., Krueger, E. W., Orth, J. D., Cao, H., and Wong, T. W. (2000). Regulated interactions between dynamin and the actin-binding protein cortactin modulate cell shape. *J. Cell Biol.* **151,** 187–198.

Meyer, H. W., Westermann, M., Stumpf, M., Richter, W., Ulrich, A. S., and Hoischen, C. (1998). Minimal radius of curvature of lipid bilayers in the gel phase state corresponds to the dimension of biomembrane structures "caveolae." *J. Struct. Biol.* **124,** 77–87.

Minshall, R. D., Sessa, W. C., Stan, R. V., Anderson, R. G., and Malik, A. B. (2003). Caveolin regulation of endothelial function. *Am J. Physiol. Lung Cell Mol. Physiol.* **285,** L1179–L1183.

Minshall, R. D., Tiruppathi, C., Vogel, S. M., Niles, W. D., Gilchrist, A., Hamm, H. E., and Malik, A. B. (2000). Endothelial cell-surface gp60 activates vesicle formation and trafficking via G(i)-coupled Src kinase signaling pathway. *J. Cell Biol.* **150,** 1057–1070.

Monier, S., Parton, R. G., Vogel, F., Behlke, J., Henske, A., and Kurzchalia, T. V. (1995). VIP21-caveolin, a membrane protein constituent of the caveolar coat, oligomerizes *in vivo* and *in vitro*. *Mol. Biol. Cell* **6,** 911–927.

Muhlberg, A. B., Warnock, D. E., and Schmid, S. L. (1997). Domain structure and intramolecular regulation of dynamin GTPase. *EMBO J.* **16,** 6676–6683.

O'Bryan, J. P., Mohney, R. P., and Oldham, C. E. (2001). Mitogenesis and endocytosis: What's at the intersection? *Oncogene* **20,** 6300–6308.

Oh, P., McIntosh, D. P., and Schnitzer, J. E. (1998). Dynamin at the neck of caveolae mediates their budding to form transport vesicles by GTP-driven fission from the plasma membrane of endothelium. *J. Cell Biol.* **141,** 101–114.

Okamoto, M., Schoch, S., and Sudhof, T. C. (1999a). EHSH1/intersectin, a protein that contains EH and SH3 domains and binds to dynamin and SNAP-25. A protein connection between exocytosis and endocytosis? *J. Biol. Chem.* **274,** 18446–18454.

Okamoto, P. M., Tripet, B., Litowski, J., Hodges, R. S., and Vallee, R. B. (1999b). Multiple distinct coiled-coils are involved in dynamin self-assembly. *J. Biol. Chem.* **274,** 10277–10286.

Parton, R. G., Joggerst, B., and Simons, K. (1994). Regulated internalization of caveolae. *J. Cell Biol.* **127,** 1199–1215.

Parton, R. G., Way, M., Zorzi, N., and Stang, E. (1997). Caveolin-3 associates with developing T-tubules during muscle differentiation. *J. Cell Biol.* **136,** 137–154.

Pawson, A. J., Maudsley, S. R., Lopes, J., Katz, A. A., Sun, Y. M., Davidson, J. S., and Millar, R. P. (2003). Multiple determinants for rapid agonist-induced internalization of a nonmammalian gonadotropin-releasing hormone receptor: A putative palmitoylation site and threonine doublet within the carboxyl-terminal tail are critical. *Endocrinology* **144,** 3860–3871.

Pelkmans, L., Puntener, D., and Helenius, A. (2002). Local actin polymerization and dynamin recruitment in SV40-induced internalization of caveolae. *Science* **296,** 535–539.

Powell, K. A. and Robinson, P. J. (1995). Dephosphin/dynamin is a neuronal phosphoprotein concentrated in nerve terminals: Evidence from rat cerebellum. *Neuroscience* **64,** 821–833.

Predescu, S. A., Predescu, D. N., Timblin, B. K., Stan, R. V., and Malik, A. B. (2003). Intersectin regulates fission and internalization of caveolae in endothelial cells. *Mol. Biol. Cell* **14,** 4997–5010.

Pucharcos, C., Casas, C., Nadal, M., Estivill, X., and de la Luna, S. (2001). The human intersectin genes and their spliced variants are differentially expressed. *Biochim. Biophys. Acta* **1521,** 1–11.

Qualmann, B., Roos, J., DiGregorio, P. J., and Kelly, R. B. (1999). Syndapin I, a synaptic dynamin-binding protein that associates with the neural Wiskott-Aldrich syndrome protein. *Mol. Biol. Cell* **10,** 501–513.

Ramjaun, A. R., Micheva, K. D., Bouchelet, I., and McPherson, P. S. (1997). Identification and characterization of a nerve terminal-enriched amphiphysin isoform. *J. Biol. Chem.* **272,** 16700–16706.

Remacle, A., Murphy, G., and Roghi, C. (2003). Membrane type I-matrix metalloproteinase (MT1-MMP) is internalised by two different pathways and is recycled to the cell surface. *J. Cell Sci.* **116,** 3905–3916.

Richterova, Z., Liebl, D., Horak, M., Palkova, Z., Stokrova, J., Hozak, P., Korb, J., and Forstova, J. (2001). Caveolae are involved in the trafficking of mouse polyomavirus virions and artificial VP1 pseudocapsids toward cell nuclei. *J. Virol.* **75,** 10880–10891.

Roy, S., Luetterforst, R., Harding, A., Apolloni, A., Etheridge, M., Stang, E., Rolls, B., Hancock, J. F., and Parton, R. G. (1999). Dominant-negative caveolin inhibits H-Ras function by disrupting cholesterol-rich plasma membrane domains. *Nat. Cell Biol.* **1,** 98–105.

Sargiacomo, M., Sudol, M., Tang, Z., and Lisanti, M. P. (1993). Signal transducing molecules and glycosyl-phosphatidylinositol-linked proteins form a caveolin-rich insoluble complex in MDCK cells. *J. Cell Biol.* **122,** 789–807.

Scaife, R. and Margolis, R. L. (1990). Biochemical and immunochemical analysis of rat brain dynamin interaction with microtubules and organelles *in vivo* and *in vitro*. *J. Cell Biol.* **111,** 3023–3033.

Schafer, D. A., Weed, S. A., Binns, D., Karginov, A. V., Parsons, J. T., and Cooper, J. A. (2002). Dynamin2 and cortactin regulate actin assembly and filament organization. *Curr. Biol.* **12,** 1852–1857.

Scherer, P. E., Tang, Z., Chun, M., Sargiacomo, M., Lodish, H. F., and Lisanti, M. P. (1995). Caveolin isoforms differ in their N-terminal protein sequence and subcellular distribution. Identification and epitope mapping of an isoform-specific monoclonal antibody probe. *J. Biol. Chem.* **270,** 16395–16401.

Schlegel, A., Arvan, P., and Lisanti, M. P. (2001). Caveolin-1 binding to endoplasmic reticulum membranes and entry into the regulated secretory pathway are regulated by serine phosphorylation. Protein sorting at the level of the endoplasmic reticulum. *J. Biol. Chem.* **276,** 4398–4408.

Schlegel, A., Schwab, R. B., Scherer, P. E., and Lisanti, M. P. (1999). A role for the caveolin scaffolding domain in mediating the membrane attachment of caveolin-1. The caveolin scaffolding domain is both necessary and sufficient for membrane binding *in vitro*. *J. Biol. Chem.* **274,** 22660–22667.

Schnitzer, J. E., Oh, P., and McIntosh, D. P. (1996). Role of GTP hydrolysis in fission of caveolae directly from plasma membranes. *Science* **274,** 239–242.

Schnitzer, J. E., Oh, P., Pinney, E., and Allard, J. (1994). Filipin-sensitive caveolae-mediated transport in endothelium: Reduced transcytosis, scavenger endocytosis, and capillary permeability of select macro-molecules. *J. Cell Biol.* **127,** 1217–1232.

Sekimata, M., Kabuyama, Y., Emori, Y., and Homma, Y. (1999). Morphological changes and detachment of adherent cells induced by p122, a GTPase-activating protein for Rho. *J. Biol. Chem.* **274,** 17757–17762.

Shajahan, A. N., Timblin, B. K., Sandoval, R., Tiruppathi, C., Malik, A. B., and Minshall, R. D. (2004). Role of Src-induced dynamin-2 phosphorylation in caveolae-mediated endocytosis in endothelial cells. *J. Biol. Chem.* **279,** 20392–20400.

Sharma, D. K., Brown, J. C., Choudhury, A., Peterson, T. E., Holicky, E., Marks, D. L., Simari, R., Parton, R. G., and Pagano, R. E. (2004). Selective stimulation of caveolar endocytosis by glycosphingolipids and cholesterol. *Mol. Biol. Cell* **15,** 3114–3122.

Smart, E. J., Ying, Y. S., and Anderson, R. G. (1995). Hormonal regulation of caveolae internalization. *J. Cell Biol.* **131,** 929–938.

Smirnova, E., Shurland, D. L., Newman-Smith, E. D., Pishvaee, B., and van der Bliek, A. M. (1999). A model for dynamin self-assembly based on binding between three different protein domains. *J. Biol. Chem.* **274,** 14942–14947.

Sowa, G., Pypaert, M., Fulton, D., and Sessa, W. C. (2003). The phosphorylation of caveolin-2 on serines 23 and 36 modulates caveolin-1-dependent caveolae formation. *Proc. Natl. Acad. Sci. USA* **100,** 6511–6516.

Stahlhut, M. and van Deurs, B. (2000). Identification of filamin as a novel ligand for caveolin-1, evidence for the organization of caveolin-1-associated membrane domains by the actin cytoskeleton. *Mol. Biol. Cell* **11,** 325–337.

Staneva, G., Angelova, M. I., and Koumanov, K. (2004). Phospholipase A2 promotes raft budding and fission from giant liposomes. *Chem. Phys. Lipids* **129,** 53–62.

Staples, R. R. and Ramaswami, M. (1999). Functional analysis of dynamin isoforms in *Drosophila melano-gaster*. *J. Neurogenet.* **13,** 119–143.

Szaszak, M., Gaborik, Z., Turu, G., McPherson, P. S., Clark, A. J., Catt, K. J., and Hunyady, L. (2002). Role of the proline-rich domain of dynamin-2 and its interactions with Src homology 3 domains during endocytosis of the AT1 angiotensin receptor. *J. Biol. Chem.* **277,** 21650–21656.

Thompson, H. M., Skop, A. R., Euteneuer, U., Meyer, B. J., and McNiven, M. A. (2002). The large GTPase dynamin associates with the spindle midzone and is required for cytokinesis. *Curr. Biol.* **12,** 2111–2117.

Thomsen, P., Roepstorff, K., Stahlhut, M., and van Deurs, B. (2002). Caveolae are highly immobile plasma membrane microdomains, which are not involved in constitutive endocytic trafficking. *Mol. Biol. Cell* **13,** 238–250.

Tigges, U., Koch, B., Wissing, J., Jockusch, B. M., and Ziegler, W. H. (2003). The F-actin cross-linking and focal adhesion protein filamin A is a ligand and in vivo substrate for protein kinase C α. *J. Biol. Chem.* **278,** 23561–23569.

Tiruppathi, C., Song, W., Bergenfeldt, M., Sass, P., and Malik, A. B. (1997). Gp60 activation mediates albumin transcytosis in endothelial cells by tyrosine kinase-dependent pathway. *J. Biol. Chem.* **272,** 25968–25975.

Upla, P., Marjomaki, V., Kankaanpaa, P., Ivaska, J., Hyypia, T., Van Der Goot, F. G., and Heino, J. (2004). Clustering induces a lateral redistribution of α 2 β 1 integrin from membrane rafts to caveolae and subsequent protein kinase C-dependent internalization. *Mol. Biol. Cell* **15,** 625–636.

van der Bliek, A. M. and Meyerowitz, E. M. (1991). Dynamin-like protein encoded by the Drosophila shibire gene associated with vesicular traffic. *Nature* **351,** 411–414.

Wigge, P. and McMahon, H. T. (1998). The amphiphysin family of proteins and their role in endocytosis at the synapse. *Trends Neurosci.* **21,** 339–344.

Yamabhai, M., Hoffman, N. G., Hardison, N. L., McPherson, P. S., Castagnoli, L., Cesareni, G., and Kay, B. K. (1998). Intersectin, a novel adaptor protein with two Eps15 homology and five Src homology 3 domains. *J. Biol. Chem.* **273,** 31401–31407.

Yamaga, M., Sekimata, M., Fujii, M., Kawai, K., Kamata, H., Hirata, H., Homma, Y., and Yagisawa, H. (2004). A PLCdelta1-binding protein, p122/RhoGAP, is localized in caveolin-enriched membrane domains and regulates caveolin internalization. *Genes Cells* **9,** 25–37.

Yao, Q., Chen, J., Cao, H., Orth, J. D., McCaffery, J. M., Stan, R. V., and McNiven, M. A. (2005). Caveolin-1 interacts directly with dynamin-2. *J. Mol. Biol.* **348,** 491–501.

Zhang, P. and Hinshaw, J. E. (2001). Three-dimensional reconstruction of dynamin in the constricted state. *Nat. Cell Biol.* **3,** 922–926.

Zhao, Y. Y., Liu, Y., Stan, R. V., Fan, L., Gu, Y., Dalton, N., Chu, P. H., Peterson, K., Ross, J., Jr., and Chien, K. R. (2002). Defects in caveolin-1 cause dilated cardiomyopathy and pulmonary hypertension in knockout mice. *Proc. Natl. Acad. Sci. USA* **99,** 11375–11380.

Chapter 4

Lipid Raft-Mediated Entry of Bacteria into Host Cells

Matthew J. Duncan[1] and Soman N. Abraham[2]

[1]Department of Molecular Genetics and Microbiology, Duke University Medical Center, Durham, North Carolina 27710; [2]Department of Pathology, Duke University Medical Center, North Carolina 27710

I. INTRODUCTION

Many bacterial pathogens are known to enter into host cells, either as a means of avoiding the host immune system or as an integral part of their replicative cycle. One major hurdle intracellular pathogens must overcome is degradation in the classical endosomal/lysosomal fusion pathway. Some pathogens avoid this by escaping from their phagosomes before lysosomal fusion occurs, others actively modify their compartment to prevent lysosomal fusion, while still others have evolved defenses that allow them to survive in the harsh lysosomal environment (Small *et al.*, 1994). Recent work has demonstrated that some pathogens have evolved a means of entering cells in a way that completely sidesteps the endosomal/lysosomal pathway altogether by utilizing discrete plasma membrane microdomains located on the host cell surface. These microdomains, commonly referred to as caveolae or lipid rafts, are enriched in cholesterol, glycosphingolipids, and glycosylphosphatidylinosotol (GPI)-anchored molecules (Anderson, 1998; Kurzchalia and Parton, 1999). Caveolae were initially described as cave-like invaginations of the plasma membrane, 50–100 nm in diameter, which contained a distinctive protein, caveolin. Since the protein caveolin is not found in all microdomains with the biochemical properties of caveolae, and not all domains with these properties have the distinctive shape of caveolae, we will define caveolae as pleomorphic lateral assemblies containing the protein caveolin-1 that are enriched in cholesterol, glycosphingolipids, and GPI-anchored molecules and that have the

Advances in Molecular and Cell Biology, Vol. 36

ISBN: 0-444-51500-3
DOI: 10.1016/S0079-6603(05)36004-1

distinct morphological appearance of cave-like invaginations by electron microscopy (EM). Microdomains with the same biochemical properties that do not exhibit the cave-like structure of caveolae will be referred to as lipid rafts, regardless of the presence or absence of caveolin-1 within these microdomains (Harder and Simons, 1997; Simons and Ikonen, 1997; Kurzchalia and Parton, 1999). For ease of reading, except when it is necessary to make a clear distinction between caveolae and lipid rafts, the term lipid raft will be used.

Most of the microbes that are utilizing lipid rafts for entry into cells bind distinct receptors on the host cell surface, which could lead one to envision multiple endocytic mechanisms for the various microbes. Another explanation for this conserved mechanism of entry is that the various receptors can all lead to the same endocytic pathway. This would imply that the various microbial receptors are located within lipid rafts or move into lipid rafts after microbial contact, as has been found to be the case for the receptors of many of the microbes and bacterial toxins that use lipid rafts (Lencer *et al.*, 1999; Shin *et al.*, 2000; Duncan *et al.*, 2004). Many different signaling molecules are concentrated within lipid rafts (Anderson, 1998; Okamoto *et al.*, 1998), including all the machinery necessary for vesicle budding, docking, and fusion (Schnitzer *et al.*, 1995), making this an ideal pathway for microbial entry. Lipid rafts also seem to be linked to the actin cytoskeleton based on the localization of actin mobilizing/modulating molecules to lipid rafts and the recent report that the F-actin cross-linking protein filamin is a ligand for the protein caveolin-1 (Rozelle *et al.*, 2000; Stahlhut and van Deurs, 2000).

The idea that such vastly different bacterial pathogens such as FimH-expressing *Escherichia coli* (Baorto *et al.*, 1997; Shin *et al.*, 2000; Duncan *et al.*, 2004), *Salmonella typhimurium* (Catron *et al.*, 2002; Garner *et al.*, 2002), and *Chlamydia trachomatis* (Norkin *et al.*, 2001) can gain entry into host cells through a pathway involving lipid rafts, or use these structures to aid in intracellular survival, could be thought to indicate a high degree of similarity in their mechanism of entry. The fact that FimH-expressing *E. coli* and the infectious form of *C. trachomatis* (the metabolically inert elementary body) enter in a passive manner while *S. typhimurium* enters by actively secreting effector proteins into the cytoplasm of the host cell via a type III secretion system, thereby inducing uptake (Collazo and Galan, 1997; Suarez and Russmann, 1998), indicates that, though similarities in their mechanism of entry exist, there are distinct differences as well. Regardless, it is becoming clear that several pathogenic bacteria have recognized the endocytic functions of lipid rafts as a means of gaining entry into host cells in a manner that aids in the avoidance of lysosomal fusion and have evolved to fully exploit this mechanism of entry.

II. LIPID RAFT-MEDIATED ENTRY OF BACTERIA INTO HOST CELLS

Typically, the major obstacle to a bacterial infection is the host's immune system. Many pathogenic bacteria have evolved mechanisms that allow avoidance of host defenses, and one of the most studied is the entry of bacteria into host cells. In the intracellular environment, the bacteria are given protection from the humoral arm of the immune system, giving them a somewhat sheltered niche in which to survive. A simplified way to look at bacterial entry is to separate host cell entry into two groups, active entry and passive entry. A paradigm of active entry is that of *S. typhimurium*. The bacterium is known to actively secrete various effector proteins into the host cell cytoplasm through its type III secretion machinery and thereby induce bacterial uptake (Collazo and Galan, 1997; Suarez and Russmann, 1998). An example of passive bacterial entry is that of *C. trachomatis*, in which the infectious chlamydial elementary body (EB) is metabolically inert and yet is able to gain entry (Norkin *et al.*, 2001).

Foremost among the dangers of the intracellular lifestyle are the potentially lethal effects of fusion of the bacteria-containing endosome with the lysosomal compartment. Many bacteria have evolved mechanisms to avoid this fate. For example, *S. typhimurium* actively modifies its compartment to avoid fusion with lysosomes (referred to as active persistence), while *Listeria monocytogenes* escapes from its phagosome and survives in the host cell's cytoplasm (Small *et al.*, 1994). An additional, newly discovered mechanism to avoid the endosomal/lysosomal compartment is to gain entry via lipid rafts, which are believed to naturally avoid fusion with lysosomes (Table I) (Fig. 1). As is discussed below, some bacteria may merge the active avoidance of lysosomal fusion with entry through lipid rafts.

One of the best-characterized interactions between bacteria and lipid rafts is the entry of the normally noninvasive type 1 fimbriated *E. coli* into murine bone marrow derived mast cells (BMMCs) (Shin *et al.*, 2000). The interaction is mediated by the type 1 fimbrial adhesin, FimH, and its cognate receptor on BMMCs, the GPI-anchored protein CD48 (Malaviya *et al.*, 1999). *E. coli* is taken up into BMMCs in a tight-fitting compartment that is positive for various markers of lipid rafts such as cholesterol, the ganglioside GM1, and the protein caveolin-1 (Shin *et al.*, 2000). In a more definitive experiment, FimH-mediated bacterial entry was found to be susceptible to the lipid raft-disrupting agents filipin and methyl β-cyclodextrin, unlike opsonin-mediated bacterial entry, which was not inhibitable by these substances. As a result of utilizing the lipid raft pathway to gain entry, the bacteria were able to avoid the potent bacteriocidal functions of the BMMCs and survive intracellularly. It is believed that entry through lipid rafts is responsible for this avoidance of lysosomal fusion, as the laboratory strain of *E. coli* used in these experiments is not thought to modify its compartment after entry (referred to as passive persistence). Though a single caveolae or lipid raft could be large enough to internalize a small viral particle, such as SV40 (Anderson *et al.*, 1996; Stang *et al.*, 1997), it would be far too small to encapsulate and internalize a large bacterium. Therefore, it has been hypothesized that many plasmalemmal and vesicular lipid rafts are recruited to sites of bacterial attachment to form large bacteria-encapsulating chambers (Shin *et al.*, 2000; Shin and Abraham, 2001a,b).

FimH-expressing *E. coli* have also recently been shown to invade human and mouse bladder epithelial cells (BECs) via a mechanism requiring lipid rafts (Duncan *et al.*, 2004), an event that is thought to play an important role in the pathogenesis of urinary tract infections (UTIs) (Mulvey *et al.*, 1998, 2000). In contrast to *E. coli*

Table I
Characteristics of Bacterial Entry Through Lipid Rafts

Pathogen	Site of entry	Nature of entry and persistence	Receptor	References
FimH-expressing *E. coli*	Caveolae/ Raft (caveolin-1 required for BEC entry)	Passive entry and passive persistence	CD48, Uroplakin Ia (MΦA/BMMC, BEC)	Baorto *et al.*, 1997; Shin *et al.*, 2000; Duncan *et al.*, 2004
C. trachomatis (certain serovars)	Raft/Caveolae (caveolin not required)	Passive entry and active persistence	Unknown	Norkin *et al.*, 2001; Stuart *et al.*, 2003
SbfI-expressing *S. pyogenes*	Caveolae	Passive entry and unknown persistence	SbfI-fibronectin-$\alpha5\beta1$ integrin interaction	Rohde *et al.*, 2003
C. jejuni	Lipid raft (no caveolin association)	Active entry and active/passive persistence	Unknown	Wooldridge *et al.*, 1996
Mycobacterium	Lipid raft associated with cholesterol after entry	Passive entry and active persistence	CR3 and an unknown GPI anchored molecule	Gatfield and Pieters, 2000; Peyron *et al.*, 2000
Brucella spp.	Lipid raft associated with cholesterol and GPI-anchored proteins	Active/passive entry and active persistence	GPI-anchored protein?	Naroeni and Porte, 2002; Watarai *et al.*, 2002
S. typhimurium	Lipid raft associated with cholesterol and GPI-anchored proteins	Active entry and active persistence	Unknown	Catron *et al.*, 2002; Garner *et al.*, 2002
S. flexneri	Lipid raft associated with cholesterol and GPI-anchored proteins	Active entry and active persistence	Unknown CD44-IpaB interaction may be required for optimal binding and invasion	Lafont *et al.*, 2002

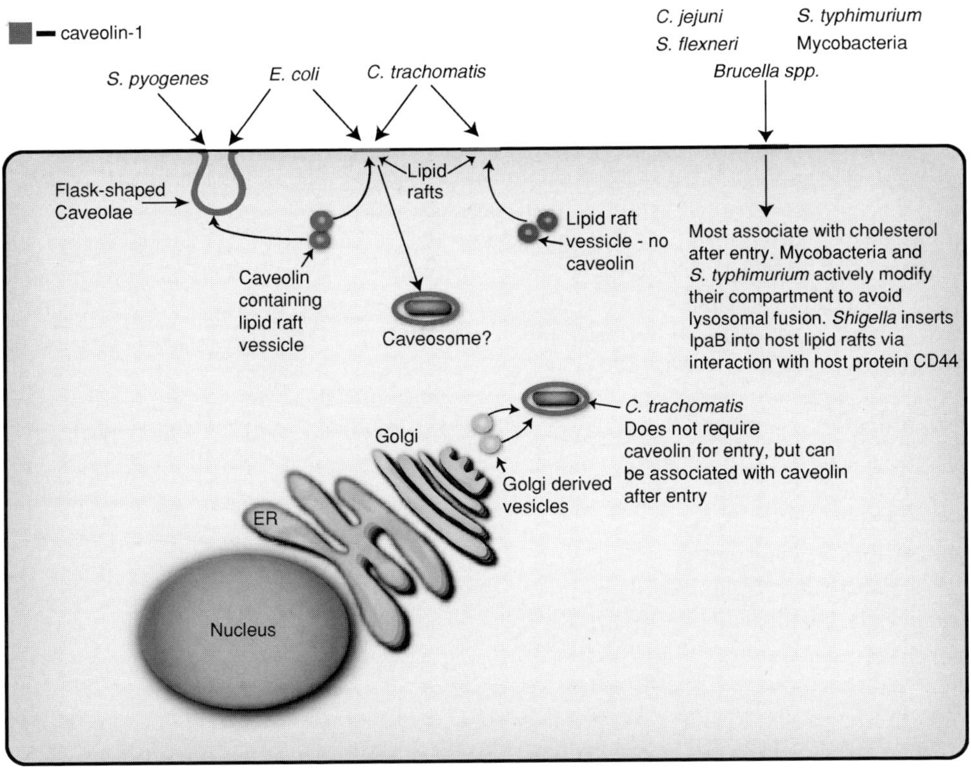

Figure 1. Model depicting the major defining characteristics of entry of various bacteria into most cells via lipid rafts. (See Color Insert.)

infecting BMMCs, *E. coli* infecting BECs induced the formation of flask-shaped caveolae at the cell surface, which intimately associated with the bacteria and remained associated with the bacteria after entry had occurred (Duncan *et al.*, 2004). In addition, the *in vivo* host BEC receptor for FimH-expressing *E. coli* was shown to be located in BEC lipid rafts, and methyl β-cyclodextrin treatment during infection in a mouse model of UTI was shown to greatly inhibit *E. coli* invasion of BEC (Duncan *et al.*, 2004). Caveolar induction has also been observed on endothelial cells infected by SfbI-expressing *Streptococcus pyogenes* (Rohde *et al.*, 2003). This physical association between flask-shaped caveolae and *E. coli* or *S. pyogenes* is reminiscent of the association between various viruses (Simian virus 40, polyoma virus, Echovirus 1) and caveolae on the cell surface (Stang *et al.*, 1997; Richterova *et al.*, 2001; Marjomaki *et al.*, 2002). The absolute requirement for caveolae formation in *E. coli* or *S. pyogenes* invasion is not known, but the protein caveolin-1, which is required for caveolae formation, was found to be necessary for optimal *E. coli* invasion of BEC (Vogel *et al.*, 1998; Razani *et al.*, 2001; Rohde *et al.*, 2003; Duncan *et al.*, 2004).

Some serovars of *C. trachomatis* are also thought to exploit lipid rafts as a means of entry into both phagocytic and nonphagocytic cells (Norkin *et al.*, 2001; Jutras *et al.*, 2003; Stuart *et al.*, 2003). Like the entry of *E. coli* into mast cells, *C. trachomatis* entry is inhibited by lipid raft-disrupting agents, and the bacteria-containing compartments react with antisera against the protein caveolin-1, though caveolin-1 is not thought to be essential for entry to occur (Stuart *et al.*, 2003). A significant difference is that *C. trachomatis* plays a much more active role in avoiding lysosomal fusion at later stages than *E. coli*, as the chlamydial vesicle traffics to the Golgi region and actively intercepts sphingolipid-, and presumably caveolin-1-containing vesicles from the Golgi (Hackstadt *et al.*, 1995, 1996), and some serovars of *C. trachomatis* are believed to actively modify their compartment to avoid lysosomal fusion (Scidmore *et al.*, 1996). Not all serovars of *C. trachomatis* appear to utilize lipid rafts for entry, and some may, in fact, utilize a clathrin-dependent mechanism of entry (Hodinka *et al.*, 1988; Wyrick *et al.*, 1989; Stuart *et al.*, 2003). Which pathway is utilized could depend on the particular receptor(s) recognized by different chlamydial strains on the host cell surface (Stuart *et al.*, 2003), where lipid raft-associated molecules lead to lipid raft-mediated entry, and coated-pit-associated molecules direct entry via clathrin-coated pits.

Other bacteria that have been linked to entry via lipid rafts include *Campylobacter jejuni* (Wooldridge *et al.*, 1996), *Mycobacterium bovis* and *Mycobacterium kansasii* (Gatfield and Pieters, 2000; Peyron *et al.*, 2000), *Brucella suis* and *Brucella abortus* (Naroeni and Porte, 2002; Watarai *et al.*, 2002), *S. typhimurium* (Catron *et al.*, 2002; Garner *et al.*, 2002), and *Shigella flexneri* (Lafont *et al.*, 2002). The entry of these bacteria into various cell types has been shown to require cholesterol, and in the cases of *M. bovis, B. abortus, S. typhimurium,* and *S. flexneri*, cholesterol has been seen to accumulate at sites of bacterial entry. *S. flexneri* seems to utilize lipid rafts for adherence to host cells as well as invasion, since host cell cholesterol depletion inhibited both events (Lafont *et al.*, 2002), unlike the interaction of type 1 fimbriated *E. coli* with BEC, where lipid raft integrity was only required for entry (Duncan *et al.*, 2004). It is noteworthy that secreted effector proteins of the type III secretion systems of both *S. typhimurium* and *S. flexneri* associate with lipid rafts during host cell infection (Lafont *et al.*, 2002; Knodler *et al.*, 2003). This points to a sophisticated interaction between pathogen and host, in which the bacteria could actually be "remodeling" host cell lipid rafts to initiate bacterial entry and survival. In addition, both Mycobacteria and Salmonella are thought to also actively modulate their compartments to avoid lysosomal fusion, and lipid rafts may also play a role in these remodeling events (Garcia-del Portillo, 1996; Ferrari *et al.*, 1999; Knodler *et al.*, 2003).

The differences that exist in the entry of various bacteria into host cells through lipid rafts could be mediated by a number of factors. One possibility is the receptor(s) on the host cells recognized by the bacteria, at least in cases where a specific receptor is known to mediate entry. For example, type 1 fimbriated *E. coli* recognize uroplakin Ia on BEC while *M. bovis* entry into macrophages has been shown to involve CR3 and an, as yet, unknown GPI-anchored molecule (Peyron *et al.*, 2000; Zhou *et al.*, 2001; Duncan *et al.*, 2004). Another possibility is the extent to which the bacteria play an active role in their entry into host cells. The entry of *E. coli* into BEC is entirely dependent on activity derived from the BEC, as even FimH-coated latex beads can gain entry (Martinez *et al.*, 2000), while *S. typhimurium* is known to secrete effector proteins directly into the host cytoplasm to induce bacterial uptake (Collazo and Galan, 1997; Suarez and Russmann, 1998). The possibility arises that the specific way in which bacteria interact with lipid rafts could lead to different levels of dependence upon these structures to gain entry into host cells, and could potentially affect the level to which the bacteria must actively resist lysosomal fusion.

III. CONCLUSION

Lipid rafts are an increasingly recognized portal of entry for a surprising variety of pathogenic microorganisms. These structures have been linked to the entry and/or intracellular survival of viruses, which have been associated with flask-shaped caveolae (Stang *et al.*, 1997; Richterova *et al.*, 2001; Marjomaki *et al.*, 2002), bacteria, whether they gain entry and persist in a passive (*E. coli*) or active (*S. typhimurium*) manner, and even large parasites (*P. falciparum*) (Lauer *et al.*, 2000; Shin *et al.*, 2000; Catron *et al.*, 2002; Garner *et al.*, 2002; Duncan *et al.*, 2004). Though these various microbes have all co-opted lipid rafts, they seem to not interact with them in the same manner. Trafficking after entry, and even the level to which the microbes depend on lipid rafts for entry and persistence, may vary from microbe to microbe. These differences could be mediated by the specific host cell receptor(s) each bacteria recognizes on the cell surface, the type or shape of the microdomain involved in entry (lipid raft vs flask-shaped caveolae), or the particular nature of entry/persistence (passive vs active) each microbe expresses. Microbes are not the only organisms to utilize the flexibility of lipid rafts for targeting and gaining entry into mammalian cells. This was demonstrated by McIntosh *et al.,* who used an antibody that specifically targets lung caveolae to achieve specific delivery and transendothelial transport of the antibody to underlying tissue cells (McIntosh *et al.*, 2002). Indeed, further study of the nature of lipid rafts, potentially using microbes as the tools of study, could allow new and exciting methods of specifically delivering therapeutic agents to treat a variety of illnesses, including microbial infections themselves.

REFERENCES

Anderson, H. A., Chen, Y., and Norkin, L. C. (1996). Bound simian virus 40 translocates to caveolin-enriched membrane domains, and its entry is inhibited by drugs that selectively disrupt caveolae. *Mol. Biol. Cell* **7,** 1825–1834.

Anderson, R. G. (1998). The caveolae membrane system. *Annu. Rev. Biochem.* **67,** 199–225.

Baorto, D. M., Gao, Z., Malaviya, R., Dustin, M. L., van der Merwe, A., Lublin, D. M., and Abraham, S. N. (1997). Survival of FimH-expressing enterobacteria in macrophages relies on glycolipid traffic. *Nature* **389,** 636–639.

Catron, D. M., Sylvester, M. D., Lange, Y., Kadekoppala, M., Jones, B. D., Monack, D. M., Falkow, S., and Haldar, K. (2002). The Salmonella-containing vacuole is a major site of intracellular cholesterol accumulation and recruits the GPI-anchored protein CD55. *Cell Microbiol.* **4,** 315–328.

Collazo, C. M. and Galan, J. E. (1997). The invasion-associated type-III protein secretion system in Salmonella—a review. *Gene* **192,** 51–59.

Duncan, M. J., Li, G., Shin, J. S., Carson, J. L., and Abraham, S. N. (2004). Bacterial penetration of bladder epithelium through lipid rafts. *J. Biol. Chem.* **279,** 18944–18951.

Ferrari, G., Langen, H., Naito, M., and Pieters, J. (1999). A coat protein on phagosomes involved in the intracellular survival of mycobacteria. *Cell* **97,** 435–447.

Garcia-del Portillo, F. (1996). Interaction of Salmonella with lysosomes of eukaryotic cells. *Microbiologia* **12,** 259–266.

Garner, M. J., Hayward, R. D., and Koronakis, V. (2002). The Salmonella pathogenicity island 1 secretion system directs cellular cholesterol redistribution during mammalian cell entry and intracellular trafficking. *Cell Microbiol.* **4,** 153–165.

Gatfield, J. and Pieters, J. (2000). Essential role for cholesterol in entry of mycobacteria into macrophages. *Science* **288,** 1647–1650.

Hackstadt, T., Rockey, D. D., Heinzen, R. A., and Scidmore, M. A. (1996). *Chlamydia trachomatis* interrupts an exocytic pathway to acquire endogenously synthesized sphingomyelin in transit from the Golgi apparatus to the plasma membrane. *EMBO J.* **15,** 964–977.

Hackstadt, T., Scidmore, M. A., and Rockey, D. D. (1995). Lipid metabolism in *Chlamydia trachomatis*-infected cells: Directed trafficking of Golgi-derived sphingolipids to the chlamydial inclusion. *Proc. Natl. Acad. Sci. USA* **92,** 4877–4881.

Harder, T. and Simons, K. (1997). Caveolae, DIGs, and the dynamics of sphingolipid-cholesterol microdomains. *Curr. Opin. Cell Biol.* **9,** 534–542.

Hodinka, R. L., Davis, C. H., Choong, J., and Wyrick, P. B. (1988). Ultrastructural study of endocytosis of *Chlamydia trachomatis* by McCoy cells. *Infect. Immun.* **56,** 1456–1463.

Jutras, I., Abrami, L., and Dautry-Varsat, A. (2003). Entry of the lymphogranuloma venereum strain of *Chlamydia trachomatis* into host cells involves cholesterol-rich membrane domains. *Infect. Immun.* **71,** 260–266.

Knodler, L. A., Vallance, B. A., Hensel, M., Jackel, D., Finlay, B. B., and Steele-Mortimer, O. (2003). Salmonella type III effectors PipB and PipB2 are targeted to detergent-resistant microdomains on internal host cell membranes. *Mol. Microbiol.* **49,** 685–704.

Kurzchalia, T. V. and Parton, R. G. (1999). Membrane microdomains and caveolae. *Curr. Opin. Cell Biol.* **11,** 424–431.

Lafont, F., Tran Van Nhieu, G., Hanada, K., Sansonetti, P., and van der Goot, F. G. (2002). Initial steps of Shigella infection depend on the cholesterol/sphingolipid raft-mediated CD44-IpaB interaction. *EMBO J.* **21,** 4449–4457.

Lauer, S., VanWye, J., Harrison, T., McManus, H., Samuel, B. U., Hiller, N. L., Mohandas, N., and Haldar, K. (2000). Vacuolar uptake of host components, and a role for cholesterol and sphingomyelin in malarial infection. *EMBO J.* **19,** 3556–3564.

Lencer, W. I., Hirst, T. R., and Holmes, R. K. (1999). Membrane traffic and the cellular uptake of cholera toxin. *Biochim. Biophys. Acta* **1450,** 177–190.

Malaviya, R., Gao, Z., Thankavel, K., van der Merwe, P. A., and Abraham, S. N. (1999). The mast cell tumor necrosis factor alpha response to FimH-expressing *Escherichia coli* is mediated by the glycosylphosphatidylinositol-anchored molecule CD48. *Proc. Natl. Acad. Sci. USA* **96,** 8110–8115.

Marjomaki, V., Pietiainen, V., Matilainen, H., Upla, P., Ivaska, J., Nissinen, L., Reunanen, H., Huttunen, P., Hyypia, T., and Heino, J. (2002). Internalization of echovirus 1 in caveolae. *J. Virol.* **76,** 1856–1865.

Martinez, J. J., Mulvey, M. A., Schilling, J. D., Pinkner, J. S., and Hultgren, S. J. (2000). Type 1 pilus-mediated bacterial invasion of bladder epithelial cells. *EMBO J.* **19,** 2803–2812.

McIntosh, D. P., Tan, X. Y., Oh, P., and Schnitzer, J. E. (2002). Targeting endothelium and its dynamic caveolae for tissue-specific transcytosis *in vivo:* A pathway to overcome cell barriers to drug and gene delivery. *Proc. Natl. Acad. Sci. USA* **99,** 1996–2001.

Mulvey, M. A., Lopez-Boado, Y. S., Wilson, C. L., Roth, R., Parks, W. C., Heuser, J., and Hultgren, S. J. (1998). Induction and evasion of host defenses by type 1-piliated uropathogenic *Escherichia coli*. *Science* **282,** 1494–1497.

Mulvey, M. A., Schilling, J. D., Martinez, J. J., and Hultgren, S. J. (2000). Bad bugs and beleaguered bladders: Interplay between uropathogenic *Escherichia coli* and innate host defenses. *Proc. Natl. Acad. Sci. USA* **97,** 8829–8835.

Naroeni, A. and Porte, F. (2002). Role of cholesterol and the ganglioside GM(1) in entry and short-term survival of Brucella suis in murine macrophages. *Infect. Immun.* **70,** 1640–1644.

Norkin, L. C., Wolfrom, S. A., and Stuart, E. S. (2001). Association of caveolin with *Chlamydia trachomatis* inclusions at early and late stages of infection. *Exp. Cell Res.* **266,** 229–238.

Okamoto, T., Schlegel, A., Scherer, P. E., and Lisanti, M. P. (1998). Caveolins, a family of scaffolding proteins for organizing "preassembled signaling complexes" at the plasma membrane. *J. Biol. Chem.* **273,** 5419–5422.

Peyron, P., Bordier, C., N'Diaye, E. N., and Maridonneau-Parini, I. (2000). Nonopsonic phagocytosis of Mycobacterium kansasii by human neutrophils depends on cholesterol and is mediated by CR3 associated with glycosylphosphatidylinositol-anchored proteins. *J. Immunol.* **165,** 5186–5191.

Razani, B., Engelman, J. A., Wang, X. B., Schubert, W., Zhang, X. L., Marks, C. B., Macaluso, F., Russell, R. G., Li, M., Pestell, R. G., Di Vizio, D., Hou, H., Jr., Kneitz, B., Lagaud, G., Christ, G. J., Edelmann, W., and Lisanti, M. P. (2001). Caveolin-1 null mice are viable but show evidence of hyperproliferative and vascular abnormalities. *J. Biol. Chem.* **276,** 38121–38138.

Richterova, Z., Liebl, D., Horak, M., Palkova, Z., Stokrova, J., Hozak, P., Korb, J., and Forstova, J. (2001). Caveolae are involved in the trafficking of mouse polyomavirus virions and artificial VP1 pseudocapsids toward cell nuclei. *J. Virol.* **75,** 10880–10891.

Rohde, M., Muller, E., Chhatwal, G. S., and Talay, S. R. (2003). Host cell caveolae act as an entry-port for group A streptococci. *Cell Microbiol.* **5,** 323–342.

Rozelle, A. L., Machesky, L. M., Yamamoto, M., Driessens, M. H., Insall, R. H., Roth, M. G., Luby-Phelps, K., Marriott, G., Hall, A., and Yin, H. L. (2000). Phosphatidylinositol 4,5-bisphosphate induces actin-based movement of raft-enriched vesicles through WASP-Arp2/3. *Curr. Biol.* **10,** 311–320.

Schnitzer, J. E., Liu, J., and Oh, P. (1995). Endothelial caveolae have the molecular transport machinery for vesicle budding, docking, and fusion including VAMP, NSF, SNAP, annexins, and GTPases. *J. Biol. Chem.* **270,** 14399–14404.

Scidmore, M. A., Rockey, D. D., Fischer, E. R., Heinzen, R. A., and Hackstadt, T. (1996). Vesicular interactions of the *Chlamydia trachomatis* inclusion are determined by chlamydial early protein synthesis rather than route of entry. *Infect. Immun.* **64,** 5366–5372.

Shin, J. S. and Abraham, S. N. (2001a). Caveolae as portals of entry for microbes. *Microbes Infect.* **3,** 755–761.

Shin, J. S. and Abraham, S. N. (2001b). Cell biology. Caveolae–not just craters in the cellular landscape. *Science* **293,** 1447–1448.

Shin, J. S., Gao, Z., and Abraham, S. N. (2000). Involvement of cellular caveolae in bacterial entry into mast cells. *Science* **289,** 785–788.

Simons, K. and Ikonen, E. (1997). Functional rafts in cell membranes. *Nature* **387,** 569–572.

Small, P. L., Ramakrishnan, L., and Falkow, S. (1994). Remodeling schemes of intracellular pathogens. *Science* **263,** 637–639.

Stahlhut, M. and van Deurs, B. (2000). Identification of filamin as a novel ligand for caveolin-1: Evidence for the organization of caveolin-1-associated membrane domains by the actin cytoskeleton. *Mol. Biol. Cell* **11,** 325–337.

Stang, E., Kartenbeck, J., and Parton, R. G. (1997). Major histocompatibility complex class I molecules mediate association of SV40 with caveolae. *Mol. Biol. Cell* **8,** 47–57.

Stuart, E. S., Webley, W. C., and Norkin, L. C. (2003). Lipid rafts, caveolae, caveolin-1, and entry by Chlamydiae into host cells. *Exp. Cell Res.* **287,** 67–78.

Suarez, M. and Russmann, H. (1998). Molecular mechanisms of Salmonella invasion: The type III secretion system of the pathogenicity island 1. *Int. Microbiol.* **1,** 197–204.

Vogel, U., Sandvig, K., and van Deurs, B. (1998). Expression of caveolin-1 and polarized formation of invaginated caveolae in Caco-2 and MDCK II cells. *J. Cell Sci.* **111,** 825–832.

Watarai, M., Makino, S., Fujii, Y., Okamoto, K., and Shirahata, T. (2002). Modulation of Brucella-induced macropinocytosis by lipid rafts mediates intracellular replication. *Cell Microbiol.* **4,** 341–355.

Wooldridge, K. G., Williams, P. H., and Ketley, J. M. (1996). Host signal transduction and endocytosis of *Campylobacter jejuni. Microb. Pathog.* **21,** 299–305.

Wyrick, P. B., Choong, J., Davis, C. H., Knight, S. T., Royal, M. O., Maslow, A. S., and Bagnell, C. R. (1989). Entry of genital *Chlamydia trachomatis* into polarized human epithelial cells. *Infect. Immun.* **57,** 2378–2389.

Zhou, G., Mo, W. J., Sebbel, P., Min, G., Neubert, T. A., Glockshuber, R., Wu, X. R., Sun, T. T., and Kong, X. P. (2001). Uroplakin Ia is the urothelial receptor for uropathogenic *Escherichia coli*: Evidence from *in vitro* FimH binding. *J. Cell Sci.* **114,** 4095–4103.

Section III

Examples of the Role of Caveolins in Cell Signalling

Chapter 5

The Caveolin Interaction with Endothelial Nitric Oxide Synthase (eNOS)

Olivier Feron

University of Louvain Medical School, Unit of Pharmacology and Therapeutics, B-1200 Brussels, Belgium

I. INTRODUCTION

Nitric oxide (NO) is an important signaling molecule involved in many biological processes and is synthesized by a family of calmodulin-activated NO synthase (NOS). The endothelial NOS (eNOS) is the NOS isoform responsible for cardiovascular homeostasis, including regulation of blood pressure, permeability, and angiogenesis but also anti-thrombotic activity, vessel remodeling, and cardiac contractility. Indeed, eNOS is not only expressed in the vascular endothelium but also in platelets, smooth muscle cells, and cardiac myocytes. Of the NOS isoforms, eNOS is unique in that it is dually acylated by the fatty acids myristate and palmitate, thereby leading to the membrane location of this enzyme. Since eNOS activity is acutely regulated by multiple extracellular stimuli and the NO produced is a very labile messenger molecule with primarily paracrine function, the site of NO synthesis has a major influence on its biological activity. These different observations (i.e., acylation, extracellular stimulus sensing, paracrine influence) prompted the initial examinations of eNOS in

ISBN: 0-444-51500-3
DOI: 10.1016/S0079-6603(05)36005-3

caveolae (Garcia-Cardena *et al.*, 1996b; Shaul *et al.*, 1996). Accordingly, using cultured endothelial cells, Shaul *et al.* showed that eNOS was preferentially located in caveolae (versus the rest of the plasma membrane), with each acylation process enhancing the caveolar enrichment by 10-fold (Shaul *et al.*, 1996).

In addition to the requirement that NO be produced in a site adapted to its gaseous nature and function, the cytotoxicity of high levels of NO implies a tight regulation of its production. Accordingly, the discovery of the eNOS targeting to plasmalemmal caveolae rapidly led to the identification of such a control mechanism regulating NO release. Indeed, in 1996, we reported that in endothelial cells and cardiac myocytes, eNOS was quantitatively associated with caveolin-1 and caveolin-3, respectively (Feron *et al.*, 1996). One year later, several independent groups demonstrated that this interaction was inhibitory (Garcia-Cardena *et al.*, 1997; Ju *et al.*, 1997; Michel *et al.*, 1997a).

II. THE CAVEOLIN–eNOS INTERACTIONS

A. Molecular Determinants

Consecutively to the original observation of the caveolin interaction with eNOS, several laboratories investigated the molecular determinants of the caveolin–eNOS interaction using *in vitro* binding assay systems with glutathione S-transferase (GST) fusion proteins (including deletion mutants) and *in vivo* yeast two-hybrid system (Garcia-Cardena *et al.*, 1997; Ju *et al.*, 1997; Michel *et al.*, 1997a,b; Feron *et al.*, 1998a,b). The major conclusions from these studies were that eNOS and caveolin-1 interact directly rather than indirectly and that this interaction involves multiple sites within both proteins (Fig. 1). Indeed, two cytoplasmic domains of caveolin-1 were found to interact with eNOS: the N-terminal scaffolding and dimerization domain (amino acids 61–101) and to a lesser extent the C-terminal tail (amino acids 135–178). In the eNOS sequence, the sites of caveolin binding were localized both in the oxygenase and reductase domains (Fig. 1).

As mentioned above, N-terminal myristoylation and thiopalmitoylation of the eNOS participate in the targeting of the enzyme to caveolae (Feron *et al.*, 1998a). By exploring the differential effects of detergents (CHAPS vs octyl glucoside), we have shown that although eNOS palmitoylation is not absolutely required to induce the formation of the caveolin–eNOS complex, acylation largely facilitates the interaction between both proteins. Moreover, two groups of investigators, using either cell lines expressing caveolin but lacking caveolae (Sowa *et al.*, 2001) or siRNA to knock down caveolin-1 in cells normally expressing caveolin (Gonzalez *et al.*, 2004), reported that caveolin-1 itself was not required for the targeting of eNOS to caveolae/lipid rafts. Finally, tyrosine phosphorylation of eNOS was shown to parallel the enzyme interaction with caveolin, but whether both phenomena were causally related is still an open question (Garcia-Cardena *et al.*, 1996a; Venema *et al.*, 1997).

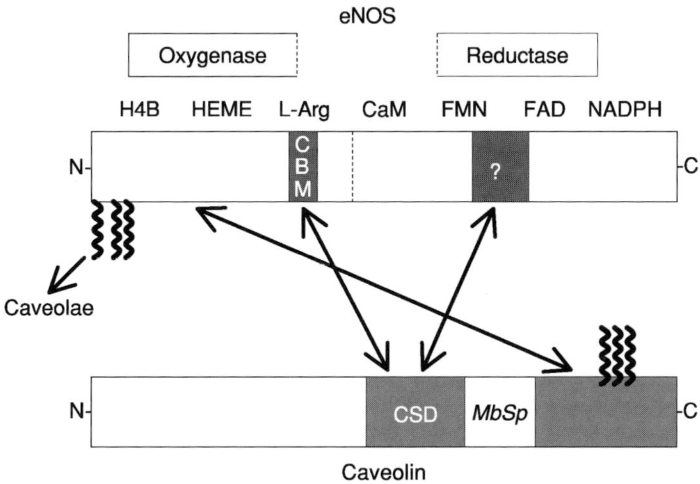

Figure 1. Scheme of the caveolin–eNOS interactions. Both the oxygenase and the reductase domains of eNOS interact with caveolin. The caveolin-binding motif (CBM) corresponds to the consensus sequence FPAAPFSGW that recognizes the caveolin scaffolding domain (CSD) within the caveolin sequence. At the other side of the membrane-spanning region of caveolin (MBSp), the 40 C-terminal residues of caveolin (involved in caveolin oligomer–oligomer interaction) also interact with the oxygenase domain of eNOS. Both proteins are acylated, and in the case of eNOS, myristoylation on the glycine and palmitoylation on two cysteines are key for facilitating the targeting of eNOS to lipid rafts/caveolae. Note that eNOS is active as a dimer and that the size of the two proteins is not to scale (human sequences: eNOS, 1205 residues and caveolin-1, 179 residues). (See Color Insert.)

B. Catalytic Regulation

Like all known nitric oxide synthases, eNOS enzyme activity is dependent on calmodulin (CaM) binding, the activation of which requires an increase in intracellular calcium (Michel and Feron, 1997). With the discovery of the caveolin–eNOS interaction, it appeared that calmodulin acted, in fact, as a direct allosteric competitor promoting the disruption of the heteromeric complex formed between eNOS and caveolin in a Ca^{2+}-dependent fashion (Michel *et al.*, 1997a,b). This inhibitory effect is mostly mediated by a specific region identified within the caveolin sequence (amino acids 82–101), the so-called caveolin scaffolding domain (CSD) (Fig. 1). Peptides corresponding to this domain interact directly with the enzyme and markedly inhibit NOS activity in endothelial cells (Michel *et al.*, 1997b).

The prototypical caveolin-binding motif (CBM) (e.g., a consensus sequence $\phi X \phi XXXX \phi XX \phi$, where ϕ is an aromatic residue (Trp, Phe, or Tyr) (Couet *et al.*, 1997)) was found within the eNOS sequence (i.e., FPAAPFSGW) (Fig. 1). Site-directed mutagenesis of this predicted motif within eNOS sequence (amino acids 350–358) blocked the ability of caveolin-1 to suppress NO release in co-transfection experiments (Garcia-Cardena *et al.*, 1997). Furthermore, incubation of pure eNOS with peptides derived from the scaffolding domains of caveolin-1 and -3, but not the analogous regions from caveolin-2, resulted in inhibition of eNOS, inducible NOS (iNOS), and neuronal NOS (nNOS) activities, suggesting a common mechanism and site of inhibition for all the NOS isoforms.

eNOS is a two-domain enzyme consisting of a N-terminal oxygenase domain (amino acids 1–491) that contains binding sites for heme, L-arginine (L-Arg), and tetrahydrobiopterin (H_4B) and a reductase domain (amino acids 492–1205) containing binding sites for flavin mononucleotide (FMN), flavin adenine dinucleotide (FAD), nicotinamide (NADPH), and CaM (Fig. 1). During NO synthesis, NADPH-derived electrons pass into the reductase domain flavins and then must be transferred to the heme located in the oxygenase domain so that the heme iron can bind O_2 and catalyze stepwise NO synthesis from L-Arg. CaM binding to its site (amino acids 493–512) activates NO synthesis by enabling the reductase domain to transfer electrons to the eNOS oxygenase domain.

The cav-1 scaffolding domain peptide (amino acids 82–101) was reported to inhibit NADPH oxidation in a manner reversible by CaM but did not affect NADPH-independent NO synthesis by full-length eNOS or its oxygenase domain, indicating that inhibition required the reductase domain (Ghosh *et al.*, 1998). Altogether, these studies suggest a model according to which caveolin interaction with the oxygenase domain only helps targeting the eNOS-caveolin complex to caveolae, whereas caveolin interaction with the reductase domain is primarily responsible for antagonizing CaM binding and for slowing electron transfer from the reductase, thus inhibiting heme iron reduction and NO synthesis (Fig. 1).

Interestingly, the caveolin binding motif in eNOS lies between the heme and the calmodulin-binding domains adjacent to a glutamate residue (Glu-361) necessary for the binding of L-arginine, suggesting that caveolin may interfere with heme iron reduction, similar to L-arginine-based NOS inhibitors.

III. CAVEOLIN, eNOS, AND CHOLESTEROL

The focus on the relationship between cholesterol and the caveolin–eNOS interaction naturally emerged from the observations by Fielding and colleagues (Bist *et al.*, 1997; Fielding *et al.*, 1997) according to which in human fibroblasts, the level of cellular free cholesterol (FC) regulates caveolin gene transcription. Because of the inhibitory interaction between eNOS and caveolin, a hypothesis was formulated stating that an increase in caveolin abundance in endothelial cells exposed to extracellular cholesterol could account for the reduction in NO bioavailability usually reported to be associated with hypercholesterolemia. Indeed, before the appearance of any ultrastructural change in the vessel wall, chronic elevations in serum cholesterol are usually associated with an impaired endothelium-dependent vasodilation due to a defect in the L-arginine/NO pathway. This primitive hallmark of endothelial dysfunction is thought to lead to the unopposed influence of thrombogenic and proliferative factors on the vessel wall and thereby to initiate the atherosclerotic process and lead to coronary and peripheral ischemic diseases.

As anticipated, exposure of endothelial cells to serum from hypercholesterolemic patients led to the upregulation of caveolin abundance and was paralleled by an increase in caveolin–eNOS complex formation (Feron *et al.*, 1999) (Fig. 2). Endothelial cell exposure to the low-density lipoprotein (LDL) fraction alone dose-dependently reproduced the upregulation of caveolin expression and its heterocomplex formation with eNOS. Importantly, both the basal NO release and agonist-stimulated NO production were repressed by high LDL cholesterol levels. Together, these data established a new mechanism that participates, through the modulation of caveolin abundance, in the pathogenesis of endothelial dysfunction. This mechanism is reversible, since treatment of endothelial cells with statins (HMGCoA reductase inhibitors) was shown to reduce caveolin abundance and to restore the basal and agonist-stimulated eNOS activity (Feron *et al.*, 2001) (Fig. 2). This finding was further validated in genetically dyslipidemic, apolipoprotein E (apoE)$^{-/-}$ mice, in which statins decreased caveolin-1 expression and corrected the altered heart rate and blood pressure variabilities, two parameters correlated with adverse prognosis in cardiovascular diseases (Pelat *et al.*, 2003).

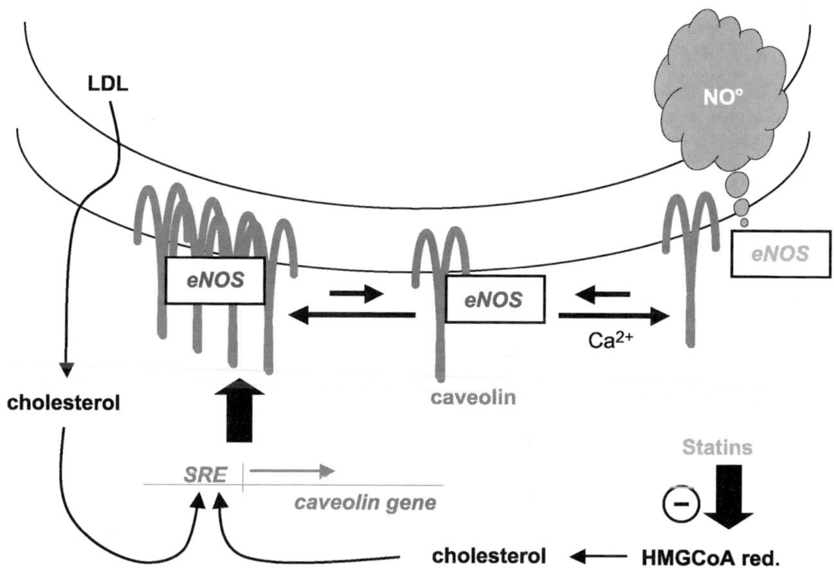

Figure 2. Regulation of the caveolin–eNOS interaction by cholesterol levels. The presence of sterol regulatory elements (SRE) and other regulatory sequences in the promoter of the caveolin gene accounts for the stimulatory transcription of caveolin in the presence of excess cholesterol, as encountered when endothelial cells are exposed to high levels of LDL cholesterol. The consecutive increase in caveolin blocks the basal and agonist-stimulated eNOS activity in endothelial cells. Statins, by inhibiting the endogenous synthesis of cholesterol in (peripheral) endothelial cells (as well as indirectly by reducing circulatory LDL cholesterol), negatively affect caveolin expression. A reduction in caveolin abundance (or a restoration if first exposed to high LDL levels) facilitates the calcium-calmodulin-mediated disruption of the caveolin–eNOS complex (e.g., displaces the equilibrium toward the caveolin-free eNOS state), thereby leading to NO release. (See Color Insert.)

Interestingly, we found that statins affect caveolin abundance in endothelial cells (EC) exposed to elevated but also normal amounts of LDL cholesterol. This finding highlighted the therapeutic potential of the statins' administration to decrease caveolin abundance in endothelial cells and thereby to correct NO-dependent endothelial dysfunction secondary to diseases other than hypercholesterolemia, such as hypertension or heart failure. Notably, changes in caveolin abundance also appeared to be sufficient to modify the ability of eNOS to interact with other modulators such as the chaperone Hsp90, resulting in substantial increases in eNOS activity (Feron *et al.*, 2001).

In parallel to the above observations that cholesterol through the modulation of caveolin gene transcription could affect eNOS activity, another paradigm dealing with the eNOS regulation by lipoproteins through the alteration of caveolae homeostasis was reported by the groups of P. Shaul and E. Smart. Accordingly, in a first set of studies, Blair *et al.* (1999) documented that oxidized low density lipoprotein (oxLDL) caused the displacement of both eNOS and caveolin from caveolae to an internal membrane fraction containing endoplasmic reticulum and Golgi apparatus (Fig. 3). In fact, oxLDL was shown to act as acceptors of cholesterol, leading to marked

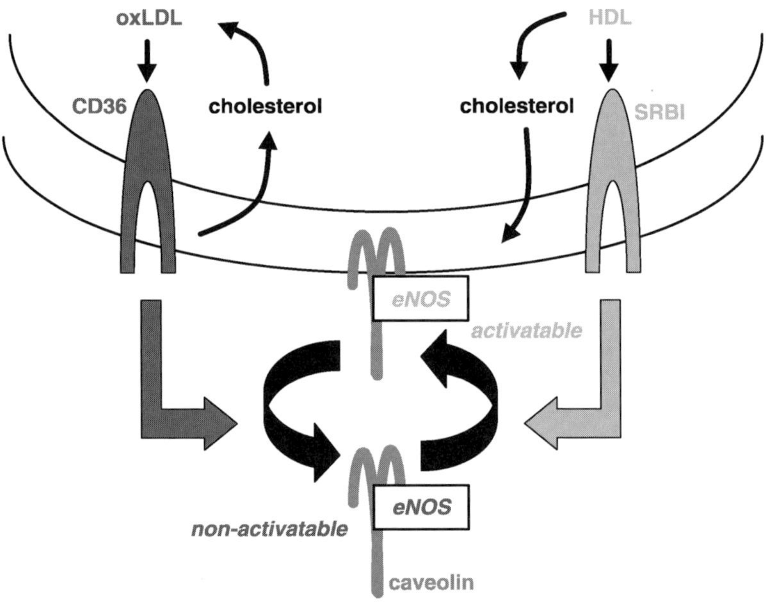

Figure 3. Regulation of the caveolin–eNOS interaction by oxLDL and HDL. The lipoprotein receptors CD36 and scavenger receptor class B member 1 (SRB1) are both enriched in caveolae. Oxidized LDL binds to CD36 and loads itself with cholesterol, leading to a marked depletion of caveolar cholesterol. Consecutively, caveolin and eNOS are translocated to intracellular compartments wherein basal and agonist-stimulated NO productions are dramatically reduced. HDL binds to SRBI and provisions cholesterol esters to the cell, thereby maintaining the integrity of caveolae and even reversing the deleterious effects of oxLDL. The normal subcellular location of caveolin and eNOS allows eNOS to be activated when needed and NO to exert its major function of vascular homeostasis control. (See Color Insert.)

depletion of caveolae cholesterol and consecutive redistribution of caveolin and eNOS (but not of other caveolar proteins). Importantly, the intracellular eNOS translocation led to the attenuation of eNOS activation upon exposure to acetylcholine, yielding a shift in the dose–response curve to the right by 100-fold; this process was shown to be independent of co- or posttranslational modifications of eNOS such as myristoylation, palmitoylation, or phosphorylation.

Further experiments using antibody blockade revealed that these effects were mediated through the binding of oxLDL to the class B CD36 receptor (Uittenbogaard *et al.*, 2000) (see Fig. 3). Other authors further reported that although caveolae isolated from apo$E^{-/-}$ mouse vessels do not contain eNOS and are not responsive to acetylcholine, the eNOS localization to caveolae and acetylcholine-evoked reduction in blood pressure are conserved in apoE/CD36 double knockout mice (Kincer *et al.*, 2002).

Because of the well-documented negative correlation between high-density lipoproteins (HDL) and the atherosclerosis risks induced by oxidative stress, the same authors tested whether HDL could prevent the adverse effects of oxLDL described previously. Amazingly, they found that the addition of HDL to medium containing oxLDL not only prevented the depletion of caveolar cholesterol and the eNOS displacement from caveolae but also restored the Ach-induced activation of the enzyme (Fig. 3). They further documented that the ability of HDL to maintain the concentration of caveolae-associated cholesterol was due to the provision of cholesterol esters and not to the inhibition of cholesterol transfer from caveolae to oxLDL (Kincer *et al.*, 2002). Interestingly, the scavenger receptor BI, the principal HDL receptor, was found to be highly enriched in endothelial cell caveolae and the use of SR-BI blocking antibody prevented the restoration of both the eNOS localization and activation by HDL (Yuhanna *et al.*, 2001) (see Fig. 3). Also, HDL was documented to induce NO-dependent dilation and enhanced Ach-induced relaxation of aorta from wild-type mice but not from SR-BI knockout mice. More recently, Mineo *et al.* (2003) reported that HDL stimulates eNOS through common upstream, Src-mediated signaling, which leads to parallel activation of Akt and mitogen-activated protein (MAP) kinases and their resultant independent activation of eNOS.

IV. CAVEOLIN, eNOS, AND ANGIOGENESIS

Angiogenesis is a complex process that involves a series of discrete phases, including endothelial cell migration, proliferation, and differentiation. In 1999, Liu *et al.* documented that, consistently with previous evidence that caveolin-1 is a negative regulator of cell proliferation, treatment of endothelial cells with a variety of growth factors including vascular endothelial growth factor (VEGF) and basic fibroblast growth factor (bFGF) resulted in the downregulation of caveolin-1 and disappearance of caveolae from cell surface. In good agreement with this study, Griffoni *et al.* (2000) reported that active (but not the scrambled) caveolin antisense oligodeoxynucleotides dramatically reduced vessel formation in the chorio-allantoic membrane assay. More recently, Liu and colleagues, by using superconfluent endothelial cells, showed that caveolin-1 levels were actually transiently upregulated during the endothelial cell differentiation process (e.g., just prior to endothelial tube formation) (Liu *et al.*, 2002). After these first studies, the temptation was great for many investigators to explore how nitric oxide, a key angiogenic mediator acting downstream of VEGF, was related to these changes in caveolin abundance considering the exquisite regulation of eNOS activity by the scaffold protein.

Evidence of this link came from a study using statins or HMGCoA reductase inhibitors. As mentioned above, these drugs exert direct beneficial effects on the endothelium through an increase in eNOS activity partly mediated by a decrease in caveolin abundance (Feron *et al.*, 2001), thereby eventually correcting the deleterious effects of an increased caveolin expression triggered by high LDL-cholesterol levels (Feron *et al.*, 1999). Brouet *et al.* (2001b) found that statins could stimulate tube or precapillary formation from endothelial cells cultured on reconstituted basement membrane matrix (Matrigel). Interestingly, while in macrovascular endothelial cells the pro-angiogenic effects of statins could be attributed to a decrease in caveolin abundance (e.g., in its inhibitory interaction with eNOS) (see Fig. 4), the role of caveolin was different in microvascular endothelial cells. In these cells, statins did not influence the abundance of caveolin protein. More exactly, the much larger pool of caveolin in microvascular endothelial cells (e.g., up to 10-fold higher than that of macrovascular endothelial cells) was not significantly reduced by the statin treatment to affect the regulation of eNOS activity. In fact, in microvascular endothelial cells, statins stimulated eNOS activity and angiogenesis through the recruitment of the chaperone protein hsp90 and consecutively of Akt that phosphorylates eNOS on Serine 1177. Still, Brouet *et al.* (2001b) showed that in the presence of an excess of the caveolin scaffolding domain-derived peptide, statins failed to induce the formation of the eNOS/Hsp90/Akt multicomplex and the associated endothelial tube formation on Matrigel. The calcium chelator BAPTA exerted similar blocking effects, indicating that statins first need to disrupt the caveolin–eNOS interaction by the classical increase in calcium–calmodulin complex to be active and stimulate angiogenesis. In a parallel study Brouet and colleagues documented that this sequence of events leading to eNOS activation was also true for VEGF signaling, making caveolin an important control point for angiogenesis regulation (Brouet *et al.*, 2001a).

Very recently, several groups of investigators directly explored this latter concept in *in vivo* experiments. Accordingly, the role of caveolin in tumor angiogenesis was examined. Gratton *et al.* (2003) reported that the caveolin scaffolding domain-derived peptide was able to block the NO-dependent increase in vascular permeability in tumors and thereby to block endothelial cell migration and exert anti-angiogenic effects by preventing the formation of a provisional matrix from the extravasated plasma protein (see below). In the same framework, Sonveaux *et al.* (2002) showed that the exposure of endothelial cells to ionizing radiations, as those administered during cancer radiotherapy, led to a decrease in caveolin abundance. This event, together with a direct increase in eNOS abundance, accounted for an increase in NO production, mediating stimulation of endothelial cell migration and re-organization in tubes (Sonveaux *et al.*, 2003). This study indicates that irradiation may be endowed

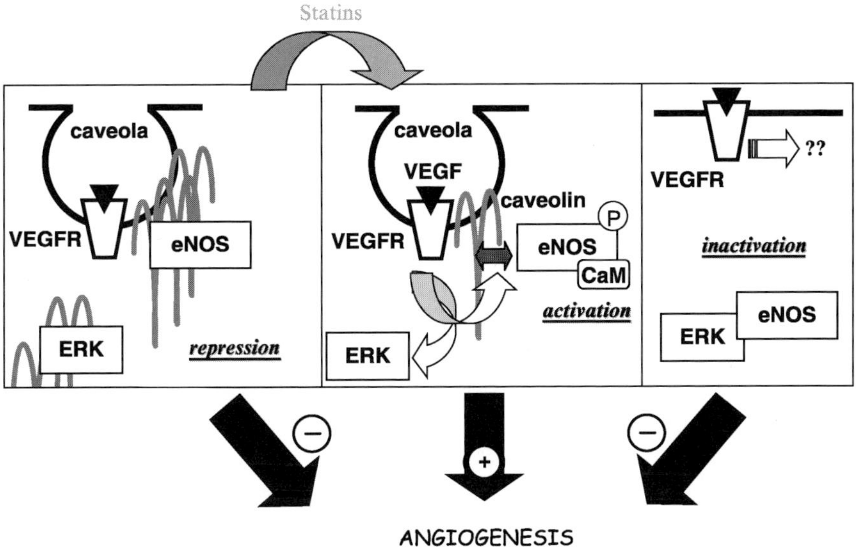

Figure 4. Model of the apparent paradoxical regulation of NO-mediated angiogenesis by caveolin. This three-panel scheme represents the relationship between the abundance of caveolin and the NO-mediated angiogenesis. Middle panel: the VEGF binding to the VEGFR-2 receptor leads to the activation of eNOS through the local increase in intracellular calcium (the endoplasmic reticulum has been located near to caveolae) and the activation of the PI3K/Akt pathway leading to the phosphorylation of eNOS on Serine 1177. Activation of ERK is also induced upon VEGF exposure and, together with the increase in NO, leads to endothelial cell migration, proliferation, and reorganization in a new vascular network. Left panel: when caveolin is increased in the cells, such as after exposure to high levels of LDL cholesterol or expression of recombinant caveolin, eNOS and ERK are maintained inactivated, preventing full activation on VEGF exposure and leading to a defect in angiogenesis. Such a negative effect may be desired to inhibit tumor neovascularization and thereby block tumor growth. Right panel: in the absence of caveolin, as encountered in mice deficient for the caveolin gene, the compartmentation of the different actors of the VEGF/NO signaling cascade is lacking and leads to a decrease in the efficacy of VEGF to stimulate angiogenesis. (See Color Insert.)

with collateral effects conferring resistance to tumors through the induction of angiogenesis that may participate in tumor regrowth after treatment.

Paradoxically, we and others found that angiogenesis was defective in mice deficient for caveolin ($Cav^{-/-}$). Accordingly, Woodman *et al.* (2003) reported that in $Cav^{-/-}$ mice, implanted Matrigel plugs and subcutaneously injected melanoma resulted in a dramatic reduction in both vessel infiltration and density, as compared with wild-type $Cav^{+/+}$ control mice. We reached similar conclusions by examining the impact of caveolae suppression in a model of adaptive angiogenesis obtained after femoral artery resection (Sonveaux *et al.*, 2004). Evaluation of the ischemic tissue perfusion and histochemical analyses revealed that contrary to $Cav^{+/+}$ mice, $Cav^{-/-}$ and $Cav^{+/-}$ mice failed to recover a functional vasculature and actually lost part of the ligated limbs, thereby recapitulating the effects of the NOS inhibitor L-NAME administered to operated $Cav^{+/+}$ mice. Isolation of endothelial cells from $Cav^{-/-}$ aorta allowed them to document that upon VEGF stimulation, NO production and endothelial tube formation were dramatically abrogated when compared with $Cav^{+/+}$ endothelial cells; the presence of a single allele of caveolin ($Cav^{+/-}$) was also insufficient to restore normal eNOS activation. The Ser1177 eNOS phosphorylation and Thr495 dephosphorylation but also the ERK phosphorylation were similarly altered in VEGF-treated $Cav^{-/-}$ endothelial cells. Caveolin transfection in $Cav^{-/-}$ endothelial cells redirected the VEGFR-2 in caveolar membranes and restored the VEGF-induced extracellular signal-regulated kinase (ERK) and eNOS activation (Fig. 4). However, when high levels of recombinant caveolin were reached, VEGF exposure failed to activate ERK and eNOS, thereby emphasizing the critical role of caveolae in ensuring the coupling between VEGFR-2 stimulation and downstream mediators of angiogenesis (Fig. 4).

More generally, these studies emphasize the complexity of the regulation of the multiple pathways supported by NO production in endothelial cells. Although endothelial cells completely deprived of caveolae is an extreme condition very unlikely to be reached in any cardiovascular diseases, the 50% reduction in caveolin and the still-associated functional alterations as observed in $Cav^{+/-}$ animals make these finding particularly relevant in the context of cardiovascular diseases. Indeed, important changes in the absolute or subcellular expression levels of caveolin-1 and -3 have been reported in hypertension (Piech *et al.*, 2003), hypercholesterolemia (Blair *et al.*, 1999; Feron *et al.*, 1999; Pelat *et al.*, 2003; Zhu *et al.*, 2003), and cardiomyopathies (Hare *et al.*, 2000; Piech *et al.*, 2002; Ratajczak *et al.*, 2003; Uray *et al.*, 2003). In all these disease states, both the "inhibitory" hypothesis (e.g., inhibition proportional to caveolin levels) and the "compartmentalizing" effect of caveolin (e.g., receptor/effector coupling is either prevented or promoted when/where caveolin is down- or upregulated) need to be integrated to determine the resulting effects of changes in caveolin abundance.

V. CAVEOLIN, eNOS, AND ENDOTHELIAL PERMEABILITY

Evidence of a direct link between NO, caveolin, and permeability was reported by Caldwell and colleagues. In the first of two papers, Feng *et al.* (1999a) reported that VEGF-R2 and eNOS colocalize with caveolin-1 in plasma membrane caveolae in retinal microvascular and aortic endothelial cells and that VEGF increases endothelial cell permeability by an eNOS-dependent mechanism of transcytosis in caveolae. In a second paper, Feng *et al.* (1999b) showed by using confocal microscopy that VEGF stimulated the translocation of eNOS, caveolin-1, and the VEGF receptor into the nucleus. They further confirmed by cell fractionation studies that levels of all three proteins within the caveolar compartment declined in a time-dependent fashion following VEGF treatment. This first set of studies suggested a double role for caveolae in the context of endothelial permeability: a direct structural implication in transcytosis but also the regulation through caveolin and NO of the transcription of genes potentially involved in the associated signaling process.

More recently, Chen *et al.* (2001) transfected endothelial cells with a green fluorescent protein (GFP)-caveolin vector to enable intravital microscopy of the dynamics of VEGF-induced caveolin-mediated increase in permeability. The most striking finding from this study was that upon VEGF exposure, caveolin appeared organized into elongated cell-spanning structures in endothelial cells. Electron microscopic studies further illustrated that this transcytotic pathway resulted from an enrichment in caveolae structures followed by their fission and finally their fusion (see Fig. 5). These authors also reported that diaphragmned fenestrae were also observed at later time points after the application of VEGF. Similarly, Yokomori and colleagues had previously documented by electron microscopy the presence of eNOS and caveolin-1 on the plasma membrane of sinusoidal endothelial fenestrae in liver tissue as well as in isolated sinusoidal endothelial cells (Yokomori *et al.*, 2001). These observations support the idea that fusion of caveolae eventually leads to the biogenesis of fenestrae or at least to clusters of vesicles allowing direct communication between opposite poles of endothelial cells. This later paradigm has been extensively described by Dvorak and

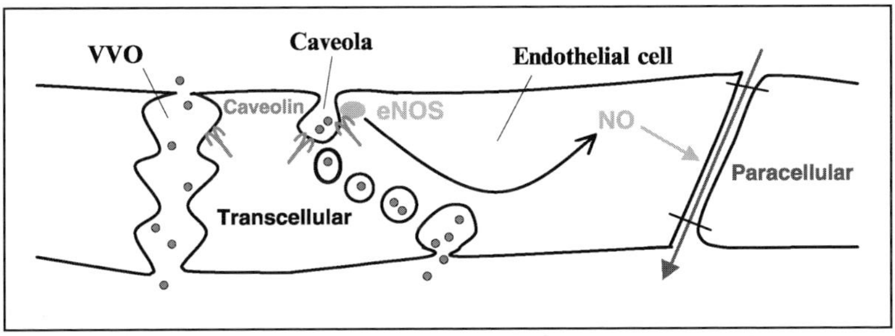

Permeability mode	Transcytosis (CAVEOLAE)	Paracellular (GAPS)
Wild-type	+	+
eNOS$^{-/-}$	+	-
Cav$^{-/-}$	-	++
Cav$^{-/-}$ + L-NAME	-	-

Figure 5. Implication of caveolin and eNOS in endothelial cell permeability. NO is known to regulate the paracellular permeability (e.g., the transport through the intercellular junctions), but the location of eNOS in caveolae makes the role of NO more complex. Indeed, the other major mode of permeability, transcytosis either through the fission of caveolae and transport from one pole of the cell to the other or through fusion of several caveolae to form vesicular-vacuolar organelles (VVO) or channels connecting each pole of the cell, requires caveolae and caveolin. It is therefore likely that alteration in transcytotic permeability will affect caveolin function and thereby the eNOS locale and regulation. In the bottom of the figure, the phenotypes observed in mice deficient for eNOS or caveolin gene are illustrated. In eNOS$^{-/-}$ mice, transcytosis is the major mode of permeability in endothelial cells, whereas in Cav$^{-/-}$ mice, paracellular transport is mostly involved in macromolecules transport and becomes exquisitely sensitive to the NOS inhibitor L-NAME. The efficacy of this pathway is even increased when compared to wild-type mice (in order to compensate for the lack of caveolae). (See Color Insert.)

Feng (2001) in venules in which cytoplasmic vesicles (caveolae) and vacuoles are proposed to organize into prominent clusters called vesiculo-vacuolar organelles (VVOs) (see Fig. 5). Whether fenestration results from the fusion of caveolae is, however, still under debate (Esser *et al.*, 1998; Vasile *et al.*, 1999), but recent studies have shown that at least for glomerular endothelial cells, fenestrae are still formed in caveolin-1-deficient mice (Sorensson *et al.*, 2002).

A series of studies either dealing with caveolin-derived peptides or using caveolin-deficient mice followed these initial reports and allowed the better understanding of the multiple roles of caveolae in the regulation of NO-mediated increase in endothelial cell permeability in physiological conditions but also, interestingly, in the context of inflammation and tumor angiogenesis. Bucci *et al.* (2000) examined the effects of the caveolin scaffolding domain in two experimental inflammatory models: subplantar administration of carrageenan and ear application of mustard oil. The caveolin-1 scaffolding domain peptide fused with a cellular internalization sequence (derived from the Antennapedia homeodomain) was systemically administered and led to the suppression of acute inflammation and vascular leak to the same extent as the NOS inhibitor L-NAME. More recently, Gratton *et al.* (2003) showed that the same fusion peptide derived from caveolin-1 inhibited eNOS-dependent vascular leakage in established tumors and consecutively delayed tumor progression in mice. The tumor vessel hyperpermeability is, indeed, thought to contribute to the extravasation of plasma proteins that provide a provisional matrix for the migration of endothelial cells in order to initiate the new vascular network formation. Through the inhibition of NO-dependent permeability, the caveolin-derived peptide was therefore proposed to block angiogenesis. These data emphasize the key role of caveolin and caveolae in regulating endothelial permeability. Surprisingly, with the dosage regimen that suppressed edema formation and tumor growth, the caveolin fusion peptide did not influence systemic blood pressure, administered intraperitoneally but also intravenously. This observation suggests that the peptide may preferentially inhibit eNOS in the endothelium of inflamed tissues or tumors (although hemodynamic experiments on non-anesthetized animals are awaiting).

Caveolin-deficient mice ($Cav^{-/-}$) also contributed to the understanding of the permeability process in the endothelium. Amazingly, the development of a compensatory mechanism to palliate the lack of caveolin and caveolae led the group of M. Lisanti to very informative insights on the relationship between the paracellular and transcellular modes of permeability (see Fig. 5, lower panel). In a first study, Schubert *et al.* reported that both fibroblasts and endothelial cells derived from caveolin-deficient mice were incapable of endocytosing albumin, a protein normally transported via transcytosis (Schubert *et al.*, 2001). In a consecutive paper, the same authors showed that despite the absence of caveolae, caveolin-deficient mice exhibited an important increase in microvascular permeability (Schubert *et al.*, 2002). In this study, injected radio-iodinated albumin was cleared from the circulation in $Cav^{-/-}$ mice at a faster rate than in control mice. Defects in tight junction morphology and in attachment to the basement membranes led the authors to conclude that the paracellular movement of macromolecules was increased in $Cav^{-/-}$ mice. Interestingly, they further showed that the use of the NOS-inhibitor L-NAME reversed the microvascular permeability phenotype of $Cav^{-/-}$ mice (Fig. 5). Some questions, however, about the level of implication of NO remain: did a constitutive activation of eNOS due to the lack of caveolin directly contribute to the observed alterations in capillaries' morphology or was the sensitivity to L-NAME in $Cav^{-/-}$ mice observed because of the compensatory development of the paracellular permeability that is intrinsically more NO dependent than the transcytotic pathway?

More work is needed, but from the studies described above, it is clear that both transcytosis (because of the implication of caveolae to form channels or act as shuttling vesicles) and paracellular permeability (because of the NO dependence of the phenomenon) can be influenced by the abundance of caveolin and the extent of its inhibitory interaction with eNOS. Finally, it should be kept in mind that although NO is usually described as a mediator of increased permeability in endothelial cells, there are reports claiming that the use of NOS inhibitor increases extravasation of albumin and macromolecules (Draijer *et al.*, 1995; Kurose *et al.*, 1995). A consensual view would be that basal NO generation by endothelial cells is necessary for the maintenance of the barrier functions of these cells, whereas high levels of NO stimulate vascular permeability. From animals deficient in the eNOS gene, it is however clear that the VEGF-induced increase in vascular permeability is blunted in these mice (Fukumura *et al.*, 2001).

VI. CAVEOLIN, eNOS, AND VASODILATION

It has been known for a long time that acute changes in pressure or shear stress induce the rapid release of NO from the vascular endothelium, resulting in vasodilation. Rizzo *et al.* (1998) reported that increasing vascular flow and pressure *in situ* rapidly activate caveolar eNOS with apparent eNOS dissociation from caveolin and association with calmodulin. Similarly, Omura *et al.* (2001) documented that eicosapentaenoic acid stimulated NO production in endothelial cells *in situ* and induced endothelium-dependent relaxation of precontracted coronary arteries by inducing the dissociation of eNOS from caveolin.

The paradigm of the caveolin abundance being critical to account for the dilatory capacity of vessels is also illustrated in pathological models. In a model of portal hypertension, Shah *et al.* (2001) found that caveolin protein levels from detergent-soluble liver lysates are significantly increased within sinusoids and venules and are associated with a significant reduction in NOS catalytic activity. Also, Pelligrino *et al.* (2000), by comparing ovariectomized rats and ovariectomized rats treated with 17beta-estradiol for 2 weeks, found that *in vivo* acetylcholine-induced vasodilation of the pial artery was absent in the former and restored in the latter. They further documented that these results were in part due to the estradiol-associated diminution in endothelial caveolin-1 (CAV-1) expression.

Using isolated tumor arterioles mounted on a pressure myograph, Sonveaux *et al.* (2002) documented that local tumor irradiation induced NO-mediated vasorelaxation through an increase in the abundance of the endothelial NO synthase but also a dramatic decrease in caveolin-1 abundance. The potentiation of the NO-dependent pathway was further shown to induce a marked increase in tumor blood flow and oxygenation as well as the delivery of drugs into the tumor.

In the two original papers reporting the phenotype of caveolin-deficient mice by the groups of T. Kurzchalia and M. Lisanti, the role of caveolae in endothelium-dependent and NO-mediated vascular relaxation was studied in isolated aortic rings (Drab *et al.*, 2001; Razani *et al.*, 2001). For this purpose, they employed phenylephrine (an α_1-adrenergic receptor agonist) as a vasoconstrictor and acetylcholine (Ach) to induce NO-dependent relaxation. Upon achieving a contractile tone with phenylephrine (that was oscillating and less important in $Cav^{-/-}$ mice), relaxation was induced by adding Ach in gradually increasing doses. A significantly greater relaxation was observed in Cav-1 null aortic rings at all acetylcholine concentrations examined. The two groups used different approaches to clarify whether this difference was dependent on the NO pathway. Drab and colleagues directly measured eNOS activity in primary cell culture of aortic vascular smooth muscle cells (Drab *et al.*, 2001). Accordingly, the basal release of NO, as measured with a selective microelectrode, was one-third higher than in wild-type cells, and the content of cyclic guanosine monophosphate (cGMP, the major mediator of NO signaling) was about threefold higher in knockout animals. Razani *et al.* (2001) chose to expose aorta rings to the NOS inhibitor L-NAME and found that the increase in the contractile response to phenylephrine was significantly greater for the Cav-1 null mice and reached a steady-state indistinguishable from that observed in the wild-type mice. Thus, both studies led to the conclusion that eNOS become hyperactivated in the absence of caveolin-1.

VII. CONCLUSION

Together with the other studies reviewed in this chapter and although caveolin-deficient mice are viable, the above described dissection of the vascular phenotype of the caveolin-1-deficient mice provides clear evidence that the structural protein caveolin-1 and caveolae are essential for the control of NO production and the many functions mediated by this signal-transducing molecule in the cardiovascular system. More generally, among the responses to the many incoming pathways integrated in the caveolar organelles, NO appears today as one of the major signal emanating from these signaling platforms.

REFERENCES

Bist, A., Fielding, P. E., and Fielding, C. J. (1997). Two sterol regulatory element-like sequences mediate up-regulation of caveolin gene transcription in response to low density lipoprotein free cholesterol. *Proc. Natl. Acad. Sci. USA* **94,** 10693–10698.

Blair, A., Shaul, P. W., Yuhanna, I. S., Conrad, P. A., and Smart, E. J. (1999). Oxidized low density lipoprotein displaces endothelial nitric-oxide synthase (eNOS) from plasmalemmal caveolae and impairs eNOS activation. *J. Biol. Chem.* **274,** 32512–32519.

Brouet, A., Sonveaux, P., Dessy, C., Balligand, J. L., and Feron, O. (2001a). Hsp90 ensures the transition from the early Ca^{2+}-dependent to the late phosphorylation-dependent activation of the endothelial nitric-oxide synthase in vascular endothelial growth factor-exposed endothelial cells. *J. Biol. Chem.* **276,** 32663–32669.

Brouet, A., Sonveaux, P., Dessy, C., Moniotte, S., Balligand, J. L., and Feron, O. (2001b). Hsp90 and caveolin are key targets for the proangiogenic nitric oxide-mediated effects of statins. *Circ. Res.* **89,** 866–873.

Bucci, M., Gratton, J. P., Rudic, R. D., Acevedo, L., Roviezzo, F., Cirino, G., and Sessa, W. C. (2000). *In vivo* delivery of the caveolin-1 scaffolding domain inhibits nitric oxide synthesis and reduces inflammation. *Nat. Med.* **6,** 1362–1367.

Chen, D., Zangl, A. L., Zhao, Q., Markley, J. L., Zheng, J., Bird, I. M., and Magness, R. R. (2001). Ovine caveolin-1: cDNA cloning, *E. coli* expression, and association with endothelial nitric oxide synthase. *Mol. Cell Endocrinol.* **175,** 41–56.

Couet, J., Li, S., Okamoto, T., Ikezu, T., and Lisanti, M. P. (1997). Identification of peptide and protein ligands for the caveolin-scaffolding domain. Implications for the interaction of caveolin with caveolae-associated proteins. *J. Biol. Chem.* **272,** 6525–6533.

Drab, M., Verkade, P., Elger, M., Kasper, M., Lohn, M., Lauterbach, B., Menne, J., Lindschau, C., Mende, F., Luft, F. C., Schedl, A., Haller, H., and Kurzchalia, T. V. (2001). Loss of caveolae, vascular dysfunction, and pulmonary defects in caveolin-1 gene-disrupted mice. *Science* **293,** 2449–2452.

Draijer, R., Atsma, D. E., van der, L. A., and van Hinsbergh, V. W. (1995). cGMP and nitric oxide modulate thrombin-induced endothelial permeability. Regulation via different pathways in human aortic and umbilical vein endothelial cells. *Circ. Res.* **76,** 199–208.

Dvorak, A. M. and Feng, D. (2001). The vesiculo-vacuolar organelle (VVO). A new endothelial cell permeability organelle. *J. Histochem. Cytochem.* **49,** 419–432.

Esser, S., Wolburg, K., Wolburg, H., Breier, G., Kurzchalia, T., and Risau, W. (1998). Vascular endothelial growth factor induces endothelial fenestrations *in vitro*. *J. Cell Biol.* **140,** 947–959.

Feng, Y., Venema, V. J., Venema, R. C., Tsai, N., Behzadian, M. A., and Caldwell, R. B. (1999a). VEGF-induced permeability increase is mediated by caveolae. *Invest. Ophthalmol. Vis. Sci.* **40,** 157–167.

Feng, Y., Venema, V. J., Venema, R. C., Tsai, N., and Caldwell, R. B. (1999b). VEGF induces nuclear translocation of Flk-1/KDR, endothelial nitric oxide synthase, and caveolin-1 in vascular endothelial cells. *Biochem. Biophys. Res. Commun.* **256,** 192–197.

Feron, O., Belhassen, L., Kobzik, L., Smith, T. W., Kelly, R. A., and Michel, T. (1996). Endothelial nitric oxide synthase targeting to caveolae. Specific interactions with caveolin isoforms in cardiac myocytes and endothelial cells. *J. Biol. Chem.* **271,** 22810–22814.

Feron, O., Dessy, C., Desager, J. P., and Balligand, J. L. (2001). Hydroxy-methylglutaryl-coenzyme A reductase inhibition promotes endothelial nitric oxide synthase activation through a decrease in caveolin abundance. *Circulation* **103,** 113–118.

Feron, O., Dessy, C., Moniotte, S., Desager, J. P., and Balligand, J. L. (1999). Hypercholesterolemia decreases nitric oxide production by promoting the interaction of caveolin and endothelial nitric oxide synthase. *J. Clin. Invest.* **103,** 897–905.

Feron, O., Michel, J. B., Sase, K., and Michel, T. (1998a). Dynamic regulation of endothelial nitric oxide synthase: Complementary roles of dual acylation and caveolin interactions. *Biochemistry* **37,** 193–200.

Feron, O., Saldana, F., Michel, J. B., and Michel, T. (1998b). The endothelial nitric-oxide synthase-caveolin regulatory cycle. *J. Biol. Chem.* **273,** 3125–3128.

Fielding, C. J., Bist, A., and Fielding, P. E. (1997). Caveolin mRNA levels are up-regulated by free cholesterol and down-regulated by oxysterols in fibroblast monolayers. *Proc. Natl. Acad. Sci. USA* **94,** 3753–3758.

Fukumura, D., Gohongi, T., Kadambi, A., Izumi, Y., Ang, J., Yun, C. O., Buerk, D. G., Huang, P. L., and Jain, R. K. (2001). Predominant role of endothelial nitric oxide synthase in vascular endothelial growth factor-induced angiogenesis and vascular permeability. *Proc. Natl. Acad. Sci. USA* **98,** 2604–2609.

Garcia-Cardena, G., Fan, R., Stern, D. F., Liu, J., and Sessa, W. C. (1996a). Endothelial nitric oxide synthase is regulated by tyrosine phosphorylation and interacts with caveolin-1. *J. Biol. Chem.* **271,** 27237–27240.

Garcia-Cardena, G., Martasek, P., Masters, B. S., Skidd, P. M., Couet, J., Li, S., Lisanti, M. P., and Sessa, W. C. (1997). Dissecting the interaction between nitric oxide synthase (NOS) and caveolin. Functional significance of the nos caveolin binding domain *in vivo*. *J. Biol. Chem.* **272,** 25437–25440.

Garcia-Cardena, G., Oh, P., Liu, J., Schnitzer, J. E., and Sessa, W. C. (1996b). Targeting of nitric oxide synthase to endothelial cell caveolae via palmitoylation: Implications for nitric oxide signaling. *Proc. Natl. Acad. Sci. USA* **93,** 6448–6453.

Ghosh, S., Gachhui, R., Crooks, C., Wu, C., Lisanti, M. P., and Stuehr, D. J. (1998). Interaction between caveolin-1 and the reductase domain of endothelial nitric-oxide synthase. Consequences for catalysis. *J. Biol. Chem.* **273,** 22267–22271.

Gonzalez, E., Nagiel, A., Lin, A. J., Golan, D. E., and Michel, T. (2004). siRNA-mediated down-regulation of caveolin-1 differentially modulates signaling pathways in endothelial cells. *J. Biol. Chem.* **279,** 40659–40669.

Gratton, J. P., Lin, M. I., Yu, J., Weiss, E. D., Jiang, Z. L., Fairchild, T. A., Iwakiri, Y., Groszmann, R., Claffey, K. P., Cheng, Y. C., and Sessa, W. C. (2003). Selective inhibition of tumor microvascular permeability by cavtratin blocks tumor progression in mice. *Cancer Cell* **4,** 31–39.

Griffoni, C., Spisni, E., Santi, S., Riccio, M., Guarnieri, T., and Tomasi, V. (2000). Knockdown of caveolin-1 by antisense oligonucleotides impairs angiogenesis *in vitro* and *in vivo*. *Biochem. Biophys. Res. Commun.* **276,** 756–761.

Hare, J. M., Lofthouse, R. A., Juang, G. J., Colman, L., Ricker, K. M., Kim, B., Senzaki, H., Cao, S., Tunin, R. S., and Kass, D. A. (2000). Contribution of caveolin protein abundance to augmented nitric oxide signaling in conscious dogs with pacing-induced heart failure. *Circ. Res.* **86,** 1085–1092.

Ju, H., Zou, R., Venema, V. J., and Venema, R. C. (1997). Direct interaction of endothelial nitric-oxide synthase and caveolin-1 inhibits synthase activity. *J. Biol. Chem.* **272,** 18522–18525.

Kincer, J. F., Uittenbogaard, A., Dressman, J., Guerin, T. M., Febbraio, M., Guo, L., and Smart, E. J. (2002). Hypercholesterolemia promotes a CD36-dependent and endothelial nitric-oxide synthase-mediated vascular dysfunction. *J. Biol. Chem.* **277,** 23525–23533.

Kurose, I., Wolf, R., Grisham, M. B., Aw, T. Y., Specian, R. D., and Granger, D. N. (1995). Microvascular responses to inhibition of nitric oxide production. Role of active oxidants. *Circ. Res.* **76,** 30–39.

Liu, J., Razani, B., Tang, S., Terman, B. I., Ware, J. A., and Lisanti, M. P. (1999). Angiogenesis activators and inhibitors differentially regulate caveolin-1 expression and caveolae formation in vascular endothelial cells. Angiogenesis inhibitors block vascular endothelial growth factor-induced down-regulation of caveolin-1. *J. Biol. Chem.* **274,** 15781–15785.

Liu, J., Wang, X. B., Park, D. S., and Lisanti, M. P. (2002). Caveolin-1 expression enhances endothelial capillary tubule formation. *J. Biol. Chem.* **277,** 10661–10668.

Michel, J. B., Feron, O., Sacks, D., and Michel, T. (1997a). Reciprocal regulation of endothelial nitric-oxide synthase by Ca2+-calmodulin and caveolin. *J. Biol. Chem.* **272,** 15583–15586.

Michel, J. B., Feron, O., Sase, K., Prabhakar, P., and Michel, T. (1997b). Caveolin versus calmodulin. Counterbalancing allosteric modulators of endothelial nitric oxide synthase. *J. Biol. Chem.* **272,** 25907–25912.

Michel, T. and Feron, O. (1997). Nitric oxide synthases: Which, where, how, and why? *J. Clin. Invest.* **100,** 2146–2152.

Mineo, C., Yuhanna, I. S., Quon, M. J., and Shaul, P. W. (2003). High density lipoprotein-induced endothelial nitric-oxide synthase activation is mediated by Akt and MAP kinases. *J. Biol. Chem.* **278,** 9142–9149.

Omura, M., Kobayashi, S., Mizukami, Y., Mogami, K., Todoroki-Ikeda, N., Miyake, T., and Matsuzaki, M. (2001). Eicosapentaenoic acid (EPA) induces Ca(2+)-independent activation and translocation of endothelial nitric oxide synthase and endothelium-dependent vasorelaxation. *FEBS Lett.* **487,** 361–366.

Pelat, M., Dessy, C., Massion, P., Desager, J. P., Feron, O., and Balligand, J. L. (2003). Rosuvastatin decreases caveolin-1 and improves nitric oxide-dependent heart rate and blood pressure variability in apolipoprotein E−/− mice *in vivo*. *Circulation* **107,** 2480–2486.

Pelligrino, D. A., Ye, S., Tan, F., Santizo, R. A., Feinstein, D. L., and Wang, Q. (2000). Nitric-oxide-dependent pial arteriolar dilation in the female rat: Effects of chronic estrogen depletion and repletion. *Biochem. Biophys. Res. Commun.* **269,** 165–171.

Piech, A., Dessy, C., Havaux, X., Feron, O., and Balligand, J. L. (2003). Differential regulation of nitric oxide synthases and their allosteric regulators in heart and vessels of hypertensive rats. *Cardiovasc. Res.* **57,** 456–467.

Piech, A., Massart, P. E., Dessy, C., Feron, O., Havaux, X., Morel, N., Vanoverschelde, J. L., Donckier, J., and Balligand, J. L. (2002). Decreased expression of myocardial eNOS and caveolin in dogs with hypertrophic cardiomyopathy. *Am J. Physiol. Heart Circ. Physiol.* **282,** H219–H231.

Ratajczak, P., Damy, T., Heymes, C., Oliviero, P., Marotte, F., Robidel, E., Sercombe, R., Boczkowski, J., Rappaport, L., and Samuel, J. L. (2003). Caveolin-1 and -3 dissociations from caveolae to cytosol in the heart during aging and after myocardial infarction in rat. *Cardiovasc. Res.* **57,** 358–369.

Razani, B., Engelman, J. A., Wang, X. B., Schubert, W., Zhang, X. L., Marks, C. B., Macaluso, F., Russell, R. G., Li, M., Pestell, R. G., Di Vizio, D., Hou, H., Jr., Kneitz, B., Lagaud, G., Christ, G. J., Edelmann, W., and Lisanti, M. P. (2001). Caveolin-1 null mice are viable but show evidence of hyperproliferative and vascular abnormalities. *J. Biol. Chem.* **276,** 38121–38138.

Rizzo, V., McIntosh, D. P., Oh, P., and Schnitzer, J. E. (1998). *In situ* flow activates endothelial nitric oxide synthase in luminal caveolae of endothelium with rapid caveolin dissociation and calmodulin association. *J. Biol. Chem.* **273,** 34724–34729.

Schubert, W., Frank, P. G., Razani, B., Park, D. S., Chow, C. W., and Lisanti, M. P. (2001). Caveolae-deficient endothelial cells show defects in the uptake and transport of albumin *in vivo*. *J. Biol. Chem.* **276,** 48619–48622.

Schubert, W., Frank, P. G., Woodman, S. E., Hyogo, H., Cohen, D. E., Chow, C. W., and Lisanti, M. P. (2002). Microvascular hyperpermeability in caveolin-1 (−/−) knock-out mice. Treatment with a specific nitric-oxide synthase inhibitor, L-name, restores normal microvascular permeability in Cav-1 null mice. *J. Biol. Chem.* **277,** 40091–40098.

Shah, V., Cao, S., Hendrickson, H., Yao, J., and Katusic, Z. S. (2001). Regulation of hepatic eNOS by caveolin and calmodulin after bile duct ligation in rats. *Am J. Physiol. Gastrointest. Liver Physiol.* **280,** G1209–G1216.

Shaul, P. W., Smart, E. J., Robinson, L. J., German, Z., Yuhanna, I. S., Ying, Y., Anderson, R. G., and Michel, T. (1996). Acylation targets emdothelial nitric-oxide synthase to plasmalemmal caveolae. *J. Biol. Chem.* **271,** 6518–6522.

Sonveaux, P., Brouet, A., Havaux, X., Gregoire, V., Dessy, C., Balligand, J. L., and Feron, O. (2003). Irradiation-induced angiogenesis through the up-regulation of the nitric oxide pathway: Implications for tumor radiotherapy. *Cancer Res.* **63,** 1012–1019.

Sonveaux, P., Dessy, C., Brouet, A., Jordan, B. F., Gregoire, V., Gallez, B., Balligand, J. L., and Feron, O. (2002). Modulation of the tumor vasculature functionality by ionizing radiation accounts for tumor radiosensitization and promotes gene delivery. *FASEB J.* **16,** 1979–1981.

Sonveaux, P., Martinive, P., DeWever, J., Batova, Z., Daneau, G., Pelat, M., Ghisdal, P., Gregoire, V., Dessy, C., Balligand, J. L., and Feron, O. (2004). Caveolin-1 expression is critical for vascular endothelial growth factor-induced ischemic hind limb collateralization and nitric oxide-mediated angiogenesis. *Circ. Res.* **95,** 154–161.

Sorensson, J., Fierlbeck, W., Heider, T., Schwarz, K., Park, D. S., Mundel, P., Lisanti, M., and Ballermann, B. J. (2002). Glomerular endothelial fenestrae *in vivo* are not formed from caveolae. *J. Am. Soc. Nephrol.* **13,** 2639–2647.

Sowa, G., Pypaert, M., and Sessa, W. C. (2001). Distinction between signaling mechanisms in lipid rafts vs. caveolae. *Proc. Natl. Acad. Sci. USA* **98,** 14072–14077.

Uittenbogaard, A., Shaul, P. W., Yuhanna, I. S., Blair, A., and Smart, E. J. (2000). High density lipoprotein prevents oxidized low density lipoprotein-induced inhibition of endothelial nitric-oxide synthase localization and activation in caveolae. *J. Biol. Chem.* **275,** 11278–11283.

Uray, I. P., Connelly, J. H., Frazier, O. H., Taegtmeyer, H., and Davies, P. J. (2003). Mechanical unloading increases caveolin expression in the failing human heart. *Cardiovasc. Res.* **59,** 57–66.

Vasile, E., Qu, H., Dvorak, H. F., and Dvorak, A. M. (1999). Caveolae and vesiculo-vacuolar organelles in bovine capillary endothelial cells cultured with VPF/VEGF on floating Matrigel-collagen gels. *J. Histochem. Cytochem.* **47,** 159–167.

Venema, V. J., Zou, R., Ju, H., Marrero, M. B., and Venema, R. C. (1997). Caveolin-1 detergent solubility and association with endothelial nitric oxide synthase is modulated by tyrosine phosphorylation. *Biochem. Biophys. Res. Commun.* **236,** 155–161.

Woodman, S. E., Ashton, A. W., Schubert, W., Lee, H., Williams, T. M., Medina, F. A., Wyckoff, J. B., Combs, T. P., and Lisanti, M. P. (2003). Caveolin-1 knockout mice show an impaired angiogenic response to exogenous stimuli. *Am J. Pathol.* **162,** 2059–2068.

Yokomori, H., Oda, M., Ogi, M., Kamegaya, Y., Tsukada, N., and Ishii, H. (2001). Endothelial nitric oxide synthase and caveolin-1 are co-localized in sinusoidal endothelial fenestrae. *Liver* **21,** 198–206.

Yuhanna, I. S., Zhu, Y., Cox, B. E., Hahner, L. D., Osborne-Lawrence, S., Lu, P., Marcel, Y. L., Anderson, R. G., Mendelsohn, M. E., Hobbs, H. H., and Shaul, P. W. (2001). High-density lipoprotein binding to scavenger receptor-BI activates endothelial nitric oxide synthase. *Nat. Med.* **7,** 853–857.

Zhu, Y., Liao, H. L., Niu, X. L., Yuan, Y., Lin, T., Verna, L., and Stemerman, M. B. (2003). Low density lipoprotein induces eNOS translocation to membrane caveolae: The role of RhoA activation and stress fiber formation. *Biochim. Biophys. Acta* **1635,** 117–126.

Chapter 6

Caveolae and Estrogen Receptor Signaling

Philip W. Shaul and Ken L. Chambliss

Department of Pediatrics, University of Texas Southwestern Medical Center at Dallas, Dallas, Texas 75390

I. INTRODUCTION

The hormone estrogen has both slow genomic actions and rapid nongenomic actions in a variety of cell types. Recent work indicates that the rapid responses are mediated by a subpopulation of plasma membrane-associated estrogen receptors (ERs). The mechanisms of action of plasma membrane-associated ERs have been elucidated in considerable detail in endothelial cells, in which estrogen causes rapid stimulation of nitric oxide (NO) production by endothelial NO synthase (eNOS) (Caulin-Glaser *et al.*, 1997; Lantin-Hermoso *et al.*, 1997; Chen *et al.*, 1999). This chapter reviews studies that localized eNOS activation by estrogen to the plasma membrane of endothelial cells, and then addresses additional experiments that implicated a subpopulation of plasma membrane-associated ERα. The further localization of ERα–eNOS coupling to endothelial cell caveolae is then discussed, followed by a review of comparable localization and action of a subpopulation of caveolae-associated ERβ in endothelium. The role of G-proteins in ERα coupling to eNOS is also considered. Last, additional intricacies about the identity of ERs in caveolae

ISBN: 0-444-51500-3
DOI: 10.1016/S0079-6603(05)36006-5

are addressed. The cumulative observations that will be reviewed offer compelling evidence that caveolae provide a signaling module in which plasma membrane-initiated ER signaling is organized.

II. ESTROGEN AND eNOS ACTIVATION

Estrogen has important atheroprotective properties that are at least partially related to its capacity to enhance the bioavailability of NO (Farhat *et al.*, 1996; Mendelsohn and Karas, 1999; Miller, 1999). NO is a potent regulator of blood pressure, platelet aggregation, leukocyte adhesion, and vascular smooth muscle mitogenesis that is produced in the vascular wall primarily by eNOS upon the conversion of the substrate L-arginine to L-citrulline (Moncada and Higgs, 1993). The function of the L-arginine/eNOS system is altered in a variety of vascular disorders (Harrison, 1997).

Experiments in cultured endothelial cells directly demonstrated that estradiol (E_2) rapidly stimulates eNOS activity, that the response is attenuated by ER antagonism but not by inhibiting gene transcription, and that ERα is expressed in endothelium (Lantin-Hermoso *et al.*, 1997; Chen *et al.*, 1999). In addition, the overexpression of ERα in endothelial cells causes enhancement of the acute response to E_2 that is blocked by ER antagonism, specific to E_2 versus other agonists, and dependent on the ERα hormone-binding domain. Furthermore, in COS-7 cells that do not constitutively express ER or eNOS, the acute stimulation of eNOS by E_2 can be demonstrated in a reconstitution paradigm (Chen *et al.*, 1999). There is evidence that the tyrosine kinase-mitogen-activated protein (MAP) kinase signaling pathway is involved (Chen *et al.*, 1999) and that there is recruitment of the phosphatidylinositol 3 (PI3) kinase–Akt pathway that may entail direct interaction between ERα and the p85 subunit of PI3 kinase to ultimately cause Akt-mediated eNOS phosphorylation (Haynes *et al.*, 2000; Simoncini *et al.*, 2000). Furthermore, the latter processes are mediated by Src kinase (Haynes *et al.*, 2003). Thus, the short-term effects of estrogen on eNOS that are central to cardiovascular physiology are mediated by ERα functioning in a novel, nongenomic manner, and multiple signal transduction events are likely to participate.

III. LOCALIZATION OF eNOS ACTIVATION BY E_2 TO PLASMA MEMBRANE

Studies employing immunoidentification or conjugated estrogen have suggested that a subpopulation of ER may be associated with the cell surface in certain cell types (Watson and Gametchu, 1999; Russell *et al.*, 2000). There is strong evidence that eNOS is targeted to the endothelial plasma membrane, particularly to caveolae, which are specialized, cholesterol-rich domains that compartmentalize signal transduction (Garcia-Cardena *et al.*, 1996; Shaul *et al.*, 1996; Shaul and Anderson, 1998). To begin to determine the subcellular site of interaction between ERα and eNOS, stimulation of the enzyme by E_2 was evaluated in isolated plasma membranes from ovine endothelial cells by measuring ^{3}H-L-arginine conversion to ^{3}H-L citrulline (Chambliss *et al.*, 2000). In the absence of added calcium, calmodulin, or cofactors, NOS activity was detectable in quiescent membranes, yielding 2.4 ± 0.1 pmol citrulline/mG-protein/min ($n = 4$). This basal level of NOS activity was inhibited by calcium chelation. In the absence of added calcium, calmodulin, or cofactors, 10^{-8} M E_2 (15 min) caused a 92% increase in NOS activity compared to basal levels (Fig. 1A). In contrast to 17β-estradiol (E_2), 17α-estradiol had no effect on NOS activity (data not shown).

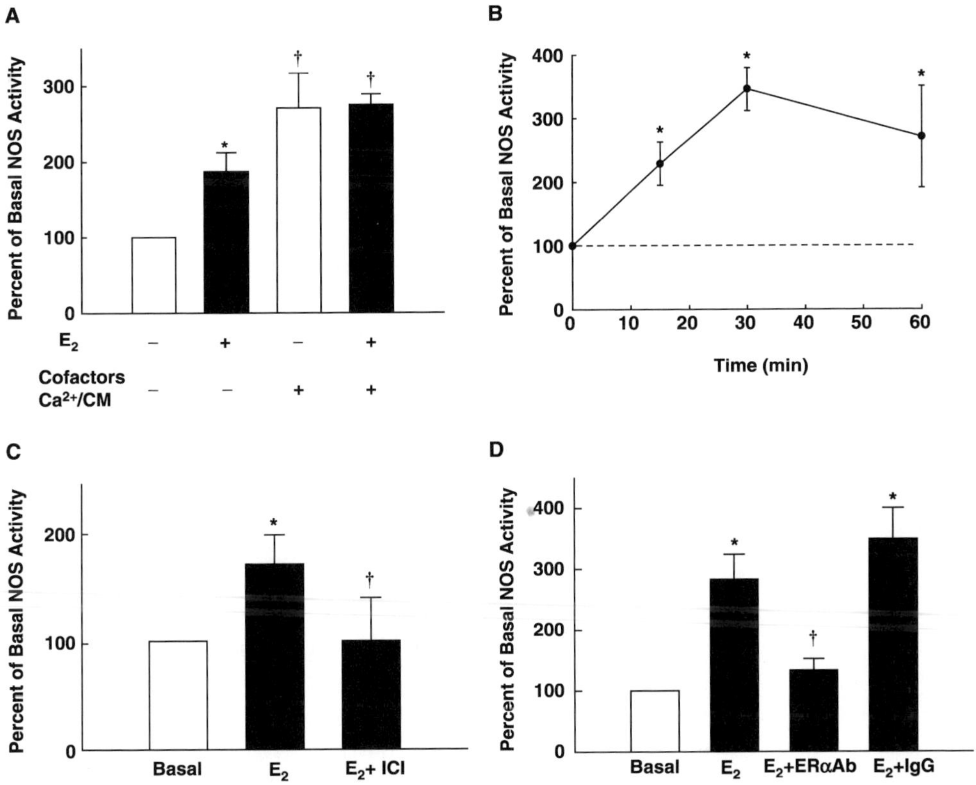

Figure 1. Activation of eNOS by ER in endothelial cell plasma membranes. (A) Effect of E_2 on NOS activity. ^{3}H-L-arginine conversion to ^{3}H-L-citrulline was measured in isolated plasma membranes in the absence (basal) or presence of 10^{-8} M E_2 and in the absence or presence of the exogenous eNOS cofactors, Ca^{2+} and calmodulin (CM). (B) Time course of the effect of E_2 on NOS activity. Incubations were performed without added cofactors, Ca^{2+} and CM, in the absence (basal) or presence of 10^{-8} M E_2 for 15 to 60 min. (C) Effect of ER antagonism on response to E_2. NOS activation was measured without added cofactors, Ca^{2+} or CM, in the absence (basal) or presence of 10^{-8} M E_2, with or without 10^{-5} M ICI 182,780 added. (D) Effect of ERα antibody on response to E_2. Incubations were performed without added cofactors, Ca^{2+} or CM, in the absence (basal) or presence of 10^{-8} M E_2 with or without antibody to ERα (TE111) or unrelated IgG added. Values are mean ± SEM, $n = 4$–6, $*p < 0.05$ vs basal, $^{\dagger}p < 0.05$ vs E_2 alone. Reprinted with permission from Chambliss *et al.* (2000).

Maximal NOS activity was assessed by replacing E_2 with a mixture of calcium, calmodulin, and cofactors, yielding a 170% rise in activity compared to no additives; E_2 (10^{-8} M) did not enhance this activity. These observations indicate that all of the signal transduction machinery required for eNOS stimulation by E_2 is associated with the plasma membrane. Since E_2 alone activates the enzyme to approximately half of maximal levels, the response is quite robust. Time course experiments with E_2 alone revealed a progressive, linear increase in NOS activity during the first 30 min of incubation, followed by a plateau (Fig. 1B). This contrasts with NOS activity with added calcium, calmodulin, and cofactors, which displayed linearity with time for at least 120 min (data not shown). As such, the availability of one or more of these molecules may be limited in the isolated membranes. In intact cells there are most likely mechanisms that replenish these factors in the locale of the plasma membrane.

The role of ER in E_2 activation of plasma membrane eNOS was examined using the ER antagonist ICI 182,780 (Fig.1C). E_2-stimulated NOS activity was prevented by ICI 182,780. In addition, antibody to the ligand binding domain of ERα (TE111), the first cloned ER subtype, blocked E_2-stimulated NOS activation, whereas unrelated IgG had no effect (Fig. 1D). These findings indicate that the response to E_2 is mediated by an ER or ER-like protein associated with the endothelial cell plasma membrane.

IV. CHARACTERIZATION OF PLASMA MEMBRANE ER IN ENDOTHELIAL CELLS

The plasma membrane-associated ER in endothelial cells was characterized in immunoidentification experiments that compared the plasma membrane ER with cytosolic and nuclear ER using antibodies directed against three different ERα epitopes. Antibodies directed against amino acids 495–595 (AER320), 302–553 (TE111), or 120–170 of human ERα (AER304) all detected a single 67-kDa protein species in endothelial cell plasma membranes that was identical in size to the protein detected in nuclear and cytosolic fractions (Fig. 2A). To confirm these observations, it was determined that epitope-tagged ERα (ERα-myc) introduced into COS-7 cells is targeted to plasma membrane (Fig. 2B). Whereas antibody to ERα revealed no signal in sham-transfected cells, there was positive signal for ERα of comparable size in nucleus, cytosol, and plasma membrane from cells transfected with ERα-myc. In parallel, immunoblot analysis with antibody to the myc tag revealed no signal in sham-transfected cells, but a similarly sized protein was detected in the nucleus, cytosol, and plasma membranes of cells expressing the tagged receptor. These cumulative findings indicate that E_2-stimulated eNOS activity is mediated by a subpopulation of ERα that is associated with the endothelial plasma membrane.

The importance of plasma membrane colocalization for ERα-stimulated eNOS activity was evaluated in a reconstitution paradigm in COS-7 cells. Plasma membranes

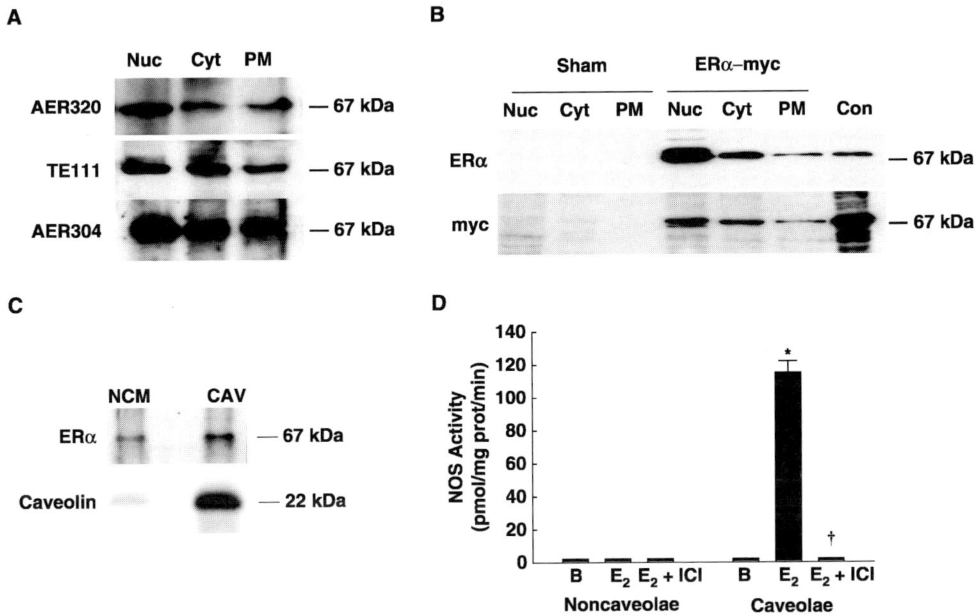

Figure 2. Characterization of plasma membrane-associated ER in endothelium. (A) Immunoblot analysis for ERα in endothelial cell nucleus (Nuc), cytosol (Cyt), and plasma membrane (PM). The monoclonal antibodies employed were directed against amino acids 495–595 (AER320), 302–553 (TE111), or 120–170 (AER304) of human ERα. (B) Targeting of epitope-tagged ERα to plasma membranes. Following transient transfection of COS-7 cells with myc-tagged ERα, immunoblot analysis was performed for ERα and myc on cell fractions. Whole cell lysate from a prior COS-7 cell transfection was used as a positive control (Con). Results are representative of three independent experiments. (C) Localization of ERα–eNOS interaction to caveolae. Immunoblot analyses were performed for ERα and caveolin-1 in noncaveolae membranes (NCM) and caveolae membranes (CAV) obtained from endothelial cell whole plasma membranes. Results are representative of three independent experiments. (D) E_2-mediated activation of eNOS in endothelial cell caveolae membranes. ^{3}H-L-arginine conversion to ^{3}H-L-citrulline was measured in noncaveolae and caveolae membranes obtained from endothelial cell plasma membranes. Membrane incubations were performed without added eNOS cofactors, Ca^{2+} or calmodulin, in the absence (basal, B) or presence of 10^{-8} M E_2 with or without 10^{-5} M ICI 182,780 added. NOS activity was undetectable in noncaveolae fractions in all groups, and it was also not detected in caveolae under basal conditions. Values are mean $\pm$ SEM, $n = 4$–6, $*p < 0.05$ vs basal, $^{\dagger}p < 0.05$ vs E_2 alone. Reprinted with permission from Chambliss *et al.* (2000).

from cells expressing eNOS and ERα displayed rapid ER-mediated NOS stimulation, whereas membranes from cells expressing eNOS alone or ERα plus myristoylation-deficient mutant eNOS were insensitive. In fact, membranes from cells expressing myristoylation-deficient mutant eNOS and ERα displayed a decline in NOS activity with E_2 that was partially reversed by ICI 182,780. Because myristoylation-deficient mutant eNOS is minimally directed to the plasma membrane but is unaltered in enzymatic activity (Shaul *et al.*, 1996), these findings indicate that both normal plasma membrane targeting of eNOS and localization of ERα to that site are required for eNOS activation by E_2 (Chambliss *et al.*, 2000).

V. LOCALIZATION OF ERα–eNOS COUPLING TO CAVEOLAE

Because plasma membrane eNOS is exclusively localized to caveolae (Shaul *et al.*, 1996), further experiments were done to determine whether ERα protein is also associated with this subfraction of endothelial cell plasma membranes. Immunoblot analysis for caveolin-1 confirmed separation of caveolae and noncaveolae fractions (Fig. 2C). ERα protein was detected in caveolae, and it was also detected in the noncaveolae fraction, but to a lesser extent.

Experiments were then performed to evaluate the capacity of E_2 to activate eNOS in isolated caveolae and noncaveolae fractions (Fig. 2D). In the absence of added calcium, calmodulin, or cofactors, there was no measurable NOS activity in the noncaveolae fraction under basal conditions or with E_2 added. Basal NOS activity was also below detection limits in caveolae membranes. However, 10^{-8} M E_2 caused robust activation of NOS in caveolae membranes, and this effect was prevented by ICI 182,780. These data strongly indicate that ERα and all of the additional molecular machinery requisite for E_2-mediated activation of eNOS exist in a functional signaling module in endothelial caveolae. Such a proposition is consistent with the known clustering of several members of the MAP kinase pathway implicated in rapid estrogen signaling, including Src and extracellular signal-regulated kinases (ERKs), in caveolae (Okamoto *et al.*, 1998). Since ERα was found in both caveolae and noncaveolae fractions and eNOS is solely in caveolae (Shaul *et al.*, 1996), the specificity of ERα coupling to eNOS to caveolae is evidently due to the localization of the effector, and not the receptor, in the microdomain. Furthermore, the effect of E_2 on eNOS in caveolae is prevented by calcium chelation (Chambliss *et al.*, 2000), suggesting that ERα activation in caveolae modifies the local calcium environment.

VI. COMPARABLE ROLE OF ERβ IN ENDOTHELIAL CAVEOLAE

Following the demonstration of a subpopulation of ERα in endothelial caveolae, the potential nongenomic actions of the other ER subtype, ERβ, were investigated after it was determined that the endothelium expresses both isoforms of the receptor (Chambliss *et al.*, 2002). In immunoblot analyses of endothelial subcellular fractions, identically sized ERα protein was present in the nucleus, cytoplasm, and plasma membrane as seen previously (Figs. 2A and 3A). Similarly, identically sized ERβ protein was detected in the nucleus, cytoplasm, and plasma membrane (Fig. 3A). These observations were confirmed in evaluations of the subcellular localization of transfected ERα and ERβ in COS-7 cells (Fig. 3B). In experiments employing isolated endothelial cell plasma membranes, the ERβ-selective antagonist RR-tetrahydrechrysene (THC) blunted E_2 activation of eNOS (Fig. 3C). In addition, ERβ protein was detected and THC attenuated E_2 stimulation of eNOS in isolated endothelial caveolae (Fig. 3D and E), and functional ERβ–eNOS coupling was recapitulated in caveolae from transfected COS-7 cells (Fig. 3F). These findings in the ER–eNOS paradigm

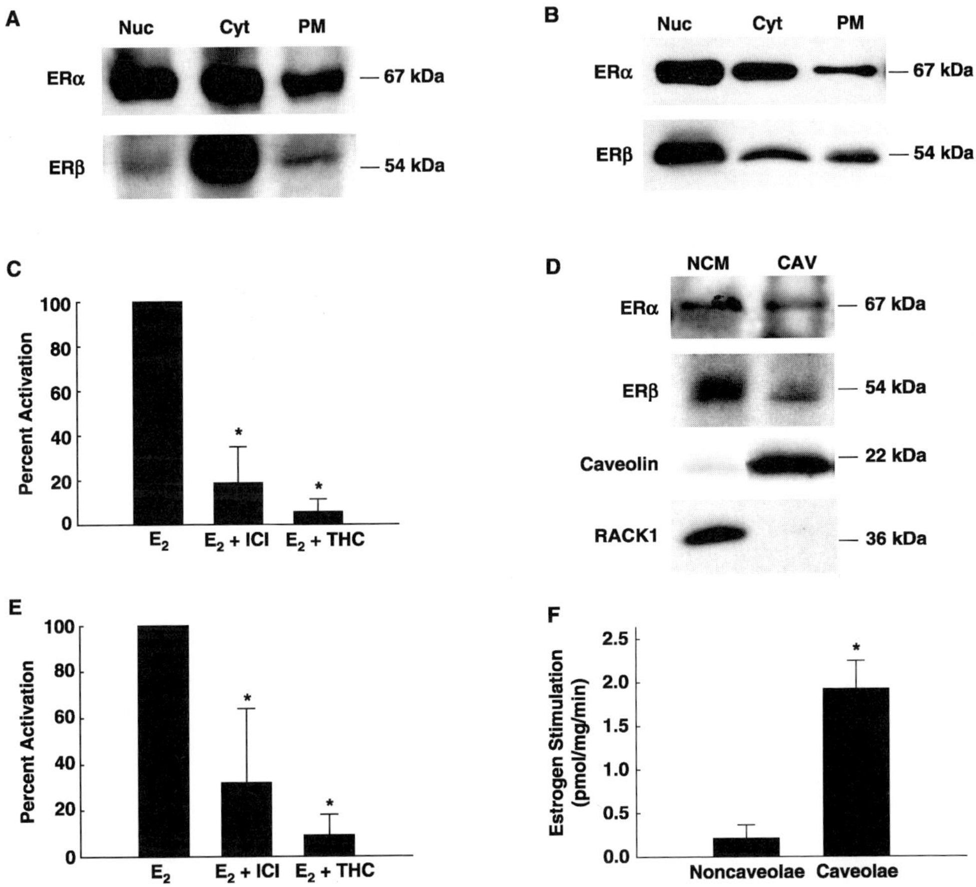

Figure 3. ERβ is localized and functional in endothelial cell caveolae. (A) Immunoblot analyses were performed on equal amounts of endothelial cell nucleus (Nuc), cytoplasm (Cyt), and plasma membrane (PM) using antibodies derived against amino acids 302–553 and 467–485, respectively, of ERα and ERβ. (B) Parallel studies were performed using subfractions of COS-7 cells transfected with either ERα or ERβ cDNA. The findings shown in (A) and (B) are representative of three independent experiments. (C) Effect of ICI 182,780 (10^{-5} M) and THC (10^{-7} M) on E_2(10^{-8} M)-mediated eNOS activation in isolated endothelial cell plasma membrane. ^{3}H-L-arginine conversion to ^{3}H-L-citrulline was measured over 60 min. Results are expressed as percent of activity induced by E_2. Values are mean ± SEM, $n = 3$, $*p < 0.05$ vs E_2 alone. (D) Immunoblot analysis for ERα and ERβ in endothelial cell noncaveolae (NCM) and caveolae (CAV) membrane fractions. Caveolin-1 and RACK1 protein abundance were also determined to assess fraction separation. Results shown are representative of three independent studies. (E) Effect of ICI 182,780 and THC on E_2-mediated eNOS activation in isolated endothelial cell caveolae membranes. ^{3}H-L-arginine conversion to ^{3}H-L-citrulline was measured over 60 min. Results are expressed as percent of activity stimulated by E_2. Values are mean ± SEM, $n = 3$, $*p < 0.05$ vs E_2 alone. (F) E_2-stimulated eNOS activity in noncaveolae and caveolae membranes from COS-7 cells expressing ERβ and eNOS. Experiments were performed as described in (E). E_2-stimulated NOS activity was not detected in caveolae from cells expressing eNOS alone (data not shown). Values are mean ± SEM, $n = 3, *p < 0.05$ vs noncaveolae. Reprinted with permission from Chambliss *et al.* (2002).

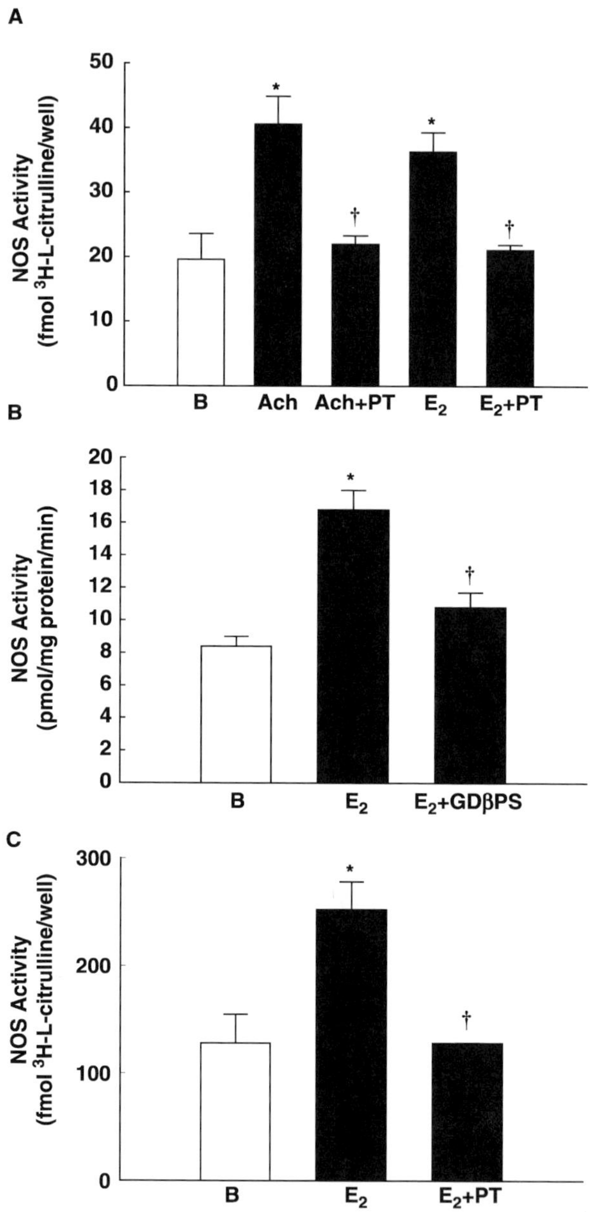

Figure 4. Role of G-proteins in E_2-stimulated eNOS activity. (A) Effect of pertussis toxin (PT) on eNOS stimulation in endothelial cells. Intact cells were pretreated with vehicle or 100 ng/ml PT, and ^{3}H-L-arginine conversion to ^{3}H-L-citrulline was assessed under basal conditions (B) or in the presence of 10^{-5} M acetylcholine (Ach) or 10^{-8} M E_2 in the continued presence of vehicle or PT. (B) Effect of exogenous GDPβS on eNOS stimulation in endothelial cell plasma membranes. The conversion of ^{3}H-L-arginine to ^{3}H-L-citrulline was measured in purified plasma membranes incubated under basal conditions (B) or in the presence of 10^{-8} M E_2, in buffer alone or buffer plus 2 mM GDPbS. (C) Effect of PT on eNOS stimulation in

indicate that the capacity for nongenomic action in caveolae is not limited to ERα but is instead also a characteristic of ERβ.

VII. G-PROTEINS AND ER COUPLING TO eNOS

The basis by which signaling events are initiated by plasma membrane ERα stimulation is poorly understood (Mendelsohn, 2000). The best-described agonists for eNOS, acetylcholine and bradykinin, activate specific plasma membrane-associated G-protein-coupled receptors (GPCR) (Liu *et al.*, 1995; Vanhoutte, 1997). G-proteins are heterotrimers of α, β, and γ subunits ($G_{\alpha\beta\gamma}$) that dissociate into G_α and $G_{\beta\gamma}$ upon GPCR stimulation; activated G_α and/or $G_{\beta\gamma}$ then modulate the activity of downstream effectors. The α subunits are divided into four subfamilies: $G_{\alpha s}$, $G_{\alpha i}$, $G_{\alpha q}$, and $G_{\alpha 12/13}$ (Gilman, 1995; Wess, 1998). To better understand the mechanisms responsible for plasma membrane-associated ERα action in the eNOS paradigm, the potential role of G-proteins was investigated (Wyckoff *et al.*, 2001). The first strategy was to evaluate E_2-stimulated eNOS activity in the absence or presence of pertussis toxin treatment, which inhibits G_α function by causing ADP ribosylation of a conserved cysteine at the fourth position (Carty, 1994). Intact endothelial cells were pretreated with vehicle or pertussis toxin and exposed to either the known GPCR agonist acetylcholine (10^{-5} M) or E_2(10^{-8} M) for 15 min (Fig. 4A). In the absence of pertussis toxin, acetylcholine and E_2 caused comparable NOS stimulation. eNOS activation by acetylcholine was fully blocked by pertussis toxin, and eNOS stimulation by E_2 was similarly affected. The inhibition of ER-mediated eNOS activation by pertussis toxin implicates $G_{\alpha i}$ subfamily members only, of which $G_{\alpha i}$ is expressed in endothelial cells, whereas $G_{\alpha o}$ is not (Carty, 1994; Gilman, 1995; Neer, 1995). In addition, exogenous GDPβS attenuated eNOS stimulation by E_2 in isolated endothelial cell plasma membranes (Fig. 4B). The plasma membrane preparations presumably contained sufficient endogenous GTP, the predominant guanine nucleotide in cells, to yield ER- and G-protein-mediated eNOS activation. The role of $G_{\alpha i}$ was further supported by the finding that pertussis toxin blocked E_2-stimulated NOS activity in COS-7 cells transfected with eNOS and ERα (Fig. 4C). These cumulative data indicate that $G_{\alpha i}$ mediates estrogen stimulation of eNOS. Interestingly, pertussis toxin also prevents E_2-mediated phosphorylation of MAP kinase in endothelial cells (data not shown), indicating that G-protein coupling occurs proximal to tyrosine kinase–MAP kinase activation in the series of events leading to eNOS stimulation.

Potential interactions between plasma membrane ERα and G_α proteins were then evaluated in coimmunoprecipitation studies using COS-7 cells transfected with ERα and $G_{\alpha i2}$, $G_{\alpha q}$, or $G_{\alpha s}$ (Fig. 5A). Immunoprecipitation was performed 48 hr after transfection

---

COS-7 cells. Cells were transfected with ERα and eNOS cDNA, and 48 hr later ^{3}H-L-arginine conversion to ^{3}H-L-citrulline was measured in intact cells under basal conditions (B) or in the presence of 10^{-8} M E_2, with or without PT pretreatment. Values are mean $\pm$ SEM, $n = 3$. $^*p < 0.05$ vs basal, $^\dagger p < 0.05$ vs no PT or no GDPβS. Reprinted with permission from Wyckoff *et al.* (2001).

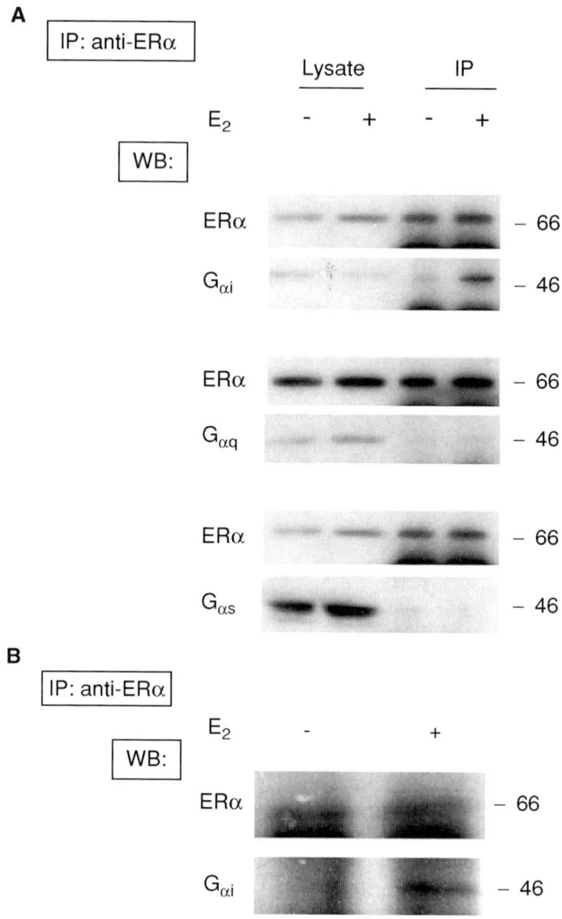

Figure 5. Interaction of plasma membrane ERα and G_α proteins. (A) Coimmunoprecipitation of ERα and $G_{\alpha i}$, $G_{\alpha q}$, or $G_{\alpha s}$ in COS-7 cell plasma membranes. Cells were transfected with ERα and $G_{\alpha i}$, $G_{\alpha q}$, or $G_{\alpha s}$ cDNAs, and 48 hr later cells were treated with vehicle or 10^{-8} M E_2 for 20 min. Plasma membranes were isolated and immunoprecipitation (IP) was done with ERα antibody. Western blot (WB) analyses were performed on whole cell lysates and plasma membrane immunoprecipitates for ERα and $G_{\alpha i}$, $G_{\alpha q}$, or $G_{\alpha s}$. The band below the ERα or G_α plain band in the IP samples is IgG heavy chain. (B) Coimmunoprecipitation of ERα and $G_{\alpha i}$ in endothelial cell plasma membranes. Endothelial cells were treated with vehicle or 10^{-8} M E_2 for 20 min, plasma membranes were isolated, and IP was done with ERα antibody. WB analyses were performed on immunoprecipitates for ERα and $G_{\alpha i}$. Results shown are representative of three independent experiments. Reprinted with permission from Wyckoff *et al.* (2001).

with ERα antibody on plasma membranes from cells treated with vehicle or 10^{-8} M E_2 for 20 min. Equivalent expression of transfected ERα and G_α proteins between study groups was confirmed by Western blot analysis of whole cell lysates. In plasma membranes from quiescent cells, $G_{\alpha i}$, $G_{\alpha q}$, and $G_{\alpha s}$ were minimally coimmunoprecipitated with ERα. However, the association of $G_{\alpha i}$ with ERα was markedly greater following

E_2 stimulation. In contrast, the association of $G_{\alpha q}$ and $G_{\alpha s}$ with ERα remained negligible after E_2 treatment. In studies of comparable interactions between endogenous ERα and $G_{\alpha i}$ in endothelial cells, parallel observations were made (Fig. 5B). Thus, ERα activation by agonist leads to interaction between the receptor and $G_{\alpha i}$.

Upon stimulation of classical GPCR, the activated G_α or $G_{\beta\gamma}$ subunit modulates the activity of downstream effectors (Neer, 1995). The potential roles of activated G_α or $G_{\beta\gamma}$ in E_2-induced activation of eNOS were distinguished by assessing the impact of overexpression of $G_{\alpha i2}$ in COS-7 cells expressing ERα and eNOS. Whereas basal eNOS activity was unchanged by cotransfection with $G_{\alpha i2}$, $G_{\alpha i2}$ overexpression augmented E_2-mediated eNOS stimulation threefold, and pertussis toxin completely blocked the enhanced response (Fig. 6A). The role of activated endogenous $G_{\alpha i}$ was further substantiated in experiments assessing the effect of overexpression of a protein

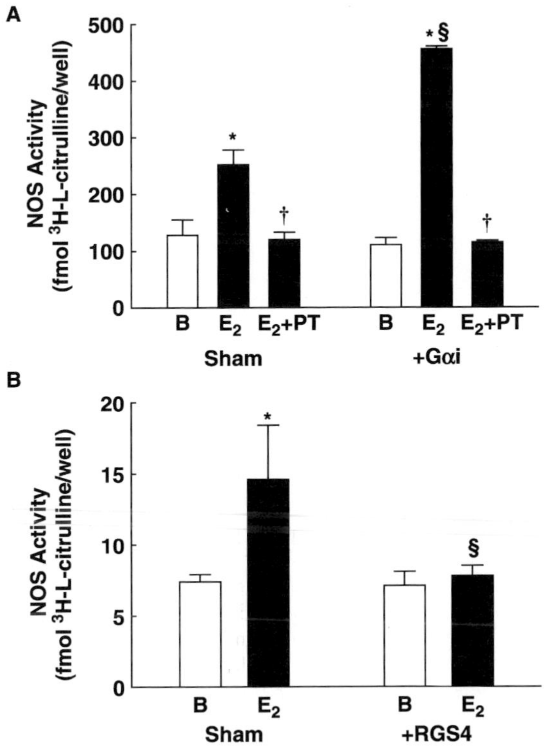

Figure 6. Activated $G_{\alpha i}$ mediates downstream signaling leading to NOS stimulation. (A) Effect of $G_{\alpha i}$ overexpression on eNOS stimulation by E_2 in COS-7 cells. Cells were transfected with cDNAs for ERα and eNOS, and either sham vector or $G_{\alpha i}$ cDNA. ^{3}H-L-arginine conversion to ^{3}H-L-citrulline was assessed in intact cells 48 hr later under basal conditions (B) or in the presence of 10^{-8} M E_2, with or without prior pertussis toxin treatment (PT) and the continued presence of vehicle or PT. (B) Effect of RGS4 overexpression on eNOS stimulation by E_2 in COS-7 cells. Cells were transfected with cDNAs for ERα and eNOS, and either sham vector or RGS4 cDNA. The conversion of ^{3}H-L-arginine to ^{3}H-L-citrulline was measured in intact cells 48 hr later under basal conditions (B) or in the presence of 10^{-8}M E_2. Values are mean ± SEM, $n = 3$. $*p < 0.05$ vs basal, $^\dagger p < 0.05$ vs no PT, $^\S\, p < 0.05$ vs sham. Reprinted with permission from Wyckoff *et al.* (2001).

regulator of G-protein signaling (RGS), RGS4. RGS4 attenuates signaling by accelerating the GTPase activity of members of the $G_{\alpha i}$ and $G_{\alpha q}$ subfamilies, but not those of $G_{\alpha s}$ or $G_{\alpha 12/13}$ (Huang *et al.*, 1997). In COS-7 cells expressing ERα and eNOS that were cotransfected with RGS4 cDNA, E_2-mediated eNOS stimulation was decreased by 90% compared to sham-transfected cells (Fig. 6B). These cumulative observations indicate that activated $G_{\alpha i}$ mediates the downstream signaling processes that ultimately couple ER to eNOS in caveolae. In addition to explaining the proximal events underlying nongenomic effects of estrogen in endothelial cells, these observations provide further support for a role for caveolae-associated ERα and G-protein coupling in nongenomic actions of estrogen in other cell types such as osteoblasts and macrophages (LeMellay *et al.*, 1997; Benten *et al.*, 2001). It is proposed that the processes by which plasma membrane ERα are localized to caveolae and coupled to partner signaling molecules such as $G_{\alpha i}$ constitute not only the basis of nongenomic estrogen action in vascular and nonvascular cells, but also the means by which other steroid hormones have important nonnuclear effects.

VIII. ADDITIONAL INTRICACIES ABOUT ER IN CAVEOLAE

Although multiple approaches in endothelium suggest that the ERs in caveolae are identical to the classical subtypes ERα and ERβ known best for their nuclear actions, other possibilities must be considered. Studies in an immortalized human endothelial cell line, EA.hy926, indicate that a 46-kDa ER (ER46), which is an N-terminal truncated product of full-length ERα (66–67 kDa), is localized to the plasma membrane. Using flow cytometry and immunofluorescence, it has been shown that the C terminus, and not theN terminus, of ER46 is antibody accessible on the plasma membrane. In addition, there is evidence of estrogen-induced, palmitoylation-dependent plasma membrane recruitment of ER46 with reorganization in caveolae (Li *et al.*, 2003). In a rat hypothalamic endothelial cell line (D12), both plasma membrane and nuclear ER that display similar affinities to E_2 are expressed, but competition assays have revealed differing affinities for certain ER ligands. Immunoblot analysis with antibodies to ERα showed that the plasma membrane-associated receptor is identical in molecular mass to classical ERα and that the protein is enriched in caveolae. Thus, in D12 cells the caveolae-associated ER appears to be similar to, but biochemically distinguishable from, classical nuclear ERα (Deecher *et al.*, 2003). In a non-endothelial cell paradigm, namely, mouse neocortex, there is evidence of a novel caveolae-associated ER. In neocortical explants both 17α-estradiol and 17β-estradiol cause rapid ERK activation, and a 62–63 kDa protein that is neither ERα nor ERβ but immunoreactive for the ligand-binding domain of ERα has been identified in caveolae. Contrasting with neocortical ERα, which is developmentally regulated and intranuclear, and neocortical ERβ, which is expressed throughout life and intranuclear, the caveolae-associated ER is found in postnatal but not adult wild-type and estrogen receptor knockout neocortical and uterine plasma membranes (Toran-Allerand *et al.*, 2002). As such, it should not be assumed that caveolae-associated ERs are exclusively either classical ERα or classical

ERβ, and detailed studies are required to clearly identify the relevant receptor protein(s) in different model systems.

IX. SUMMARY AND FUTURE DIRECTIONS

It is now apparent that a subpopulation of ERα and ERβ, or a closely related receptor protein, is coupled to eNOS in caveolae (Fig. 7), thus providing a model system for ER signaling in caveolae. In this specific setting, ligand binding to ER leads to $G_{\alpha i}$ activation, which mediates downstream events. These events include the activation of tyrosine kinase–MAP kinase signaling, PI3 kinase and Akt kinase activation causing eNOS phosphorylation, and perturbation of the local calcium environment that ultimately leads to calmodulin-dependent eNOS stimulation. The current and future challenges to our understanding of ER signaling in caveolae are numerous. Although immunocytochemical and immunofluorescence analyses suggest that the ligand-binding domain of cell surface-associated ER may be extracellular (Razandi *et al.*, 1999; Li *et al.*, 2003), the orientation of plasma membrane ER is yet to be elucidated. The mechanisms by which ER is membrane associated and the processes regulating the relative number of cell surface ERs are not well understood. The basis

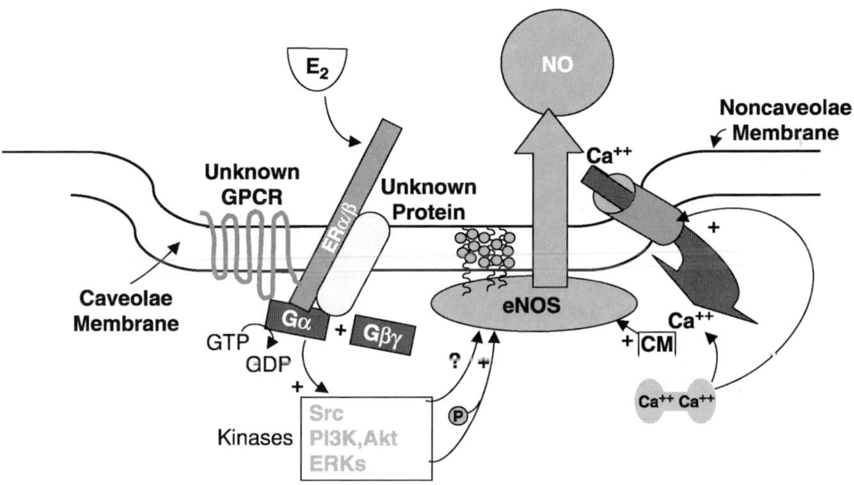

Figure 7. ER signaling to eNOS in endothelial cell caveolae. eNOS is localized to cholesterol-enriched (orange circles) caveolae by myristoylation and palmitoylation. Both ERα and ERβ have also been localized to endothelial cell caveolae. Ligand binding to ER leads to $G_{\alpha i}$ activation, which mediates downstream events. ER interaction with $G_{\alpha i}$ may be direct, or it may involve a classical *G*-protein-coupled receptor (GPCR) or an alternative unknown protein intermediate. The downstream events include the activation of Src tyrosine kinase, PI3 kinase-Akt kinase signaling causing eNOS phosphorylation, and ERK activation having an unknown impact on eNOS. In addition, there is perturbation of the local calcium environment, involving either the release of intracellular stores, which are likely to be in close proximity to caveolae membranes, or calcium influx, resulting in calmodulin (CM) binding to the enzyme. (See Color Insert.)

for ER–$G_{\alpha i}$ interaction, which may involve a classical GPCR or an alternative intermediate protein, is also entirely unknown (Fig. 7). The proximal signal transduction events following $G_{\alpha i}$ activation warrant further study. As mentioned previously, in certain systems the caveolae-associated ER may not be a classical ER but rather a variant of classical ER or possibly an entirely new receptor, and considerable additional work is needed to identify the receptor protein(s). Perhaps most importantly, the physiological and pathophysiological implications of ER localization and action in caveolae, which are currently being defined in cell or organ culture, are yet to be delineated in intact model systems. It is only through focused efforts that deeper understanding will be gained about ER localization and function in caveolae, thereby unraveling the basis for nongenomic estrogen action.

ACKNOWLEDGMENTS

The authors thank the colleagues and collaborators who have contributed to the data summarized in this chapter. They include Richard G. W. Anderson, Zohre German, Pingsheng Liu, Michael E. Mendelsohn, Chieko Mineo, Susanne M. Mumby, Todd S. Sherman, Myra H. Wyckoff, and Ivan S. Yuhanna. The authors also thank Marilyn Dixon for preparing this manuscript. This work was supported by National Institutes of Health grants HL58888, HL53546, and HD30276, the Lowe Foundation, and the Crystal Charity Ball.

REFERENCES

Benten, W. P., Stephan, C., Lieberherr, M., and Wunderlich, F. (2001). Estradiol signaling via sequestrable surface receptors. *Endocrinology* **142,** 1669–1677.

Carty, D. J. (1994). Pertussis toxin-catalyzed ADP-ribosylation of G-proteins. *Methods Enzymol.* **237,** 63–70.

Caulin-Glaser, T., Garcia-Cardena, G., Sarrel, P., Sessa, W. C., and Bender, J. R. (1997). 17 beta-estradiol regulation of human endothelial cell basal nitric oxide release, independent of cytosolic Ca2+ mobilization. *Circ. Res.* **81,** 885–892.

Chambliss, K. L., Yuhanna, I. S., Anderson, R. G., Mendelsohn, M. E., and Shaul, P. W. (2002). ERbeta has nongenomic action in caveolae. *Mol. Endocrinol.* **16,** 938–946.

Chambliss, K. L., Yuhanna, I. S., Mineo, C., Liu, P., German, Z., Sherman, T. S., Mendelsohn, M. E., Anderson, R. G., and Shaul, P. W. (2000). Estrogen receptor alpha and endothelial nitric oxide synthase are organized into a functional signaling module in caveolae. *Circ. Res.* **87,** E44–E52.

Chen, Z., Yuhanna, I. S., Galcheva-Gargova, Z., Karas, R. H., Mendelsohn, M. E., and Shaul, P. W. (1999). Estrogen receptor alpha mediates the nongenomic activation of endothelial nitric oxide synthase by estrogen. *J. Clin. Invest.* **103,** 401–406.

Deecher, D. C., Swiggard, P., Frail, D. E., and O'Connor, L. T. (2003). Characterization of a membrane-associated estrogen receptor in a rat hypothalamic cell line (D12). *Endocrine* **22,** 211–223.

Farhat, M. Y., Lavigne, M. C., and Ramwell, P. W. (1996). The vascular protective effects of estrogen. *FASEB J.* **10,** 615–624.

Garcia-Cardena, G., Oh, P., Liu, J., Schnitzer, J. E., and Sessa, W. C. (1996). Targeting of nitric oxide synthase to endothelial cell caveolae via palmitoylation: Implications for nitric oxide signaling. *Proc. Natl. Acad. Sci. USA* **93,** 6448–6453.

Gilman, A. G. (1995). Nobel Lecture. G-proteins and regulation of adenylyl cyclase. *Biosci. Rep.* **15,** 65–97.

Harrison, D. G. (1997). Cellular and molecular mechanisms of endothelial cell dysfunction. *J. Clin. Invest.* **100,** 2153–2157.

Haynes, M. P., Li, L., Sinha, D., Russell, K. S., Hisamoto, K., Baron, R., Collinge, M., Sessa, W. C., and Bender, J. R. (2003). Src kinase mediates phosphatidylinositol 3-kinase/Akt-dependent rapid endothelial nitric-oxide synthase activation by estrogen. *J. Biol. Chem.* **278,** 2118–2123.

Haynes, M. P., Sinha, D., Russell, K. S., Collinge, M., Fulton, D., Morales-Ruiz, M., Sessa, W. C., and Bender, J. R. (2000). Membrane estrogen receptor engagement activates endothelial nitric oxide synthase via the PI3-kinase-Akt pathway in human endothelial cells. *Circ. Res.* **87,** 677–682.

Huang, C., Hepler, J. R., Gilman, A. G., and Mumby, S. M. (1997). Attenuation of Gi- and Gq-mediated signaling by expression of RGS4 or GAIP in mammalian cells. *Proc. Natl. Acad. Sci. USA* **94,** 6159–6163.

Lantin-Hermoso, R. L., Rosenfeld, C. R., Yuhanna, I. S., German, Z., Chen, Z., and Shaul, P. W. (1997). Estrogen acutely stimulates nitric oxide synthase activity in fetal pulmonary artery endothelium. *Am. J. Physiol.* **273,** L119–L126.

LeMellay, V., Grosse, B., and Lieberherr, M. (1997). Phospholipase C beta and membrane action of calcitriol and estradiol. *J. Biol. Chem.* **272,** 11902–11907.

Li, L., Haynes, M. P., and Bender, J. R. (2003). Plasma membrane localization and function of the estrogen receptor alpha variant (ER46) in human endothelial cells. *Proc. Natl. Acad. Sci. USA* **100,** 4807–4812.

Liu, J., Conklin, B. R., Blin, N., Yun, J., and Wess, J. (1995). Identification of a receptor/G-protein contact site critical for signaling specificity and G-protein activation. *Proc. Natl. Acad. Sci. USA* **92,** 11642–11646.

Mendelsohn, M. E. (2000). Nongenomic, ER-mediated activation of endothelial nitric oxide synthase: How does it work? What does it mean? *Circ. Res.* **87,** 956–960.

Mendelsohn, M. E. and Karas, R. H. (1999). The protective effects of estrogen on the cardiovascular system. *N. Engl. J. Med.* **340,** 1801–1811.

Miller, V. M. (1999). Gender, estrogen, and NOS: Cautions about generalizations. *Circ. Res.* **85,** 979–981.

Moncada, S. and Higgs, A. (1993). The L-arginine-nitric oxide pathway. *N. Engl. J. Med.* **329,** 2002–2012.

Neer, E. J. (1995). Heterotrimeric G-proteins: Organizers of transmembrane signals. *Cell* **80,** 249–257.

Okamoto, T., Schlegel, A., Scherer, P. E., and Lisanti, M. P. (1998). Caveolins, a family of scaffoldinG-proteins for organizing "preassembled signaling complexes" at the plasma membrane. *J. Biol. Chem.* **273,** 5419–5422.

Razandi, M., Pedram, A., Greene, G. L., and Levin, E. R. (1999). Cell membrane and nuclear estrogen receptors (ERs) originate from a single transcript: Studies of ERalpha and ERbeta expressed in Chinese hamster ovary cells. *Mol. Endocrinol.* **13,** 307–319.

Russell, K. S., Haynes, M. P., Sinha, D., Clerisme, E., and Bender, J. R. (2000). Human vascular endothelial cells contain membrane binding sites for estradiol, which mediate rapid intracellular signaling. *Proc. Natl. Acad. Sci. USA* **97,** 5930–5935.

Shaul, P. W. and Anderson, R. G. W. (1998). Role of plasmalemmal caveolae in signal transduction. *Am. J. Physiol.* **275,** L843–L851.

Shaul, P. W., Smart, E. J., Robinson, L. J., German, Z., Yuhanna, I. S., Ying, Y., Anderson, R. G., and Michel, T. (1996). Acylation targets endothelial nitric-oxide synthase to plasmalemmal caveolae. *J. Biol. Chem.* **271,** 6518–6522.

Simoncini, T., Hafezi-Moghadam, A., Brazil, D. P., Ley, K., Chin, W. W., and Liao, J. K. (2000). Interaction of oestrogen receptor with the regulatory subunit of phosphatidylinositol-3-OH kinase. *Nature* **407,** 538–541.

Toran-Allerand, C. D., Guan, X., MacLusky, N. J., Horvath, T. L., Diano, S., Singh, M., Connolly, E. S., Jr., Nethrapalli, I. S., and Tinnikov, A. A. (2002). ER-X: A novel, plasma membrane-associated, putative estrogen receptor that is regulated during development and after ischemic brain injury. *J. Neurosci.* **22,** 8391–8401.

Vanhoutte, P. M. (1997). Endothelial dysfunction and atherosclerosis. *Eur. Heart J.* **18,** E19–E29.

Watson, C. S. and Gametchu, B. (1999). Membrane-initiated steroid actions and the proteins that mediate them. *Proc. Soc. Exp. Biol. Med.* **220,** 9–19.

Wess, J. (1998). Molecular basis of receptor/G-protein-coupling selectivity. *Pharmacol. Ther.* **80,** 231–264.

Wyckoff, M. H., Chambliss, K. L., Mineo, C., Yuhanna, I. S., Mendelsohn, M. E., Mumby, S. M., and Shaul, P. W. (2001). Plasma membrane estrogen receptors are coupled to endothelial nitric-oxide synthase through Galpha(i). *J. Biol. Chem.* **276,** 27071–27076.

Chapter 7

eNOS Regulation by Sphingosine 1-Phosphate and Caveolin

Junsuke Igarashi, M.D., Ph.D.

Department of Cardiovascular Physiology, Kagawa University Faculty of Medicine, 761-0793 Kagawa, Japan

Nitric oxide (NO), a unique and multi-potent signaling gaseous molecule, is produced by three independent isoforms of NO synthases (NOS) in a wide variety of mammalian cellular systems. This chapter focuses on the regulatory mechanisms of endothelial isoform of NOS (eNOS), especially on recent studies that explored how sphingolipids modulate eNOS activity within vascular endothelial cells. The importance of plasmalemmal caveolae and the scaffolding protein caveolin, which enable endothelial signaling machineries to efficiently communicate cell stimulation by sphingolipids to the activation of eNOS, is also discussed.

I. eNOS AS A KEY VASCULAR ENZYME

The endothelial isoform of NO synthase (eNOS), or NOS III, produces NO within vascular endothelial cells to modulate numerous essential vascular functions, including regulation of blood pressure, inhibition of platelet aggregation, and inhibition of leukocyte adhesion, among others (reviewed in Loscalzo and Welch, 1995). Additionally, recent studies have documented the roles of eNOS to promote angiogenesis, or

Advances in Molecular and Cell Biology, Vol. 36

ISBN: 0-444-51500-3

DOI: 10.1016/S0079-6603(05)36007-7

growth of new blood capillaries (for example, see Murohara *et al.*, 1998). Because angiogenesis plays key roles in the normal development as well as in the pathogenesis of a wide array of common diseases (reviewed in Carmeliet, 2003), these novel aspects of eNOS are no less important than its other classical functions. Since NO rapidly diffuses to remote target cells including vascular smooth muscle cells as well as blood platelets to exert its biological effects, eNOS activity within endothelial cells must be always governed by tight control mechanisms. eNOS activity increases in response to the elevations of intracellular calcium concentration (i[Ca^{2+}]) following various stimulation to endothelial cells. Increases in i[Ca^{2+}] enhance interaction of eNOS with calmodulin, a ubiquitous calcium regulatory protein, which is essential for the NOS catalytic reaction to produce NO. This dependence of eNOS on transient activation of i[Ca^{2+}]/calmodulin stands in contrast to that of another NOS isoform, iNOS, or NOS II, whose activity is principally modulated at its expression levels rather than by changes in i[Ca^{2+}]/calmodulin. Physiologically relevant eNOS activators comprise a very wide variety of endothelial stimuli, including hormones such as bradykinin (Mombouli and Vanhoutte, 1995), growth factors such as vascular endothelial growth factor (VEGF) (Ku *et al.*, 1993), and mechanical stimulation (Kuchan and Frangos, 1994), among others. On the other hand, derangements in eNOS regulation and resulting insufficiency of vascular NO bioavailability are associated with the pathogenesis of various cardiovascular diseases, including atherosclerosis, hypertension, and congestive heart failure (reviewed in Arnal *et al.*, 1999), although elucidation of detailed molecular mechanisms underlying eNOS dysregulation remains incomplete. It is therefore key to more fully understand how eNOS is (dys)regulated within vascular endothelial cells from the standpoint of molecular/cellular medicine.

II. eNOS ACYLATION, TARGETING TO CAVEOLAE, AND REGULATION BY CAVEOLIN

eNOS represents a unique NOS isoform to be dually acylated at its N terminus by saturated fatty acids myristate and palmitate (reviewed in Sase and Michel, 1997). eNOS myristoylation occurs cotranslationally at its glycine 2 residue (Busconi and Michel, 1993) and is irreversible, while eNOS palmitoylation occurs posttranslationally at cysteine residues 15 and 26 and is reversible (Robinson and Michel, 1995; Yeh *et al.*, 1999). Precise mechanisms of how eNOS dual acylation takes place are not fully understood. Because dual acylation by myristate and palmitate represents an important feature of many caveolae-targeted signaling proteins (Lisanti *et al.*, 1994), and eNOS activity had been detected within endothelial plasma membrane in earlier studies (Hecker *et al.*, 1994), subcellular localization of eNOS was examined in detail. A large body of biochemical as well as cell biological evidence has accumulated that establishes that a major fraction of eNOS protein is specifically enriched in endothelial caveolae, where the enzyme undergoes complex regulatory processes (reviewed in Shaul, 2002). It should be also noted, however, that some eNOS protein has been also detected in undefined intracellular locales (see Prabhakar *et al.*, 1998, for example).

Roles of eNOS protein residing in non-caveolae fractions, which may still be functional (Jobin *et al.*, 2003), are currently less well appreciated.

eNOS interacts with a scaffolding protein of caveolae, caveolin. Among the three isoforms of caveolin, caveolin-1 and caveolin-3 principally serve to interact with and regulate eNOS within vascular endothelial cells and within cardiac myocytes, respectively (Feron *et al.*, 1996). Caveolin directly interacts with eNOS to tonically inhibit the enzyme through its middle portion, termed the "scaffolding domain" (Michel *et al.*, 1997). Importantly, when endothelial cells are stimulated to elevate i[Ca^{2+}], calmodulin replaces caveolin to increase NO production and eNOS dissociates from caveolae. Once i[Ca^{2+}] returns to basal level, caveolin re-interacts with eNOS, which now re-associates with caveolae. Thus, these regulatory processes, the so-called eNOS-caveolin regulatory cycle (Michel and Feron, 1997), occurring at the plasmalemmal caveolae, play key roles to determine eNOS activity, reciprocally controlled by caveolin versus calcium/calmodulin.

III. SPHINGOSINE 1-PHOSPHATE AND REGULATION OF VASCULAR ENDOTHELIAL CELLS

Sphingolipids are derivatives of sphingosine, the core structure of this family of lipids. Caveolae are known to be specifically enriched in sphingolipids and cholesterol (Brown and Rose, 1992). It was the unique lipid composition of these micro-organelles in comparison with other plasma membrane fractions that initially prompted us to explore the relationships of sphingolipids to the regulation of eNOS, a caveolae-targeted enzyme. This unique lipid composition principally serves to create biophysical properties of caveolae membranes, i.e., the liquid-ordered phase (Brown and London, 1998). In addition to these important structural roles, sphingolipids had also started attracting much attention as signaling molecules. In the initial experiments, we tested the roles of ceramide, an acylated derivative of sphingosine, which has been best studied in relation to the regulation of apoptosis, or programmed cell death (reviewed in Pettus *et al.*, 2002). We found that exogenously added ceramide analogue is able to activate eNOS in cultured endothelial cells in a procedure appearing to be independent of the elevation of i[Ca^{2+}] (Igarashi *et al.*, 1999). Although endogenous ceramide generation following endothelial stimulation may also contribute to eNOS activation at least in some experimental conditions (Igarashi *et al.*, 1999; Li *et al.*, 2002; Florio *et al.*, 2003), molecular mechanisms whereby ceramide regulates eNOS remain rather elusive. Notably, ceramide may also modulate functions of caveolin (Zundel *et al.*, 2000; Shack *et al.*, 2003).

Other sphingolipids had started to be implicated in mammalian signal transduction, including sphingosine 1-phosphate (S1P). S1P is a phosphorylated derivative of sphingosine by an action of the enzymes termed sphingosine kinase and is found in normal human serum at the concentration ranges of several hundred nanomolars (Yatomi *et al.*, 1997a). S1P was initially supposed to exert its functions as an intracellular second messenger, which mediates platelet-derived growth factor (PDGF) stimulation to mobilization of calcium and activation of phospholipase D (Olivera and Spiegel, 1993), although its precise intracellular target molecule(s) remained less well identified. Instead, the activation of G-protein-coupled receptor pathways was implicated in the actions of this sphingolipid, because glass beads-conjugated S1P was still capable of exerting cellular activation (Yamamura *et al.*, 1997), and pertussis-toxin, which inhibits some of the heterotrimeric G-proteins, was found to markedly attenuate the effects of S1P (van Koppen *et al.*, 1996).

Receptor revelation for S1P derived from a completely independent line of investigation. The endothelial transcripts termed EDG-1 were identified in 1990 as an immediate early gene whose expression levels are markedly upregulated following endothelial stimulation with phorbol-12-myristate-13-acetate (PMA) (Hla and Maciag, 1990). Since treatment with PMA induces endothelial cells to differentiate to "angiogenic" phenotypes, this gene was named endothelial differentiation gene-1. Although EDG-1 was considered to belong to a member of the G-protein-coupled receptor superfamily due to its sequence similarity, the ligand for this "orphan" receptor remained unidentified for almost 10 years. It was 1998 when these two independent lines of research dovetailed nicely to demonstrate that S1P acts as a ligand for the EDG-1 receptors (Lee *et al.*, 1998) to modulate many important biological functions (see below). Five independent S1P receptor subtypes have since been found, and G-protein-coupled receptors for structurally similar other lysophospholipids were also identified (reviewed in Hla *et al.*, 2001); EDG receptors for S1P have recently been renamed $S1P_1$–$S1P_5$. It has now been established that most, if not all, of S1P's actions are mediated by its binding to and activating these S1P receptors via heterotrimeric G-protein pathways (reviewed in Hla *et al.*, 1999).

In vascular endothelial cells, receptor activation by S1P leads to a wide variety of responses, such as angiogenic morphogenesis, inhibition of apoptosis, chemotactic responses, and proliferation, among others (Hla, 2001). Many of these physiological responses of vascular endothelial cells to S1P overlap those mediated by eNOS (Loscalzo and Welch, 1995; Murohara *et al.*, 1998), suggesting functional associations of eNOS and S1P pathways. Enzyme activity of sphingosine kinase and amounts of S1P had been found specifically enriched in blood platelets, which may acutely release this lipid upon activation, possibly leading to high S1P concentrations within local environments surrounding activated platelets (Yatomi *et al.*, 1997b). Therefore, physical proximity of S1P with eNOS could be afforded not only intracellularly by plasmalemmal caveolae, but also extracellularly by platelet–endothelial interactions. It appeared plausible to us that S1P might participate in eNOS regulation based on these functional associations and potential physical proximity of two pathways.

IV. eNOS ACTIVATION BY S1P AND ENDOTHELIAL SIGNALING

This background led us to explore whether and how S1P regulates eNOS. In the initial experiments, we exploited heterologous expression systems in COS-7 cells, which had been known to lack endogenous expressions of both eNOS (Busconi and

Michel, 1993) and S1P receptors (Zondag *et al.*, 1998). When these cells are transiently co-transfected with plasmids encoding eNOS and those for $S1P_1$ (EDG-1) receptors, S1P induced robust NO production (Igarashi and Michel, 2000). S1P does not activate eNOS in the absence of co-transfected $S1P_1$ receptors, indicating that this action of S1P is mediated by $S1P_1$ receptors rather than by its intracellular actions. In the subsequent studies, we tested the effects of S1P for NO production in native endothelial cells. We made use of bovine aortic endothelial cells (BAEC), an archetypal endothelial cell culture, as a model. S1P induces marked increases in NO production in a dose-dependent manner with an EC_{50} value of approximately 10 nM (Igarashi *et al.*, 2001), which is in good agreement with many other endothelial responses to this lipid (Hla, 2001). It is important to note that the degree of eNOS activation by S1P is comparable to those attained by other classical eNOS agonists, including bradykinin or VEGF, suggesting quantitative importance of eNOS activation by S1P (Igarashi and Michel, 2001; Igarashi *et al.*, 2001). S1P activates eNOS in other endothelial cultures as well (Kimura *et al.*, 2001; Morales-Ruiz *et al.*, 2001). We then explored proximal signaling pathways that connect S1P stimulation to eNOS activation. S1P elevates i[Ca^{2+}] in BAEC (Lee *et al.*, 2000), and conversely, chelation of intracellular calcium abolishes eNOS activation (Igarashi and Michel, 2001; Igarashi *et al.*, 2001), indicating that eNOS regulation by S1P is essentially calcium dependent. Additionally, we found that eNOS regulation by S1P involves another set of signaling pathways to modulate the enzyme's phosphorylation at the serine 1179 residue; this phosphorylation of eNOS at its C terminus "sensitizes" the enzyme to lower levels of calcium *in vitro* (McCabe *et al.*, 2000). We found that full activation of eNOS by S1P requires a pathway comprising G-protein-coupled S1P receptors; pertussis toxin-sensitive G-proteins, especially G-protein $\beta\gamma$ subunits; the β-isoform of phosphoinositide 3-kinase (PI3-K); and a protein kinase, Akt, which phosphorylates eNOS at the serine 1179 residue (Igarashi and Michel, 2001; Igarashi *et al.*, 2001) (Fig. 1). These proximal signals evoked by S1P receptor activation exhibit significant similarities and differences in comparison with other eNOS activators (Igarashi and Michel, 2001), identifying potentially important points of control in receptor-regulated eNOS activation pathways. Although eNOS phosphorylation pathways also involve other protein kinase/phosphatase pathways at additional serine/threonine residues (Fischer *et al.*, 2001; Michell *et al.*, 2001; Boo *et al.*, 2002; Greif *et al.*, 2002, Kou *et al.*, 2002), these aspects of eNOS regulation by S1P remain to be explored.

V. ROLES OF CAVEOLAE FOR eNOS REGULATION BY S1P

Now that S1P is an eNOS activator; what are the roles of caveolae/caveolin for this novel eNOS-activating pathway? To begin exploring the relationships of the S1P/eNOS pathway with caveolae/caveolin, we tested whether $S1P_1$ (EDG-1) receptors are targeted to caveolae. We performed subcellular fractionation experiments in a detergent-free preparation, in which proteins that are attached to caveolae-enriched plasma membrane fractions are specifically recovered at the "light" fractions due to the

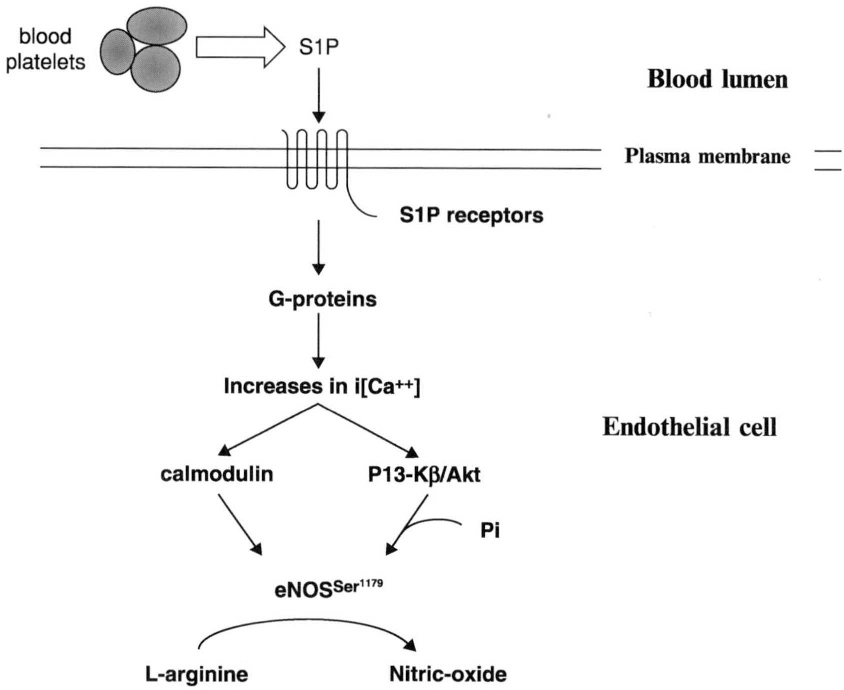

Figure 1. Signaling pathways whereby S1P mediates eNOS activation in vascular endothelial cells. Shown is a schematic presentation of how stimulation of vascular endothelial cells with sphingosine 1-phosphate (S1P) leads to the activation of endothelial NO synthase (eNOS) to produce nitric oxide. The top part represents blood lumen, at which activated blood platelets are supposed to acutely release S1P (Yatomi *et al.*, 1997a, 2000). S1P binds to and activates S1P receptors expressed on endothelial plasma membrane, specifically localized within the caveolae-enriched fractions (Igarashi and Michel, 2000). S1P receptors belong to the G-protein-coupled receptor superfamily with seven membrane-spanning regions (Hla *et al.*, 2001). Among the five independent subtypes, $S1P_1$ represents major receptors for S1P in vascular endothelial cells, coupled to pertussis toxin-sensitive G-protein pathways (Lee *et al.*, 1999; Morales-Ruiz *et al.*, 2001). S1P receptor stimulation elevates intracellular calcium concentrations ($i[Ca^{2+}]$) (Lee *et al.*, 2000; Muraki and Imaizumi, 2001), leading to the activation of eNOS in a classical manner that is dependent on calcium/calmodulin, a ubiquitous calcium regulatory protein. eNOS activation by S1P also involves another set of signaling cascade comprising the β-isoform of phosphoinositide 3-kinase (PI3-Kβ) and a downstream protein kinase, Akt, which in turn phosphorylates eNOS at its serine 1179 residue (Igarashi and Michel, 2001; Igarashi *et al.*, 2001) to sensitize the enzyme for calcium/calmodulin (McCabe *et al.*, 2000). Among the pertussis toxin-sensitive G-proteins, G-protein $\beta\gamma$ subunits appear to play a major role in activating the lipid kinase PI3-Kβ (Igarashi and Michel, 2001). Full activation of eNOS by S1P requires Akt-dependent phosphorylation at the serine 1179 residue (Igarashi *et al.*, 2001). As illustrated, calcium/calmodulin and PI3-Kβ/Akt pathways act synergistically to regulate eNOS following S1P receptor stimulation in vascular endothelial cells.

relatively low density of caveolae-enriched membranes (Song *et al.*, 1996). We exploited the transient transfection strategy of epitope-tagged $S1P_1$ (EDG-1) cDNA construct to COS-7 cells. In the resting cells, approximately half of the $S1P_1$ receptors are recovered in the caveolae-enriched fractions; conversely, overexpressed $S1P_1$ and endogenously expressed caveolin-1 of COS-7 cells are co-immunoprecipitated with each other (Igarashi and Michel, 2000). These results establish that $S1P_1$ (EDG-1) receptors are targeted to caveolae-enriched fractions to interact with caveolin-1. $S1P_1$ receptors undergo robust phosphorylation after S1P challenge (Lee *et al.*, 1998), as is the case with many other liganded G-protein-coupled receptors. Overexpression of caveolin-1 along with $S1P_1$ and eNOS in these cells leads to the attenuation of S1P-induced $S1P_1$ receptor phosphorylation as well as that of eNOS activation; treatment of these cells with S1P also markedly promotes translocation of the receptors to caveolae-enriched fractions (Igarashi and Michel, 2000). Since most of the signaling pathways following receptor stimulation are tonically inhibited by caveolin (Liu *et al.*, 2002), translocation of liganded $S1P_1$ receptors to caveolae fractions may represent a negative feedback mechanism so that caveolin-1 can more fully inhibit $S1P_1$ signaling. Translocation to caveolae fractions following ligand stimulation has been reported in several other G-protein-coupled receptors as well (de Weerd and Leeb-Lundberg, 1997; Feron *et al.*, 1997; Ishizaka *et al.*, 1998). Recently, it was shown that acylation-deficient mutants of eNOS, which have impaired targeting to caveolae, fail to undergo S1P-mediated phosphorylation at the serine 1179 residue (Gonzalez *et al.*, 2002). Together, these results suggest that caveolae may serve to facilitate S1P receptors to effectively modulate eNOS within endothelial cells, although caveolin-1 may rather serve to tonically attenuate "leakiness" of $S1P_1$ signaling pathways unless receptors are stimulated. Although protein kinase C pathways have been implicated in ligand-induced $S1P_1$ phosphorylation (Watterson *et al.*, 2002), regulatory roles of caveolin for specific protein kinase(s) in the context of $S1P_1$ phosphorylation remain unknown.

VI. eNOS REGULATION AND RECEPTOR CROSS-TALK OF S1P SIGNALING

Recent studies revealed that some of the G-protein-coupled receptors may "transactivate" receptor tyrosine kinases, and vice versa; one of the best known examples is epidermal growth factor receptor transactivation by β-adrenergic receptors in cardiac myocytes (Maudsley *et al.*, 2000), ultimately leading to hypertrophic responses of these cells (Asakura *et al.*, 2002). S1P was originally considered to act as a second messenger of fibroblasts following PDGF stimulation that activates receptor tyrosine kinases (Olivera and Spiegel, 1993); in a pathophysiological context, S1P works synergistically with VEGF to promote angiogenesis *in vivo* (Lee *et al.*, 1999), suggesting that S1P receptors may also interact with receptor tyrosine kinase pathways in endothelial cells. Since expression levels of $S1P_1$ transcripts are subject to regulation (discussed above), we sought to determine whether growth factors that modulate endothelial receptor tyrosine kinases alter $S1P_1$ receptor expressions. We found that in BAEC, VEGF, but not several other polypeptide growth factors, acutely upregulate expression of $S1P_1$ receptors at the levels of mRNA and protein, in a manner sensitive to pharmacological inhibition of protein kinase C pathways (Igarashi *et al.*, 2003). Increases in $S1P_1$ expression levels are associated with enhanced eNOS phosphorylation/activation of cultured endothelial cells as well as those of isolated blood vessels to subsequent stimulation with S1P, indicating that at least some of the newly synthesized $S1P_1$ receptor molecules are functional. It remains to be elucidated, however, whether newly produced receptors are readily targeted to plasmalemmal caveolae within such an acute time frame (~60 min). PDGF stimulation of 3T3 fibroblasts has been shown to activate sphingosine kinase within these cells, followed by a release of S1P into the culture media to stimulate overexpressed $S1P_1$ receptors, ultimately leading to the augmentation of cellular motility (Hobson *et al.*, 2001). Whether endothelial stimulation with VEGF also leads to $S1P_1$ stimulation in an autocrine/paracrine fashion remains to be established, while sphingosine kinase activation by VEGF has been recently documented (Shu *et al.*, 2002). In HEK-293 cells, S1P receptors and PDGF receptors have been found to form functional complexes to effectively activate mitogen-activated protein kinase pathways (Alderton *et al.*, 2001), making it possible that $S1P_1$ and VEGF receptors also physically interact with each other in endothelial cells. Together, these studies suggest that S1P/$S1P_1$ receptor pathways may act synergistically with receptor tyrosine kinases such as VEGF pathways to mediate eNOS responses in vascular endothelial cells.

VII. IMPLICATIONS FOR VASCULAR PHYSIOLOGY

Physiological implications of eNOS regulation by S1P have been explored in several experimental systems. We have shown that S1P induces eNOS-dependent vasorelaxation in rat mesenteric arterioles as well as in mouse thoracic arteries (Dantas *et al.*, 2003), helping to establish that treatment with S1P leads to physiologically relevant amounts of NO production in intact vessel preparations. In some other vessel preparations, however, S1P may induce vasoconstrictions (Bischoff *et al.*, 2000; Tosaka *et al.*, 2001); these differences of responses to S1P can possibly be attributed to the differences of S1P receptor subtype expression patterns in various blood vessels. For example, vascular smooth muscle cells predominantly express $S1P_2$ (EDG-5) and $S1P_3$ (EDG-3) subtypes that are coupled to pertussis toxin-insensitive G-protein pathways (Kluk and Hla, 2001; Tamama *et al.*, 2001), in contrast to endothelial cells that express $S1P_1$ as a major subtype coupled to pertussis toxin-sensitive pathways (Lee *et al.*, 1999; Morales-Ruiz *et al.*, 2001). In addition to regulation of vascular tone, eNOS activation by S1P also mediates endothelial signals leading to survival (Kwon *et al.*, 2001), proliferation, and migration (Rikitake *et al.*, 2002). S1P is found enriched in serum high-density lipoprotein (HDL) fractions (Murata *et al.*, 2000); clinically, higher concentrations of HDL are associated with favorable outcomes in the patients of cardiovascular diseases. Other lysophospholipids such as lysophosphatidylcholine and lysophosphatidic acid are higher in deleterious oxidized forms of low-density lipoproteins (Kume and Gimbrone, 1994; Maschberger *et al.*, 2000), raising a possibility that S1P and other lysophospholipids within various lipoprotein fractions may exert differential actions on vascular cells. A recent study implicated $S1P_3$ receptor subtype for eNOS activation following vascular stimulation with HDL (Nofer *et al.*, 2004), identifying a potentially important point of control whereby HDL may exert favorable actions on blood vessels. Because even normal serum contains saturating amounts of S1P, and subtype-specific agonists/antagonists for S1P receptors are not currently available, specific roles of each S1P receptor subtype for eNOS activation remain to be elucidated in more detail.

Angiogenic responses are the most prominent cardiovascular actions of S1P, especially at a level where newly formed blood vessels are maturated (Lee *et al.*, 1999; Hla, 2001), mediated principally by $S1P_1$ receptor subtype within endothelial cells (Liu *et al.*, 2000; Allende *et al.*, 2003). It has been demonstrated with a Matrigel plug assay in mice that eNOS mediates promotion of angiogenesis by S1P *in vivo* (Rikitake *et al.*, 2002), indicating that eNOS actually participates in S1P-promoted angiogenesis. In addition to eNOS and $S1P_1$ (discussed above), many of the endothelial signaling molecules that regulate angiogenesis have been determined to be localized within caveolae fractions, including receptors such as KDR (or VEGF receptor 2) and effectors such as a tyrosine kinase c-Src and several G-protein subunits (reviewed in Liu *et al.*, 2002). Downregulation of caveolin-1 has been documented to induce angiogenic differentiation of cultured endothelial cells (Liu *et al.*, 1999); the expression levels of some of these signaling molecules residing in caveolae are modulated by cell stimulation (Shen *et al.*, 1998, 1999; Liu *et al.*, 1999; Igarashi *et al.*, 2003). It is also noteworthy that overexpression of sphingosine kinase to endothelial cells leads to augmented angiogenesis in response to fibroblast growth factor in association with the extracellular export of sphingosine kinase itself (Ancellin *et al.*, 2002). It remains to be elucidated whether sphingosine kinase is enriched in endothelial caveolae to produce S1P. If this were the case, it is likely that plasmalemmal caveolae and caveolin coordinate endothelial signaling events leading to angiogenesis, communicating ligands/receptors S1P/$S1P_1$ and VEGF/KDR to their downstream effectors, including sphingosine kinase, tyrosine kinases, G-proteins, as well as eNOS (Fig. 2).

VIII. POSSIBILITIES IN NONENDOTHELIAL CELLS

Cardiac myocytes represent one of the best-studied non-endothelial cellular systems that express eNOS. eNOS of cardiac myocytes has been implicated in the regulation of contractile responses to cholinergic and adrenergic stimulation (reviewed in Kelly *et al.*, 1996). Cardiac myocytes express S1P receptors (Goetzl *et al.*, 2000), hence S1P is potentially capable of activating eNOS in these cells as well. In zebrafish, mutation of the *mil* gene, which encodes the orthologue of mammalian $S1P_2$/EDG-5 receptors, leads to derangements of cardiac morphogenesis during development (Kupperman *et al.*, 2000). It is therefore possible that S1P/eNOS signaling may play a role in mammalian cardiac morphogenesis and/or hypertrophic responses in concert with caveolin-3, a muscle-specific caveolin isoform that interacts with and regulates cardiac eNOS (Feron *et al.*, 1996). It should also be noted that in cardiac myocytes, the γ-isoform of PI3-K, rather than its β-isoform, has been shown to mediate PI3-K/Akt activation following G-protein-coupled receptor activation (Naga Prasad *et al.*, 2000,

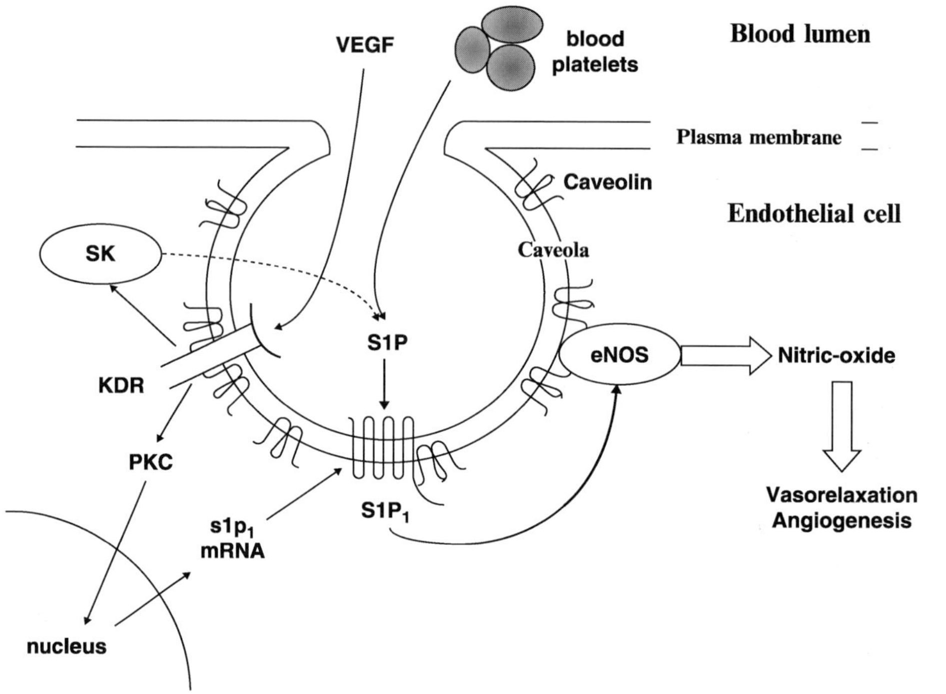

Figure 2. Possible mechanisms whereby caveolae/caveolin may coordinate cross-talks of signaling pathways evoked by VEGF and S1P receptors to regulate eNOS. Shown is a schematic presentation of how signaling pathways evoked by receptor stimulation might be coordinated within plasmalemmal caveolae by the scaffolding protein caveolin to activate endothelial nitric oxide synthase (eNOS) in vascular endothelial cells. Shown in the middle is an invagination structure representing a caveola of endothelial plasma membrane, which is coated with homo-oligomers of caveolin (illustrated as hairpin-like structures). As is eNOS, many of endothelial receptors that modulate eNOS activity are specifically enriched in caveolae (Liu *et al.*, 2002). In most cases, caveolin interacts with these signaling molecules serving to tonically inhibit receptor signaling at multiple levels both at receptors such as $S1P_1$ (Igarashi and Michel, 2000) as well as effectors such as eNOS (Feron *et al.*, 1996). Receptor tyrosine kinase KDR (or VEGF receptor 2) and G-protein-coupled receptor $S1P_1$ for sphingosine 1-phosphate (S1P) are capable of acting independently to increase the activity of eNOS. However, VEGF stimulation to endothelial cells also leads to the upregulation of $S1P_1$ mRNA/protein expression levels via protein kinase C (PKC) pathways. This upregulation of $S1P_1$ receptors is associated with enhanced responses to subsequent S1P stimulation (Igarashi *et al.*, 2003), indicating that these two pathways exhibit significant cross-talks. Although S1P has been supposed to derive from activated platelets, recent studies documented that VEGF stimulation augments the activity of endothelial sphingosine kinase (SK) that produces S1P (Shu *et al.*, 2002). Hence, it is possible that VEGF stimulation leads to S1P release into extracellular space outside the caveolae, which in turn binds to and activates $S1P_1$ receptors whose expression levels have been elevated. NO production following S1P stimulation contributes to promote not only vasorelaxation (Dantas *et al.*, 2003) but also angiogenesis (Rikitake *et al.*, 2002); VEGF and S1P can act synergistically to modulate angiogenesis (Lee *et al.*, 1999). It is therefore plausible that caveolae and caveolin serve to coordinate these multiple layers of signaling pathways evoked by VEGF/KDR and S1P/$S1P_1$ to regulate downstream effectors such as sphingosine kinase and eNOS, which ultimately lead to important vascular responses including vasorelaxation and angiogenesis.

2001; Bony *et al.*, 2001). Thus, it is plausible that this isoform of PI3-K may mediate eNOS phosphorylation/activation by S1P in the heart. Similarly, S1P might also modulate eNOS in blood platelets, which express functional S1P receptors (Yatomi *et al.*, 1997a), PI3-K isoforms (Hirsch *et al.*, 2001), Akt (Kim *et al.*, 2004), as well as eNOS (Sase and Michel, 1995). It is tempting to speculate that eNOS in platelets may exert negative feedback mechanisms for S1P-mediated platelet aggregation.

IX. CONCLUDING REMARKS

Prompted initially by the subcellular localization of eNOS and sphingolipids to the plasmalemmal caveolae, we and others have explored eNOS regulation by these lipid species. Although these studies found that S1P does activate endothelial cells to produce quantitatively relevant amounts of nitric oxide, many issues remain to be elucidated to establish the physiological consequences of this novel signaling pathway. I propose that a deeper understanding of eNOS regulation by sphingolipids, modulated at the plasmalemmal caveolae by caveolin, may lead to the identification of important points of control in mammalian cardiovascular systems.

ACKNOWLEDGEMENTS

I express sincere thanks to Professor Thomas Michel (Brigham and Women's Hospital/Harvard Medical School) for his mentorship, support, guidance, and persistent encouragement during my fellowship/instructorship in his laboratory, and am indebted to those who generously provided invaluable materials used in our experiments cited in this article. J.I. is currently supported in part by a research grant from the Ministry of Education, Culture, Sports, Science and Technology of Japan (Grant-in-Aid for Young Scientists (B) 15790119).

REFERENCES

Alderton, F., Rakhit, S., Kong, K. C., Palmer, T., Sambi, B., Pyne, S., and Pyne, N. J. (2001) Tethering of the platelet-derived growth factor beta receptor to G-protein-coupled receptors. A novel platform for integrative signaling by these receptor classes in mammalian cells. *J. Biol. Chem.* **276,** 28578–28585.

Allende, M. L., Yamashita, T., and Proia, R. L. (2003) G-protein-coupled receptor S1P1 acts within endothelial cells to regulate vascular maturation. *Blood* **102,** 3665–3667.

Ancellin, N., Colmont, C., Su, J., Li, Q., Mittereder, N., Chae, S. S., Stefansson, S., Liau, G., and Hla, T. (2002) Extracellular export of sphingosine kinase-1 enzyme. Sphingosine 1-phosphate generation and the induction of angiogenic vascular maturation. *J. Biol. Chem.* **277,** 6667–6675.

Arnal, J. F., Dinh-Xuan, A. T., Pueyo, M., Darblade, B., and Rami, J. (1999) Endothelium-derived nitric oxide and vascular physiology and pathology. *Cell Mol. Life Sci.* **55,** 1078–1087.

Asakura, M., Kitakaze, M., Takashima, S., Liao, Y., Ishikura, F., Yoshinaka, T., Ohmoto, H., Node, K., Yoshino, K., Ishiguro, H., Asanuma, H., Sanada, S., Matsumura, Y., Takeda, H., Beppu, S., Tada, M.,

Hori, M., and Higashiyama, S. (2002) Cardiac hypertrophy is inhibited by antagonism of ADAM12 processing of HB-EGF: Metalloproteinase inhibitors as a new therapy. *Nat. Med.* **8,** 35–40.

Bischoff, A., Czyborra, P., Fetscher, C., Meyer Zu Heringdorf, D., Jakobs, K. H., and Michel, M. C. (2000) Sphingosine-1-phosphate and sphingosylphosphorylcholine constrict renal and mesenteric microvessels *in vitro*. *Br. J. Pharmacol.* **130,** 1871–1877.

Bony, C., Roche, S., Shuichi, U., Sasaki, T., Crackower, M. A., Penninger, J., Mano, H., and Puceat, M. (2001) A specific role of phosphatidylinositol 3-kinase gamma. A regulation of autonomic Ca(2)+ oscillations in cardiac cells. *J. Cell Biol.* **152,** 717–728.

Boo, Y. C., Sorescu, G., Boyd, N., Shiojima, I., Walsh, K., Du, J., and Jo, H. (2002) Shear stress stimulates phosphorylation of endothelial nitric-oxide synthase at Ser1179 by Akt-independent mechanisms: Role of protein kinase A. *J. Biol. Chem.* **277,** 3388–3396.

Brown, D. A. and London, E. (1998) Structure and origin of ordered lipid domains in biological membranes. *J. Membr. Biol.* **164,** 103–114.

Brown, D. A. and Rose, J. K. (1992) Sorting of, G. P.I-anchored proteins to glycolipid-enriched membrane subdomains during transport to the apical cell surface. *Cell* **68,** 533–544.

Busconi, L. and Michel, T. (1993) Endothelial nitric oxide synthase. N-terminal myristoylation determines subcellular localization. *J. Biol. Chem.* **268,** 8410–8413.

Carmeliet, P. (2003) Angiogenesis in health and disease. *Nat. Med.* **9,** 653–660.

Dantas, A. P., Igarashi, J., and Michel, T. (2003) Sphingosine 1-phosphate and control of vascular tone. *Am. J. Physiol. Heart Circ. Physiol.* **284,** H2045–H2052.

de Weerd, W. F. C. and Leeb-Lundberg, L. M. F. (1997) Bradykinin sequesters B2 bradykinin receptors and the receptor-coupled $G\alpha$-subunits $G\alpha_q$ and $G\alpha_i$ in caveolae in DDT_1 MF-2 smooth muscle cells. *J. Biol. Chem.* **272,** 17858–17866.

Feron, O., Belhassen, L., Kobzik, L., Smith, T. W., Kelly, R. A., and Michel, T. (1996) Endothelial nitric oxide synthase targeting to caveolae. Specific interactions with caveolin isoforms in cardiac myocytes and endothelial cells. *J. Biol. Chem.* **271,** 22810–22814.

Feron, O., Smith, T. W., Michel, T., and Kelly, R. A. (1997) Dynamic targeting of the agonist-stimulated m2 muscarinic acetylcholine receptor to caveolae in cardiac myocytes. *J. Biol. Chem.* **272,** 17744–17748.

Fischer, D. J., Nusser, N., Virag, T., Yokoyama, K., Wang, D., Baker, D. L., Bautista, D., Parrill, A. L., and Tigyi, G. (2001) Short-chain phosphatidates are subtype-selective antagonists of lysophosphatidic acid receptors. *Mol. Pharmacol.* **60,** 776–784.

Florio, T., Arena, S., Pattarozzi, A., Thellung, S., Corsaro, A., Villa, V., Massa, A., Diana, F., Spoto, G., Forcella, S., Damonte, G., Filocamo, M., Benatti, U., and Schettini, G. (2003) Basic fibroblast growth factor activates endothelial nitric-oxide synthase in C.H.O-K1 cells via the activation of ceramide synthesis. *Mol. Pharmacol.* **63,** 297–310.

Goetzl, E. J., Lee, H., Azuma, T., Stossel, T. P., Turck, C. W., and Karliner, J. S. (2000) Gelsolin binding and cellular presentation of lysophosphatidic acid. *J. Biol. Chem.* **275,** 14573–14578.

Gonzalez, E., Kou, R., Lin, A. J., Golan, D. E., and Michel, T. (2002) Subcellular targeting and agonist-induced site-specific phosphorylation of endothelial nitric-oxide synthase. *J. Biol. Chem.* **277,** 39554–39560.

Greif, D. M., Kou, R., and Michel, T. (2002) Site-specific dephosphorylation of endothelial nitric oxide synthase by protein phosphatase 2A: Evidence for crosstalk between phosphorylation sites. *Biochemistry* **41,** 15845–15853.

Hecker, M., Mulsch, A., Bassenge, E., Forstermann, U., and Busse, R. (1994) Subcellular localization and characterization of nitric oxide synthase(s) in endothelial cells: Physiological implications. *Biochem. J.* **299**(Pt 1), 247–252.

Hirsch, E., Bosco, O., Tropel, P., Laffargue, M., Calvez, R., Altruda, F., Wymann, M., and Montrucchio, G. (2001) Resistance to thromboembolism in P. I.3Kgamma-deficient mice. *FASEB J.* **15,** 2019–2021.

Hla, T. (2001) Sphingosine 1-phosphate receptors. *Prostaglandins* **64,** 135–142.

Hla, T., Lee, M. J., Ancellin, N., Liu, C. H., Thangada, S., Thompson, B. D., and Kluk, M. (1999) Sphingosine-1-phosphate: Extracellular mediator or intracellular second messenger? *Biochem. Pharmacol.* **58,** 201–207.

Hla, T., Lee, M. J., Ancellin, N., Paik, J. H., and Kluk, M. J. (2001) Lysophospholipids—receptor revelations. *Science* **294,** 1875–1878.

Hla, T. and Maciag, T. (1990) An abundant transcript induced in differentiating human endothelial cells encodes a polypeptide with structural similarities to G-protein-coupled receptors. *J. Biol. Chem.* **265,** 9308–9313.

Hobson, J. P., Rosenfeldt, H. M., Barak, L. S., Olivera, A., Poulton, S., Caron, M. G., Milstien, S., and Spiegel, S. (2001) Role of the sphingosine-1-phosphate receptor EDG-1 in PDGF-induced cell motility. *Science* **291,** 1800–1803.

Igarashi, J., Bernier, S. G., and Michel, T. (2001) Sphingosine 1-phosphate and activation of endothelial nitric oxide synthase: Differential regulation of Akt and MAP kinase pathways by EDG and bradykinin receptors in vascular endothelial cells. *J. Biol. Chem.* **276,** 12420–12426.

Igarashi, J., Erwin, P. A., Dantas, A. P., Chen, H., and Michel, T. (2003) VEGF induces S1P1 receptors in endothelial cells: Implications for cross-talk between sphingolipid and growth factor receptors. *Proc. Natl. Acad. Sci. USA* **100,** 10664–10669.

Igarashi, J. and Michel, T. (2000) Agonist-modulated targeting of the EDG-1 receptor to plasmalemmal caveolae: eNOS activation by sphingosine 1-phosphate and the role of caveolin-1 in sphingolipid signal transduction. *J. Biol. Chem.* **275,** 32363–32370.

Igarashi, J. and Michel, T. (2001) Sphingosine 1-phosphate and isoform-specific activation of phosphoinositide 3-kinase beta. Evidence for divergence and convergence of receptor-regulated, endothelial nitric-oxide synthase signaling pathways. *J. Biol. Chem.* **276,** 36281–36288.

Igarashi, J., Thatte, H. S., Prabhakar, P., Golan, D. E., and Michel, T. (1999) Calcium-independent activation of endothelial nitric oxide synthase by ceramide. *Proc. Natl. Acad. Sci. USA* **96,** 12583–12588.

Ishizaka, N., Griendling, K. K., Lassegue, B., and Alexander, R. W. (1998) Angiotensin II type 1 receptor: Relationship with caveolae and caveolin after initial agonist stimulation. *Hypertension* **32,** 459–466.

Jobin, C. M., Chen, H., Lin, A. J., Yacono, P. W., Igarashi, J., Michel, T., and Golan, D. E. (2003) Receptor-regulated dynamic interaction between endothelial nitric oxide synthase and calmodulin revealed by fluorescence resonance energy transfer in living cells. *Biochemistry* **42,** 11716–11725.

Kelly, R. A., Balligand, J. L., and Smith, T. W. (1996) Nitric oxide and cardiac function. *Circ. Res.* **79,** 363–380.

Kim, S., Jin, J., and Kunapuli, S. P. (2004) Akt activation in platelets depends on gi signaling pathways. *J. Biol. Chem.* **279,** 4186–4195.

Kimura, T., Sato, K., Kuwabara, A., Tomura, H., Ishiwara, M., Kobayashi, I., Ui, M., and Okajima, F. (2001) Sphingosine 1-phosphate may be a major component of plasma lipoproteins responsible for the cytoprotective actions in human umbilical vein endothelial cells. *J. Biol. Chem.* **276,** 31780–31785.

Kluk, M. J. and Hla, T. (2001) Role of the sphingosine 1-phosphate receptor EDG-1 in vascular smooth muscle cell proliferation and migration. *Circ. Res.* **89,** 496–502.

Kou, R., Greif, D., and Michel, T. (2002) Dephosphorylation of endothelial nitric-oxide synthase by vascular endothelial growth factor implications, for the vascular responses to cyclosporin A. *J. Biol. Chem.* **277,** 29669–29673.

Ku, D. D., Zaleski, J. K., Liu, S., and Brock, T. A. (1993) Vascular endothelial growth factor induces EDRF-dependent relaxation in coronary arteries. *Am. J. Physiol.* **265,** H586–H592.

Kuchan, M. J. and Frangos, J. A. (1994) Role of calcium and calmodulin in flow-induced nitric oxide production in endothelial cells. *Am. J. Physiol.* **266,** C628–C636.

Kume, N. and Gimbrone, M. A. J. (1994) Lysophosphatidylcholine transcriptionally induces growth factor gene expression in cultured human endothelial cells. *J. Clin. Invest.* **93,** 907–911.

Kupperman, E., An, S., Osborne, N., Waldron, S., and Stainier, D. Y. (2000) A sphingosine-1-phosphate receptor regulates cell migration during vertebrate heart development. *Nature* **406,** 192–195.

Kwon, Y. G., Min, J. K., Kim, K. M., Lee, D. J., Billiar, T. R., and Kim, Y. M. (2001) Sphingosine 1-phosphate protects human umbilical vein endothelial cells from serum-deprived apoptosis by nitric oxide production. *J. Biol. Chem.* **276,** 10627–10633.

Lee, H., Goetzl, E. J., and An, S. (2000) Lysophosphatidic acid and sphingosine 1-phosphate stimulate endothelial cell wound healing. *Am. J. Physiol.* **278,** C612–C618.

Lee, M. J., Thangada, S., Claffey, K. P., Ancellin, N., Liu, C. H., Kluk, M., Volpi, M., Sha'afi, R. I., and Hla, T. (1999) Vascular endothelial cell adherens junction assembly and morphogenesis induced by sphingosine-1-phosphate. *Cell* **99,** 301–312.

Lee, M. J., Thangada, S., Liu, C. H., Thompson, B. D., and Hla, T. (1998) Lysophosphatidic acid stimulates the G-protein-coupled receptor EDG-1 as a low affinity agonist. *J. Biol. Chem.* **273,** 22105–22112.

Lee, M. J., Van Brocklyn, J. R., Thangada, S., Liu, C. H., Hand, A. R., Menzeleev, R., Spiegel, S., and Hla, T. (1998) Sphingosine-1-phosphate as a ligand for the G protein-coupled receptor EDG-1. *Science* **279,** 1552–1555.

Li, X. A., Titlow, W. B., Jackson, B. A., Giltiay, N., Nikolova-Karakashian, M., Uittenbogaard, A., and Smart, E. J. (2002) High density lipoprotein binding to scavenger receptor class B type I activates endothelial nitric-oxide synthase in a ceramide-dependent manner. *J. Biol. Chem.* **277,** 11058–11063.

Lisanti, M. P., Scherer, P. E., Tang, Z., and Sargiacomo, M. (1994) Caveolae, caveolin and caveolin-rich membrane domains: A signalling hypothesis. *Trends Cell Biol.* **4,** 231–235.

Liu, J., Razani, B., Tang, S., Terman, B. I., Ware, J. A., and Lisanti, M. P. (1999) Angiogenesis activators and inhibitors differentially regulate caveolin-1 expression and caveolae formation in vascular endothelial cells. Angiogenesis inhibitors block vascular endothelial growth factor-induced down-regulation of caveolin-1. *J. Biol. Chem.* **274,** 15781–15785.

Liu, P., Rudick, M., and Anderson, R. G. (2002) Multiple functions of caveolin-1. *J. Biol. Chem.* **277,** 41295–41298.

Liu, Y., Wada, R., Yamashita, T., Mi, Y., Deng, C. X., Hobson, J. P., Rosenfeldt, H. M., Nava, V. E., Chae, S. S., Lee, M. J., Liu, C. H., Hla, T., Spiegel, S., and Proia, R. L. (2000) Edg-1, the G protein-coupled receptor for sphingosine-1-phosphate, is essential for vascular maturation. *J. Clin. Invest.* **106,** 951–961.

Loscalzo, J. and Welch, G. (1995) Nitric oxide and its role in the cardiovascular system. *Prog. Cardiovasc. Dis.* **38,** 87–104.

Maschberger, P., Bauer, M., Baumann-Siemons, J., Zangl, K. J., Negrescu, E. V., Reininger, A. J., and Siess, W. (2000) Mildly oxidized low density lipoprotein rapidly stimulates via activation of the lysophosphatidic acid receptor Src family and Syk tyrosine kinases and Ca2+ influx in human platelets. *J. Biol. Chem.* **275,** 19159–19166.

Maudsley, S., Pierce, K. L., Zamah, A. M., Miller, W. E., Ahn, S., Daaka, Y., Lefkowitz, R. J., and Luttrell, L. M. (2000) The beta(2)-adrenergic receptor mediates extracellular signal-regulated kinase activation via assembly of a multi-receptor complex with the epidermal growth factor receptor. *J. Biol. Chem.* **275,** 9572–9580.

McCabe, T. J., Fulton, D., Roman, L. J., and Sessa, W. C. (2000) Enhanced electron flux and reduced calmodulin dissociation may explain "calcium-independent" eNOS activation by phosphorylation. *J. Biol. Chem.* **275,** 6123–6128.

Michel, J. B., Feron, O., Sase, K., Prabhakar, P., and Michel, T. (1997) Caveolin versus calmodulin. Counterbalancing allosteric modulators of endothelial nitric oxide synthase. *J. Biol. Chem.* **272,** 25907–25912.

Michel, T. and Feron, O. (1997) Nitric oxide synthases: Which, where, how, and why? *J. Clin. Invest.* **100,** 2146–2152.

Michell, B. J., Chen, Z., Tiganis, T., Stapleton, D., Katsis, F., Power, D. A., Sim, A. T., and Kemp, B. E. (2001) Coordinated control of endothelial nitric-oxide synthase phosphorylation by protein kinase C and the cAMP-dependent protein kinase. *J. Biol. Chem.* **276,** 17625–17628.

Mombouli, J. V. and Vanhoutte, P. M. (1995) Kinins and endothelial control of vascular smooth muscle. *Annu. Rev. Pharmacol. Toxicol.* **35,** 679–705.

Morales-Ruiz, M., Lee, M. J., Zoellner, S., Gratton, J. P., Scotland, R., Shiojima, I., Walsh, K., Hla, T., and Sessa, W. C. (2001) Sphingosine-1-phosphate activates Akt, nitric oxide production and chemotaxis through a Gi-protein/phosphoinositide 3-kinase pathway in endothelial cells. *J. Biol. Chem.* **276,** 19672–19677.

Muraki, K. and Imaizumi, Y. (2001) A novel function of sphingosine 1-phosphate to activate a non-selective cation channel in human endothelial cells. *J. Physiol.* **537,** 431–441.

Murata, N., Sato, K., Kon, J., Tomura, H., Yanagita, M., Kuwabara, A., Ui, M., and Okajima F. (2000). Interaction of sphingosine 1-phosphate with plasma components, including lipoproteins, regulates the lipid receptor-mediated actions. *Biochem. J.* **352**(Pt 3), 809–815.

Murohara, T., Asahara, T., Silver, M., Bauters, C., Masuda, H., Kalka, C., Kearney, M., Chen, D., Symes, J. F., Fishman, M. C., Huang, P. L., and Isner, J. M. (1998) Nitric oxide synthase modulates angiogenesis in response to tissue ischemia. *J. Clin. Invest.* **101,** 2567–2578.

Naga Prasad, S. V., Barak, L. S., Rapacciuolo, A., Caron, M. G., and Rockman, H. A. (2001) Agonist-dependent recruitment of phosphoinositide 3-kinase to the membrane by $\{\beta\}$-adrenergic receptor kinase 1. A role in receptor sequestration. *J. Biol. Chem.* **276,** 18953–18959.

Naga Prasad, S. V., Esposito, G., Mao, L., Koch, W. J., and Rockman, H. A. (2000) Gbetagamma-dependent phosphoinositide 3-kinase activation in hearts with *in vivo* pressure overload hypertrophy. *J. Biol. Chem.* **275,** 4693–4698.

Nofer, J. R., Van Der Giet, M., Tolle, M., Wolinska, I., Von Wnuck Lipinski, K., Baba, H. A., Tietge, U. J., Godecke, A., Ishii, I., Kleuser, B., Schafers, M., Fobker, M., Zidek, W., Assmann, G., Chun, J., and Levkau, B. (2004) HDL induces NO-dependent vasorelaxation via the lysophospholipid receptor S1P(3). *J. Clin. Invest.* **113,** 569–581.

Olivera, A. and Spiegel, S. (1993) Sphingosine-1-phosphate as second messenger in cell proliferation induced by PDGF and FCS mitogens. *Nature* **365,** 557–560.

Pettus, B. J., Chalfant, C. E., and Hannun, Y. A. (2002) Ceramide in apoptosis: An overview and current perspectives. *Biochim. Biophys. Acta* **1585,** 114–125.

Prabhakar, P., Thatte, H. S., Goetz, R. M., Cho, M. R., Golan, D. E., and Michel, T. (1998) Receptor-regulated translocation of endothelial nitric-oxide synthase. *J. Biol. Chem.* **273,** 27383–27388.

Rikitake, Y., Hirata, K., Kawashima, S., Ozaki, M., Takahashi, T., Ogawa, W., Inoue, N., and Yokoyama, M. (2002) Involvement of endothelial nitric oxide in sphingosine-1-phosphate-induced angiogenesis. *Arterioscler. Thromb. Vasc. Biol.* **22,** 108–114.

Robinson, L. J. and Michel, T. (1995) Mutagenesis of palmitoylation sites in endothelial nitric oxide synthase identifies a novel motif for dual acylation and subcellular targeting. *Proc. Natl. Acad. Sci. USA* **92,** 11776–11780.

Sase, K. and Michel, T. (1995) Expression of constitutive endothelial nitric oxide synthase in human blood platelets. *Life Sci.* **57,** 2049–2055.

Sase, K. and Michel, T. (1997) Expression and regulation of endothelial nitric oxide synthase. *Trends Cardiovasc. Med.* **7,** 28–37.

Shack, S., Wang, X. T., Kokkonen, G. C., Gorospe, M., Longo, D. L., and Holbrook, N. J. (2003) Caveolin-induced activation of the phosphatidylinositol 3-kinase/akt pathway increases arsenite cytotoxicity. *Mol. Cell Biol.* **23,** 2407–2414.

Shaul, P. W. (2002) Regulation of endothelial nitric oxide synthase: Location, location, location. *Annu. Rev. Physiol.* **64,** 749–774.

Shen, B. Q., Lee, D. Y., Gerber, H. P., Keyt, B. A., Ferrara, N., and Zioncheck, T. F. (1998) Homologous up-regulation of KDR/Flk-1 receptor expression by vascular endothelial growth factor *in vitro*. *J. Biol. Chem.* **273,** 29979–29985.

Shen, B. Q., Lee, D. Y., and Zioncheck, T. F. (1999) Vascular endothelial growth factor governs endothelial nitric-oxide synthase expression via a KDR/Flk-1 receptor and a protein kinase C signaling pathway. *J. Biol. Chem.* **274,** 33057–33063.

Shu, X., Wu, W., Mosteller, R. D., and Broek, D. (2002) Sphingosine kinase mediates vascular endothelial growth factor-induced activation of ras and mitogen-activated protein kinases. *Mol. Cell Biol.* **22,** 7758–7768.

Song, K. S., Li, S., Okamoto, T., Quilliam, L. A., Sargiacomo, M., and Lisanti, M. P. (1996) Co-purification and direct interaction of Ras with caveolin, an integral membrane protein of caveolae microdomains. Detergent-free purification of caveolae microdomains. *J. Biol. Chem.* **271,** 9690–9697.

Tamama, K., Kon, J., Sato, K., Tomura, H., Kuwabara, A., Kimura, T., Kanda, T., Ohta, H., Ui, M., Kobayashi, I., and Okajima, F. (2001) Extracellular mechanism through the Edg family of receptors might be responsible for sphingosine-1-phosphate-induced regulation of DNA synthesis and migration of rat aortic smooth-muscle cells. *Biochem. J.* **353,** 139–146.

Tosaka, M., Okajima, F., Hashiba, Y., Saito, N., Nagano, T., Watanabe, T., Kimura, T., and Sasaki, T. (2001) Sphingosine 1-phosphate contracts canine basilar arteries *in vitro* and *in vivo*: Possible role in pathogenesis of cerebral vasospasm. *Stroke* **32,** 2913–2919.

van Koppen, C., Meyer zu Heringdorf, M., Laser, K. T., Zhang, C., Jakobs, K. H., Bunemann, M., and Pott, L. (1996) Activation of a high affinity Gi protein-coupled plasma membrane receptor by sphingosine-1-phosphate. *J. Biol. Chem.* **271,** 2082–2087.

Watterson, K. R., Johnston, E., Chalmers, C., Pronin, A., Cook, S. J., Benovic, J. L., and Palmer, T. M. (2002) Dual regulation of EDG1/S1P(1) receptor phosphorylation and internalization by protein kinase C and G-protein-coupled receptor kinase 2. *J. Biol. Chem.* **277,** 5767–5777.

Yamamura, S., Yatomi, Y., Ruan, F., Sweeney, E. A., Hakomori, S., and Igarashi, Y. (1997) Sphingosine 1-phosphate regulates melanoma cell motility through a receptor-coupled extracellular action and in a pertussis toxin-insensitive manner. *Biochemistry* **36,** 10751–10759.

Yatomi, Y., Igarashi, Y., Yang, L., Hisano, N., Qi, R., Asazuma, N., Satoh, K., Ozaki, Y., and Kume, S. (1997a) Sphingosine 1-phosphate, a bioactive sphingolipid abundantly stored in platelets, is a normal constituent of human plasma and serum. *J. Biochem. (Tokyo)* **121,** 969–973.

Yatomi, Y., Ohmori, T., Rile, G., Kazama, F., Okamoto, H., Sano, T., Satoh, K., Kume, S., Tigyi, G., Igarashi, Y., and Ozaki, Y. (2000) Sphingosine 1-phosphate as a major bioactive lysophospholipid that is released from platelets and interacts with endothelial cells. *Blood* **96,** 3431–3438.

Yatomi, Y., Yamamura, S., Ruan, F., and Igarashi, Y. (1997b) Sphingosine 1-phosphate induces platelet activation through an extracellular action and shares a platelet surface receptor with lysophosphatidic acid. *J. Biol. Chem.* **272,** 5291–5297.

Yeh, C. D., Duncan, J. A., Yamashita, S., and Michel, T. (1999) Depalmitoylation of endothelial nitric-oxide synthase by acyl-protein thioesterase (1) is potentiated by Ca2+-calmodulin. *J. Biol. Chem.* **274,** 33148–33154.

Zondag, G. C., Postma, F. R., Etten, I. V., Verlaan, I., and Moolenaar, W. H. (1998) Sphingosine 1-phosphate signalling through the G-protein-coupled receptor Edg-1. *Biochem. J.* **330,** 605–609.

Zundel, W., Swiersz, L. M., and Giaccia, A. (2000) Caveolin 1-mediated regulation of receptor tyrosine kinase-associated phosphatidylinositol 3-kinase activity by ceramide. *Mol. Cell. Biol.* **20,** 1507–1514.

Chapter 8

Insulin Signaling and Caveolae

Peter Strålfors

Department of Cell Biology and Diabetes Research Centre, Linköping University, SE58185 Linköping, Sweden

Advances in Molecular and Cell Biology, Vol. 36

ISBN: 0-444-51500-3
DOI: 10.1016/S0079-6603(05)36008-9

I. INSULIN ACTION

A. Insulin Signal Transduction

Insulin is the most important anabolic hormone and as such has a prime role in directing nutrients to synthetic processes and storage after a meal. Although most cell types express receptors for insulin, these are found in the highest concentrations in the metabolic target cells of the hormone: skeletal and heart muscle, adipose tissue, and liver. Insulin also has crucial actions on pancreatic islet β-cells that produce the hormone and on cells in the hypothalamus that regulate feeding.

Circulating insulin can bind to and be detected by well-characterized transmembrane receptors exposed on the surface of target cells. The insulin receptor consists of two membrane-spanning β-subunits and two extracellular hormone-binding α-subunits. Insulin binding produces a conformational change, which activates the protein kinase that is inherent in the intracellular domains of the β-subunits. The receptor protein kinase is specific for phosphorylation of tyrosine residues. In a transphosphorylation reaction the β-subunits phosphorylate each other. This autophosphorylation allows the receptor to phosphorylate a number of downstream docking proteins, such as the insulin receptor substrate (IRS) family of proteins, Cbl, GAB-1, APS, and Shc. Once tyrosine is phosphorylated by the insulin receptor, such docking proteins can bind and activate a number of different signaling proteins with SH2 domains that recognize phosphotyrosine. Insulin control of intermediary metabolism is heavily dependent on members of the IRS family, particularly IRS1 and IRS2. The two proteins are highly homologous and apparently have both overlapping and specific functions downstream of the insulin receptor. Both proteins can be phosphorylated on multiple tyrosine residues by insulin, thus providing binding sites for downstream signal-mediating proteins. Phosphatidylinositol 3-kinase (PI3-kinase) is such an SH2-domain-containing enzyme that phosphorylates phosphoinositides at the 3 position of the inositol ring to produce second messengers, particularly phosphatidylinositol-(3,4,5)P_3. Since these are amphiphilic lipids they remain in the membrane where they provide specific binding sites for proteins with pleckstrin homology (PH) domains. Thus, the phosphoinositide-dependent protein kinase (PDK1) is activated to phosphorylate and activate the protein kinase B (PKB or Akt) and protein kinases $C\zeta/\lambda$. Other players may contribute under special circumstances, but the described signal chain appears to be central to insulin's metabolic control in all its target tissues and, through less-well-understood downstream signaling events, eventually leads to enhanced glycogen, triacylglycerol, and protein synthesis and inhibition of the corresponding catabolic pathways, as well as stimulation of glucose uptake via the insulin-regulated glucose transporter GLUT4.

In addition to its acute metabolic effects, insulin also controls DNA transcription and specific protein synthesis. The hormone's mitogenic and transcriptional control is at least partly mediated by the ubiquitous MAP-kinase signaling pathway involving activation of the small G-protein Ras and a protein kinase phosphorylation cascade controlling the activity of extracellular signal-regulated kinases (ERK)1 and 2.

Phosphorylated and activated ERK can then move into the nucleus to phosphorylate transcription factors. That way insulin, which is a bona fide growth factor, can initiate transcriptional programs controlling cell proliferation and differentiation.

B. Function of Insulin in Adipose Tissue

The main function of insulin in adipose tissue is to promote triacylglycerol synthesis and inhibit lipolysis. In human beings, fatty acids are obtained from hydrolysis of circulating triacylglycerol in chylomicrons or very low-density lipoprotein (VLDL) particles in the blood capillaries. By an unknown route, likely involving caveolae in the adipocytes, liberated fatty acids can enter the adipocytes. The fatty acids are rapidly re-esterified with glycerol 3-phosphate to triacylglycerol that is stored in the adipocyte oil droplet. The glycerol 3-phosphate has to be generated from glycolytic conversion of glucose or gluconeogenic synthesis from pyruvate, since adipocytes lack glycerol kinase to phosphorylate glycerol. Stimulation of glucose uptake is therefore a key function of insulin in adipocytes. Transgenic mice that overexpress the gluconeogenic enzyme phosphoenolpyruvate carboxykinase (PEPCK) in adipose tissue exhibit an increased triacylglycerol synthesis and adipose tissue mass (Franckhauser *et al.*, 2002). This indicates that the glycerol 3-phosphate supply can be rate limiting for triacylglycerol synthesis. However, glucose is not utilized for fatty acid synthesis in the adipose tissue of human beings (Goldrick, 1967), in contrast to, e.g., rats and mice. The human adipose tissue therefore quantitatively contributes to overall body glucose disposal in a minor but qualitatively critical way. Triacylglycerol hydrolysis is stimulated by β-adrenergic generation of cyclic AMP, mainly through sympathetic nerve release of noradrenalin, which mediates activation of protein kinase A and phosphorylation of the hormone-sensitive lipase (Strålfors and Belfrage, 1983; Strålfors *et al.*, 1984) and the lipid droplet-associated protein perilipin (Tansey *et al.*, 2003). Insulin inhibits lipolysis largely by counteracting these processes, reducing cyclic AMP concentration and dephosphorylating the hormone-sensitive lipase (Strålfors *et al.*, 1984; Strålfors and Honnor, 1989) and perilipin (Tansey *et al.*, 2003).

Failure of tissues and cells, particularly the adipose tissue, skeletal muscle, liver, and pancreatic islet β-cells, to properly respond to insulin is a cause of diabetes (reviewed in Saltiel and Kahn, 2001; White, 2002). The insulin resistance of the energy metabolizing tissues results in postprandial hyperglycemia and hypertriglyceridemia. It is initially compensated for by an increased insulin release from the β-cells, which gives rise to hyperinsulinemia. When insulin production, because of β-cell failure, is no longer sufficient to compensate for the insulin resistance, the patient presents with clinically overt diabetes type 2. In spite of only extracting a minor part of the blood glucose after a meal, the adipose tissue, in interplay with muscle tissues, has a central role in development of insulin resistance and diabetes. There is evidence that elevated blood levels of fatty acids, which result from insulin resistance in adipose tissue, increase muscle and liver resistance to insulin. Moreover, the adipose tissue, as described below, produces several hormones that control energy metabolism, and dysregulation of these has been described in insulin resistance. Tissue-specific knock-out of the insulin receptor or insulin-regulated glucose transporter GLUT4 has highlighted the extent and complexity of interplay between the adipose and other tissues in maintaining energy homeostasis (reviewed in Charron *et al.*, 1999; Minokoshi *et al.*, 2003).

Mice with a targeted deletion of the insulin receptor gene in skeletal and heart muscle respond normally to glucose ingestion, due to adipose tissue hyperplasia and thereby an ability to compensate for the muscle failure (Cariou *et al.*, 2004). Mice without the insulin receptor in adipose tissue have little adipose tissue mass and are protected against age-related obesity and obesity-related insulin resistance (Bluher *et al.*, 2002). Deletion of the insulin-regulated glucose transporter ($GLUT4^{-/-}$ mice) affects glucose metabolism and shortens the life span, but does not cause diabetes, while the heterozygous $GLUT4^{+/-}$ mice develop diabetes at a high rate. Targeted deletion of GLUT4 in skeletal muscle or in adipose tissue causes insulin resistance in the animals, and some proceed to develop diabetes. However, adipose tissue glucose uptake is also impaired in the muscle-GLUT4-KO mice, and in adipose tissue-GLUT4-KO mice skeletal muscle glucose uptake and suppression of hepatic glucose production by insulin are also impaired. Conversely, overexpression of GLUT4 in adipose tissue apparently results in enhanced whole-body glucose tolerance (Shepherd *et al.*, 1993).

In human beings there is a very significant concordance between insulin resistance/diabetes and obesity. More than 70% of patients with type 2 diabetes are obese. Patients who lack adipose tissue are, however, also characterized by insulin resistance and go on to develop diabetes eventually (Oral, 2003). A common denominator of this apparent paradox is poor insulin response in the adipose tissue, because of insulin resistance in the adipocytes, shortage of GLUT4, or shortage of the tissue. This in turn could force a deposition of triacylglycerol in other tissues such as muscle and liver. The examples serve to vividly illustrate the central role played by adipose tissue in the regulation of whole-body energy homeostasis.

II. CAVEOLAE AND THE ADIPOCYTE

A. Adipocytes Are Highly Specialized Cells

Adipocytes are very large cells with a diameter that can exceed 100 μm, reflecting one of their primary functions, storing energy in the form of triacylglycerol oil. A single oil droplet occupies more than 95% of the huge cell volume (Fig. 1). Squeezed between the oil and the plasma membrane is a film of cytosol <0.5 μm thick with the whole setup of intracellular membranes and organelles. The nucleus forms a distinct protrusion on the cell surface, reflecting the second primary aspect of adipocyte function—producing and releasing hormones, or adipokines, as they are sometimes referred to (Ahima and Flier, 2000; Kim and Moustaid-Moussa, 2000; Fruhbeck

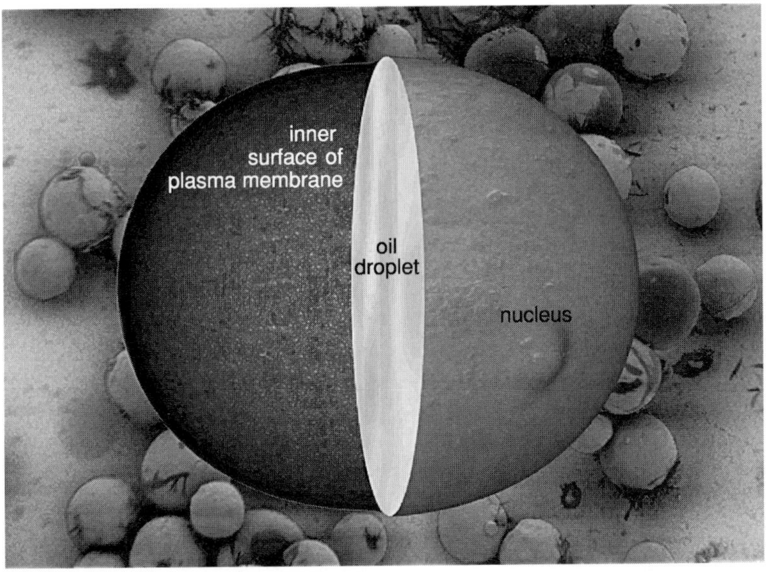

Figure 1. Scanning electron microscopy reconstruction of an adipocyte. Images from scanning electron microscopy of the surface of intact adipocytes and the inner surface of plasma membranes were used to construct this image, to scale, of an adipocyte. The inner surface of the plasma membrane is seen to be speckled with caveolae of varying size. (See Color Insert.)

et al., 2001; Rajala and Scherer, 2003). These hormones are involved in a very tight control of energy homeostasis and feeding behavior as exemplified by leptin, adiponectin, and resistin. It is worth pondering the fact that the adipose tissue is the body's largest endocrine organ and that it can be regarded as two physically distinct organs in one. Metabolic control, energy substrate transfer, conversion, and storage are confined to the lipid droplet and the surrounding cytoplasm and plasma membrane. The endocrine organ, on the other hand, should be concentrated to the nucleus, the surrounding cytosol with rough endoplasmic reticulum and Golgi.

B. Structure of Caveolae in Primary Rat Adipocytes

Some of the crucial events in insulin action take place in the specialized domains of the plasma membrane called caveolae. Caveolae are vesicular invaginations found in the plasma membrane of most cell types, but are particularly abundant in the plasma membrane of adipocytes (Fig. 1). During differentiation of 3T3-L1 fibroblasts to adipocytes they dramatically increase their expression of caveolin (Scherer *et al.*, 1994), which appears to be the primary structural protein of caveolae. Caveolin-1 and -2 of the caveolin protein family are both expressed in adipocytes, whereas caveolin-3 is restricted to muscle, as reviewed in Smart *et al.* (1999). In rat adipocytes under

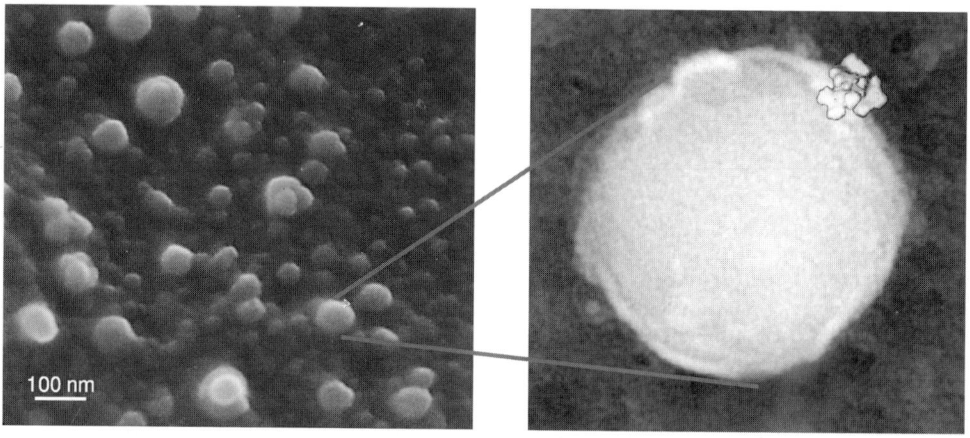

Figure 2. Scanning electron microscopy of adipocyte plasma membrane inner surface with an insulin receptor model inserted to scale. From Luo *et al.* (1999); Thorn *et al.* (2003). (See Color Insert.)

normal conditions about one-third of the plasma membrane constitutes caveolae (Thorn *et al.*, 2003) (Figs.1 and 2). Caveolae in primary rat adipocytes have been the subjects of detailed structural examination by electron microscopy (Thorn *et al.*, 2003) and examination of their lipid composition (Örtegren *et al.*, 2004). In primary adipocytes caveolae range from about 25 to 150 nm in diameter. There are two classes of caveolae. The first is a group of "ordinary" caveolae that are open to the extracellular space, ranging from 50 to 150 nm in diameter and representing about half of the total caveolae. The other half is a group of small caveolae with a diameter of less than 50 nm that are closed without access to the cell surface (Thorn *et al.*, 2003). This represents a static view and we do not yet know how these two groups of caveolae interrelate, if at all. They both, however, represent plasma membrane vesicles or invaginations that contain the protein caveolin. Of particular interest is the fact that caveolin-1 and -2 to a large extent are confined to the necks or the plasma membrane proximal regions of the caveolae, with very few in the bulbs of caveolae invaginations (Thorn *et al.*, 2003). This has obvious implications for caveolin interaction with and regulation of other caveolar proteins, such as the insulin receptor, as discussed below. By scanning electron microscopy, caveolar openings of about 20 nm were demonstrated on the surface of adipocytes (Thorn *et al.*, 2003).

Caveolae in rat adipocytes have a threefold higher concentration of cholesterol and twofold higher concentration of sphingomyelin compared to the surrounding plasma membrane (Örtegren *et al.*, 2004). Interestingly, the concentration of glycerophospholipids is more or less the same in the caveolae and in the surrounding plasma membrane. This suggests that glycerophospholipids constitute a constant membrane core, with variation in other lipids, particularly, high concentrations of cholesterol and sphingomyelin giving rise to caveolae or rafts. Of the glycosphingolipids examined, the ganglioside GM3 is found at the highest concentration in the plasma membrane and is also to some extent enriched in caveolae. GD3, which is a metabolite of GM3, is highly concentrated in caveolae. GM1 is also concentrated in caveolae.

C. Detergent-Proof Gateways for Fatty Acid Entry and Exit

During uptake and, especially, lipolytic release of fatty acids, the flow over the plasma membrane can be massive (Thorn *et al.*, 2003). Fatty acids are amphiphilic molecules and thus in effect behave as a powerful detergent. Fatty acids have to be efficiently sequestered from the plasma membrane by conversion to triacylglycerol or binding to serum albumin in the blood and transported away from the cells. If the concentration is allowed to build up in the membrane, immediate cell lysis ensues (Strålfors, 1990). In adipocytes fatty acids are taken up by a specific subclass of caveolae and are converted to triacylglycerol in the caveolae (Öst *et al.*, 2005). One function of the large number of caveolae in adipocytes may be to increase the efficiency of uptake by enlarging the plasma membrane surface area. It can be argued that fatty acids are channeled through the detergent-resistant caveolae precisely because they are relatively detergent resistant and can tolerate an otherwise lethal buildup of fatty acid detergent. Moreover, caveolin has binding sites for fatty acids (Trigatti *et al.*, 1999), and another major constituent of the plasma membrane is the caveolae localized fatty acid binding protein CD36 (Lisanti *et al.*, 1994b; Dorahy *et al.*, 1996; Ibrahimi and Abumrad, 2002; Aboulaich *et al.*, 2004). *In situ*, fatty acids inevitably have to pass the endothelial cell membranes on their way in or out of the adipocytes. It is possible that caveolae and caveolin provide a detergent-proof route for fatty acids in these cells.

III. INSULIN SIGNALING IN CAVEOLAE

A. Localization of the Insulin Receptor in Caveolae

Functional hormone signaling in cellular control hinges on the right protein at the right place at the right time. One way of achieving this is through colocalization of proteins that need to interact in a certain sequence, as in supramolecular structures such as the pyruvate dehydrogenase complex, the glycogen particle, or the mitochondrial inner membrane and the electron transport chain. Insulin signal transduction, especially to enhance glucose uptake, utilizes caveolae as scaffolds. The insulin receptor is in adipocytes localized to the caveolae (Fig. 3). This localization is not unique for the insulin receptor; to the contrary, localization to caveolae is the case for many hormone and growth factor receptors (reviewed in Anderson, 1993; Lisanti *et al.*, 1994a; Shaul and Anderson, 1998; Simons and Toomre, 2000; Razani *et al.*, 2002c); including the tyrosine kinase receptors for insulin-like growth factor-1 (IGF1) (Huo *et al.*, 2003), epidermal growth factor (EGF) (Mineo *et al.*, 1996), and platelet derived growth factor (PDGF) (Liu *et al.*, 1996). The insulin receptor in fat cells appears, however, to be largely confined to caveolae, with few receptors identified outside of

Figure 3. Immunogold electron microscopic localization of the insulin receptor in caveolae. Transmission electron microscopic image of the inner surface of human adipocyte plasma membrane immunogold labeled against the insulin receptor (large gold particles, arrowheads) and caveolin (small gold particles). Images contrast inverted. From Karlsson *et al.* (2004).

caveolae. There is no evidence for a regulated translocation of the insulin receptor in or out of caveolae in the plasma membrane, as has been reported for some hormone receptors in other cell types (Feron *et al.*, 1997; Sabourin *et al.*, 2002). However, in a hepatoma cell line that does not express caveolin or caveolae, the insulin receptor was reported to translocate into detergent-resistant rafts in response to insulin (Vainio *et al.*, 2002).

Evidence for the localization of the insulin receptor to caveolae in adipocytes is based on immunogold electron microscopy, immunofluorescence microscopy colocalization with caveolin, biochemical isolation of caveolae, co-immunoprecipitation with caveolin, interaction with caveolin scaffolding domain-based peptides, and functional assays (Strålfors, 1997; Wu *et al.*, 1997; Yamamoto *et al.*, 1998; Gustavsson *et al.*, 1999; Nystrom *et al.*, 1999; Parpal *et al.*, 2001; Karlsson *et al.*, 2004). This has been demonstrated for primary human and rat adipocytes and for 3T3-L1 adipocytes. Few proteins have met such stringent experimental criteria for localization to caveolae as has the insulin receptor. That the insulin receptor is a caveolar protein was nevertheless a new concept not easily accepted, for several reasons. The insulin receptor had earlier been found in clathrin-coated pits in different cell types by electron microscopy and autoradiography (Carpentier *et al.*, 1981; Fan *et al.*, 1982; Carpentier, 1989; Smith *et al.*, 1991b). It is quite possible that the insulin receptor is localized to both caveolae and clathrin-coated pits. Adipocytes have relatively few clathrin-coated pits, however (membrane area occupied by clathrin-coated pits is <2% of caveolae membrane area [Smith and Jarett, 1983]), whereas they are much more prevalent in, e.g., hepatocytes. Targeting of the insulin receptor: to caveolae or clathrin-coated pits may reflect two different functions of the receptor, to bind insulin and transmit signals or to bind insulin for lysosomal destruction, respectively (Gustavsson *et al.*, 1999). The first function is obviously of prime importance in adipocytes. The peripheral tissues including muscle and adipose tissues are responsible for no more than 20% of total insulin clearance from the blood; the rest is accounted for by kidney and liver (Ferrannini and Cobelli, 1987). It is conceivable that a caveolar localization of the receptor is mainly associated with signaling, whereas receptors in clathrin-coated pits mainly function to sequester insulin from the circulation and target it for proteolytic destruction in lysosomes.

Early studies found binding of insulin to pinocytotic vesicles in the plasma membrane of glutaraldehyde fixed rat adipocytes by electron microscopy (Smith and Jarett, 1983). These vesicles almost certainly represent what we now call caveolae, although the significance of the finding was not recognized at the time. It had also been reported that the insulin receptor was not present in biochemically isolated caveolae (Corley Mastick *et al.*, 1995). However, that conclusion was singularly based on detergent solubilization of adipocytes with the non-ionic detergent Triton X100 and collection of the detergent-resistant low-density residue. It has subsequently been demonstrated that the insulin receptor is completely solubilized by this treatment and that caveolae isolated without detergent extraction have a high concentration of the receptor (Gustavsson *et al.*, 1999). Caveolae fractions from rat adipocytes, isolated either with or without utilization of detergent, have been reported to contain only a minor fraction of the insulin receptor (Muller *et al.*, 2001). There is no obvious explanation for these conflicting results. Lately, a small fraction of caveolae was isolated from rat adipocytes by immunoprecipitation from the microsomal fraction of cell homogenates using antibodies against caveolin. Very few signaling proteins, including the insulin receptor, were detected in that preparation (Souto *et al.*, 2003). Suoto *et al.*, however, also subjected the microsomal fraction to density gradient centrifugation, and the small amount of insulin receptor in their microsomal fraction did cofractionate with caveolae/caveolin. When isolated from insulin-treated adipocytes, what most likely was endocytosed insulin receptor, which did not cofractionate with caveolae/caveolin, appeared in the microsomal fraction (Souto *et al.*, 2003). A recent thorough analysis of caveolae-associated proteins in human adipocytes has revealed a very high preponderance of signaling proteins (Aboulaich *et al.*, 2004).

B. Insulin Receptor May Be Contained in a Subpopulation of Caveolae

A quantitative analysis of caveolae reveals an interesting aspect of insulin receptor localization and signaling in caveolae. There are fewer insulin receptors in an adipocyte than caveolae (Thorn *et al.*, 2003). We have found a similar relationship for specific gangliosides such as GD3 or GM1, which are concentrated in caveolae, but on average there is no more than one molecule of GD3 or GM1 per caveola (Örtegren *et al.*, 2004). We often find that caveolae that are immunolabeled for insulin receptor are multiply labeled (Smith and Jarett, 1983; Gustavsson *et al.*, 1999; Karlsson *et al.*, 2004), indicating that there are several receptors per caveola in a subset of the adipocyte caveolae. Likewise, it is likely that GD3 and other glycosphingolipids present in equimolar or lower levels in caveolae are concentrated in subsets of caveolae. In fact, it is quite possible to imagine different subpopulations of caveolae, in a single cell, that host different sets of specific signaling protein and glycosphingolipid constituents and therefore serve different functions or constitute different signaling platforms. We have recently found that exogenous fatty acids are converted to triacylglycerol in a specific subclass of caveolae in the plasma membrane of adipocytes (Öst *et al.*, 2005) and that a subclass of small caveolae in the plasma membrane lack access to the cell surface (Thorn *et al.*, 2003).

C. Insulin Signaling for Metabolic Control in Caveolae

It is well known that caveolae are dependent on cholesterol for their structural integrity. Adipocyte caveolae have a threefold higher concentration of cholesterol and twofold higher concentration of sphingomyelin than the surrounding plasma membrane (Örtegren *et al.*, 2004). We have shown by electron microscopy that lowering the plasma membrane concentration of cholesterol, with β-cyclodextrin in rat or 3T3-L1 adipocytes, to half its normal level causes caveolae invaginations to flatten or disappear (Gustavsson *et al.*, 1999; Parpal *et al.*, 2001; Thorn *et al.*, 2003), There is good reason for using the word flatten, because at least acutely after cholesterol extraction with β-cyclodextrin, caveolin remains in the plasma membrane. The caveolin is not randomly dispersed but rather remains clustered as if caveolae were not removed, just flattened. Interestingly, flattening of caveolae coincides with a significant loss of insulin control in the fat cells. The insulin receptor, however, remains in the plasma membrane, binds insulin with unperturbed affinity, and autophosphorylates similarly to the way it does in normal cells (Parpal *et al.*, 2001). Insulin signaling to enhance glucose transport, however, including phosphorylation and activation of protein kinase B and IRS1, is severed. Re-addition of cholesterol to the cells restores insulin signaling to enhanced glucose transport (Gustavsson *et al.*, 1999).

Obesity is correlated to insulin resistance and hypertrophy of adipocytes. It has been shown that large rat adipocytes have about twofold lower plasma membrane concentration of cholesterol than the small adipocytes from the same fat pad (Le Lay *et al.*, 2001). How this translates into differences in amount of caveolin, caveolae structure, or number of caveolae was not examined. But a similar reduction of the plasma membrane cholesterol level, by incubation of the cells with methyl-β-cyclodextrin, severely compromised the response to insulin stimulation of glucose oxidation (Le Lay *et al.*, 2001).

The inability of the insulin receptor to signal for enhanced glucose transport in cholesterol-depleted rat adipocytes was narrowed down to an inability of the active insulin receptor to bind and phosphorylate the immediate downstream effector IRS1 (Parpal *et al.*, 2001). The results suggest that caveolae may be needed for the active insulin receptor to gain access to IRS1, which in rat epididymal adipocytes is an intracellular protein (Anai *et al.*, 1998). The insulin receptor is known to be internalized in conjunction with insulin binding and signaling (Klein *et al.*, 1987). An unresolved issue is whether internalization is part of the signaling (Khan *et al.*, 1986; Klein *et al.*, 1987), perhaps a prerequisite, or whether it is just part of the machinery to downregulate the insulin signal in response to insulin stimulation. If internalization was part of insulin receptor signaling, caveolae internalization with the receptor would provide such a mechanism. Indeed, cholesterol depletion from rat adipocytes blocks the internalization of the insulin receptor (M. Karlsson and P. Strålfors, unpublished observation). The kinin B(2) receptor is a hormone receptor that has been reported to be internalized via caveolae (Marceau *et al.*, 2002).

It is intriguing that, in contrast to rat adipocytes, human subcutaneous adipocytes have a fraction of IRS1 bound to the plasma membrane and caveolae (Karlsson *et al.*, 2004). This is analogous to the situation with the β1-adrenergic receptor in rat cardiomyocytes, which is colocalized with its downstream effector adenylate cyclase in caveolae (Rybin *et al.*, 2000; Ostrom *et al.*, 2001). This has been suggested to explain the more efficient cyclic AMP response to stimulation of the β1-compared to the β2-adrenergic receptor, which moves out of caveolae upon agonist binding, or of β1- and β2-adrenergic receptors compared with prostaglandin E2 receptor, which is not found in caveolae (Ostrom *et al.*, 2001). The sequel in human adipocytes is that caveolae destruction by cholesterol depletion neither impairs insulin receptor phosphorylation of IRS1 nor dissociates IRS1 from the plasma membrane. Instead, downstream activation of PI3-kinase or protein kinase B is impaired, which leads to poor stimulation of glucose transport by insulin (Karlsson *et al.*, 2004). It is possible that in human adipocytes both the insulin receptor and IRS1 have to be internalized with caveolae in order to transmit the downstream signal. This, however, has to be investigated.

D. Insulin Signaling for Mitogenic Control in Caveolae

In contrast to metabolic signaling, insulin control of mitogenic signaling for control of MAP kinases ERK1 and 2 is not affected by caveolae flattening/destruction in rat adipocytes (Parpal *et al.*, 2001). This indicates that IRS1 is not mediating this effect of insulin in these cells. Another candidate for this function was Shc, which in different cell lines has been demonstrated to be activated by insulin. However, short-term effects of insulin do not include phosphorylation of Shc in rat adipocytes (Liu *et al.*, 2000; Parpal *et al.*, 2001). The role of IRS proteins in insulin receptor signaling for activation of MAPkinases in rat adipocytes remains to be demonstrated.

In human subcutaneous adipocytes, on the other hand, caveolae flattening/destruction impairs both metabolic and mitogenic signaling to ERK1/2 phosphorylation by insulin (Karlsson *et al.*, 2004). This of course is compatible with IRS1 mediating only metabolic actions in the terminally differentiated adipocytes of rat and human origin, but it is entirely possible that both actions are mediated by IRS1 in human adipocytes. The findings that IRS1 is caveolar in human but not in rat adipocytes and that caveolae integrity is necessary for insulin's mitogenic control in human but not in rat adipocytes are compelling reasons to be cautious in extrapolating results from animals to human beings, and even more so from immortalized cell lines to human cells.

IV. MECHANISMS FOR TARGETING THE INSULIN RECEPTOR TO CAVEOLAE

A. Subcellular Targeting by Specific Protein or Lipid Interactions

A widely occurring mechanism for targeting proteins to a specific subcellular localization utilizes specific interaction with targeting subunits or anchor proteins, as first described for the targeting of the catalytic subunit of protein phosphatase 1 (PP1c) to the glycogen particles via a glycogen-bound subunit G_M of the phosphatase (Strålfors *et al.*, 1985). The scope has since broadened to encompass all areas of cellular regulation and signaling, and this mechanism is considered a central and ubiquitous means for the regulated targeting to a subcellular locale by specific binding to targeting proteins or scaffolding proteins, or to lipids (as reviewed in, e.g., Bayer and Schulman, 2001; Feliciello *et al.*, 2001; Hurley and Meyer, 2001; Cohen, 2002). The mechanisms for the targeting of the insulin receptor to caveolae have not yet been experimentally defined, however.

B. Fatty Acylation of the Insulin Receptor

A strong determinant for protein targeting to caveolae is the presence of saturated fatty acids in the protein (Melkonian *et al.*, 1999). Myristoylation or palmitoylation, but preferentially tandem acylation by myristate and palmitate or dual palmitoylation, have proved to be the best defined targeting signals for localization of proteins to a membrane domain in liquid-ordered phase (l_o), which is characteristic of caveolae. Interestingly, the insulin receptor in cultured human IM-9 lymphocytes has been reported to be acylated by both myristate and palmitate (Hedo *et al.*, 1987). Both the α- and β-subunits of the receptor contained both fatty acids. Since the receptor consists of two α- and two β-subunits, a minimum of eight saturated fatty acids in the receptor protein may provide a powerful signal for targeting to caveolae and an explanation for the low level of insulin receptors found outside of caveolae. It should be noted that this could provide the insulin receptor with fatty acyl anchoring in both leaflets of the caveolae membrane. It will be important to examine insulin receptor from adipocytes for the extent of fatty acylation and to identify the position of the modified amino acid residues in the protein. Likewise, it should be very interesting and informative to examine this question in the liver, which has few caveolae (Calvo *et al.*, 2001), a high density of clathrin-coated pits, and reported variances in insulin receptor localization (Carpentier *et al.*, 1985; Vainio *et al.*, 2002). Moreover, mutation of acylation sites should provide information on the importance of acylation for the caveolar localization of the receptor and perhaps of caveolar localization in insulin action.

C. Interaction between Insulin Receptor and Caveolin

The insulin receptor β-subunit carries an amino acid sequence—1220WSF GVVIW1227 (numbered according to the human insulin proreceptor sequence)—that can interact with the so-called caveolin-scaffolding domain of caveolin-1 and -3, but not caveolin-2, through the aromatic amino acid residues at positions 1220, 1222, and 1227 (Couet *et al.*, 1997a; Yamamoto *et al.*, 1998). This interaction could provide a mechanism for retaining the insulin receptor in caveolae. Precisely how this could work to retain the receptor in caveolae, however, is not clear. Caveolin-1 and -2 appear to be largely localized in the necks or in plasma membrane proximal regions of caveolae (Thorn *et al.*, 2003; Karlsson *et al.*, 2004). If only for steric considerations it can be difficult to accommodate the insulin receptor or receptors in the neck of a caveola (Fig. 2). Insulin stimulates the phosphorylation of caveolin-1 on tyrosine 14 by the insulin receptor in the 3T3-L1 adipocyte cell line (Mastick and Saltiel, 1997; Lee *et al.*, 2000; Kimura *et al.*, 2002). It was found that this tyrosine phosphorylation could specifically confer binding of caveolin-1 to the SH2-domain-containing Grb7 in Cos-7 cells (Lee *et al.*, 2000). How much of the caveolin in the cell that is phosphorylated in response to insulin has not been reported (Kimura *et al.*, 2002), but the phosphorylation of caveolin indicates that the mature insulin receptor can interact to some extent with the caveolin in cells. Expression of caveolin-Y14F, with the insulin phosphorylated tyrosine residue swapped for the non-phosphorylatable phenylalanine, in 3T3-L1 adipocytes reportedly had no effect on insulin signaling (Kimura *et al.*, 2002). Because no analysis was provided of the percentage of cells that were transfected with the caveolin mutant, the significance of this finding cannot be assessed at present. It is interesting that tyrosine phosphorylation of unidentified caveolar proteins was found to trigger caveolae-mediated internalization of the SV40 virus in cells (Pelkmans *et al.*, 2002).

It may be that the caveola and the caveolar milieu as such and not just a specific molecule, such as caveolin, function as the scaffold for targeting the insulin receptor to caveolae. It has been reported that critical sequences for localization of the EGF receptor to caveolae are in the extracellular part of the receptor (Yamabhai and Anderson, 2002), where interaction with the entirely intracellular caveolin is not possible. The EGF receptors possess a caveolin scaffolding-domain-binding motif, however (Couet *et al.*, 1997b). It was recently found that a common structural sphingolipid-binding motif in prion protein, Alzheimer β-amyloid peptide, and HIV-1 gp120 glycoprotein is responsible for caveolae/raft localization of these proteins (Mahfoud *et al.*, 2002).

A further twist is added by the ability of caveolin-1 and -3 to potentiate the activation of the insulin receptor by insulin when HEK293T cells, expressing low levels of endogenous caveolin, were transfected to overexpress caveolin-1 or -3 and the insulin-stimulated phosphorylation of IRS1 was assessed (Yamamoto *et al.*, 1998). Receptor autophosphorylation was not affected, however. The binding and direct activation of the intracellular portion of the receptor by caveolin-scaffolding domain-containing peptides were also demonstrated. This is a surprising result, since caveolin and the scaffolding domain have repeatedly been shown to inhibit enzymatic activities, including tyrosine-specific receptor protein kinases (reviewed in Razani *et al.*, 2002c). The physiological role for caveolin activation of the insulin receptor is not clear, especially since we do not know how the receptor is affected when localized in caveolae, as this was not examined in the HEK293T cells. Nystrom *et al.* expressed wild-type caveolin-1 and caveolin-1 with mutations in the scaffolding domain in Cos-7 cells or rat adipocytes, and examined the consequences on insulin signaling (Nystrom *et al.*, 1999). In Cos-7 cells, insulin-stimulated phosphorylation of the transcription factor Elk-1 was enhanced by wild-type caveolin, with no effect in the absence of insulin. In contrast, insulin stimulation of ERK2 phosphorylation was blocked. The scaffolding domain mutant of caveolin had, as expected, no effect on insulin control of Elk-1 or of ERK2. In primary rat adipocytes that express high levels of endogenous caveolin and where the insulin receptor is properly located in caveolae, overexpression of either wild-type or mutated caveolin impaired insulin stimulation of GLUT4 translocation from intracellular sites to the cell surface. Both wild-type and mutant caveolins lowered the basal level of ERK2 phosphorylation, however, without affecting the insulin stimulation. The presence of high levels of endogenous caveolin-1 in the adipocytes may have swamped effects of transfected caveolin, and interpretation of these findings is therefore difficult.

In a different approach to caveolae and insulin signaling, a synthetic caveolin-binding domain-containing peptide was reported to act as an insulin receptor-independent insulin mimetic by releasing $pp59^{Lyn}$ and $pp125^{Fak}$ from their inhibitory caveolin tethers in rat adipocytes (Muller *et al.*, 2001). Translocation out of caveolae activated these non-receptor protein tyrosine kinases to induce tyrosine phosphorylation of IRS1, activation of PI3-kinase, and glucose uptake, without engaging the insulin receptor. An excess of caveolin-scaffolding domain-containing peptide inhibited the redistribution and activation of $pp59^{Lyn}$ and $pp125^{Fak}$. The importance for normal insulin signaling is not clear, but if corroborated this signaling pathway may be useful for pharmacological intervention.

A few patients with syndrome of extreme insulin resistance have been found to have mutations of aromatic residues (W1220L, W1227S) in the caveolin-binding motif of the insulin receptor, i.e., the receptor counterpart to the caveolin-scaffolding domain (Moller *et al.*, 1990, 1991; Iwanishi *et al.*, 1993; Imamura *et al.*, 1994, 1998; Sawa *et al.*, 1996). Transfection of cells with insulin receptor harboring the W1220L mutation, but also with an E1206D mutation outside of the caveolin-binding motif, caused receptor degradation and therefore a reduced amount of cell surface-expressed receptor. The W1227S mutation, however, did not reduce the amount of surface-expressed protein. The insulin receptor with these and the third aromatic amino acid residue in this region mutated to non-aromatic residues (F1222G, W1227T, or the combined W1220G/F1222G/W1227T mutations) have been expressed in Cos-7 cells and compared to expression of wild-type receptor (Nystrom *et al.*, 1999). All mutated receptors were expressed at substantially reduced levels, without affecting the amount of 210-kDa insulin proreceptor. The receptor mutants discussed were autophosphorylation defective, so their accelerated degradation was appropriate. Albeit with inconsistencies such as the fact that W1227S did not, but the conservative swap from serine to threonine in W1227T did cause a reduced amount of receptor, the results suggest that the caveolin-binding motif may be important for the formation of a mature and active receptor, which otherwise is rapidly degraded.

Taken together, these findings suggest a role for caveolin/caveolae in folding the receptor protein during maturation of the receptor or in protecting the mature receptor from degradation, but they do not help us understand whether caveolin serves as an insulin receptor controller, of activity or localization in caveolae of the plasma membrane. It is possible that the caveolin-binding motif in the insulin receptor has a propensity for interaction with caveolin but serves other functions when the receptor is properly processed and targeted to caveolae. The interaction of caveolin with the insulin receptor and the functional consequences thereof may also be specific for different cell types and species.

V. ADIPOSE TISSUE AND INSULIN SIGNALING IN CAVEOLIN-1-DEFICIENT ANIMALS

Given the pervasive involvement of caveolae in insulin actions described here and, indeed, their role in so many other cellular processes, it is surprising that caveolin knockout mice are carried full term and survive (Drab *et al.*, 2001; Galbiati *et al.*, 2001; Razani *et al.*, 2001), albeit with a shortened life span (Park *et al.*, 2003). The physiology of these mice has been described in great detail (as reviewed in Drab *et al.*, 2001; Razani and Lisanti, 2001).

Adipocytes from caveolin-1$^{-/-}$ mice appear to lack caveolae altogether, but the mice are not overtly diabetic. They exhibit peripheral insulin resistance and, after high-fat feeding, also postprandial hyperinsulinemia (Cohen *et al.*, 2003), indicative of a malfunction in insulin response mechanisms. Lipid metabolism is deranged with very markedly elevated concentrations of triacylglycerol and fatty acids in the blood after a meal. Isolated adipocytes respond poorly to insulin, with blunted effects on insulin receptor downstream signal-mediator proteins (protein kinase B and glycogen synthase kinase-3β phosphorylation) (Cohen *et al.*, 2003). Caveolin-2$^{-/-}$ mice, on the other hand, did not display any of the abnormalities seen in energy metabolism and adipose tissue of the caveolin-1$^{-/-}$ mouse (Razani *et al.*, 2002b; Cohen *et al.*, 2003).

Resistance to diet-induced obesity of the caveolin-1$^{-/-}$ mice is conspicuous, despite a tendency to hyperphagia (Razani *et al.*, 2002a). Indeed, the mice present with a general adipose tissue derangement and atrophy, which increases with age. The fat cells are small and fail to accumulate triacylglycerol. The layer of hypodermal adipose tissue is absent and subcutaneous, and visceral adipose tissue (with possible exemption of the omental fat) is atrophied (Razani *et al.*, 2002a). Further consequences of a dysfunctional adipose tissue in the caveolin-1$^{-/-}$ mouse are reduced plasma concentrations of both leptin and adiponectin (Cohen *et al.*, 2003), which are normally produced and secreted by the adipocytes (Zhang *et al.*, 1994; Scherer *et al.*, 1995). The lack of caveolae in the caveolin-1$^{-/-}$ mice has obvious similarities with the situation, in which caveolae are flattened by cholesterol depletion, which makes the cells respond poorly to insulin. Lack of caveolin and caveolae thus compromises insulin's ability to properly control fatty acid levels, as indeed evidenced by very high levels of circulating triacylglycerol-rich lipoproteins and free fatty acids in the caveolin-1$^{-/-}$ mice (Cohen *et al.*, 2003). The lack of caveolin and caveolae, moreover, may deprive the cell membrane of its protection against fatty acid accumulation. As discussed previously, fatty acids are toxic detergents that are fatal to cells if they accumulate without proper control (Strålfors, 1990). Fatty acids are normally taken up and reacylated to triacylglycerol in the caveolae (Öst *et al.*, 2004). Insulin is a prime controller to keep fatty acids at bay. First, insulin stimulates the rapid esterification of fatty acids in the adipocytes during hydrolysis of ingested and liver-derived triacylglycerol in the circulation. Second, insulin blocks the hydrolysis of endogenous triacylglycerol by dephosphorylating the hormone-sensitive lipase. Moreover, caveolin-1 has the potential to inhibit protein kinase A-catalyzed phosphorylation and activation of the hormone-sensitive lipase (Razani *et al.*, 1999). Loss of caveolin-1/caveolae may thus remove the fatty acid proof protection that caveolae provide, disturb fatty acid reacylation to triacylglycerol, and remove insulin inhibition of lipolysis and concomitantly release protein kinase A to activate lipolysis.

In line with the dependence on caveolin-1 for maturation of the insulin receptor (described previously), the concentration of the insulin receptor is reduced by more than 90% in adipocytes from caveolin-1$^{-/-}$ mice compared to wild-type mice (Cohen *et al.*, 2003). Transfection of embryonic fibroblasts from these animals with caveolin-1 partially restored the levels of receptor, while transfection with a scaffolding domain deletion mutant of caveolin-1 (Δ61–100) failed to restore insulin receptor levels. Significantly, the proteasome inhibitor MG132 partially restored the amount of insulin receptor in the cells. However, the levels of the highly homologous IGF1 receptor remained normal in the caveolin-1$^{-/-}$ mouse adipocytes (Cohen *et al.*, 2003), in spite of the IGF1 receptor being equipped with the same caveolin-binding motif (residues 1196WSFGVVLW1203) as the insulin receptor. The IGF1 receptor has furthermore been reported to be localized in caveolae (Huo *et al.*, 2003) and to catalyze the phosphorylation of caveolin-1 on Y14 (Maggi *et al.*, 2002), similarly to the insulin receptor. These findings thus show that in the absence of caveolin-1/caveolae, the mature insulin receptor, or receptor that has failed to mature in the absence of caveolin, is rapidly hydrolyzed by the proteasome complex. Why the same effect of caveolin deficiency does not arise with the IGF1 receptor in the same cells is not clear, but it may suggest that the caveolar localization is important for insulin receptor stability. Hence, in blocking insulin receptor degradation a role of caveolin may be to form stable caveolae. It has indeed been demonstrated that caveolae actually form in the absence of or at very low levels of caveolin, but are rapidly internalized by endocytosis (Le *et al.*, 2002). Caveolin acts to stabilize the caveolae invaginations at the plasma membrane and substantially slow down endocytosis (Le *et al.*, 2002; Thomsen *et al.*, 2002). It is therefore possible that insulin receptor endocytosis is inhibited by caveolin and that lack of caveolin leads to a rapid internalization and degradation of the receptor. At the same time, the lack of caveolin/caveolae deprives the cell of its fatty acid protection, resulting in tissue degradation and malfunction. It is illustrative that mice with targeted deletion of the insulin receptor in adipose tissue, similarly to caveolin-1$^{-/-}$ mice, have a reduced adipose tissue mass and are protected from obesity, but, in contrast to caveolin-1$^{-/-}$ mice, they are also protected from glucose intolerance and insulin resistance (Bluher *et al.*, 2002). Hence, the glucose and lipid dysregulation seen in the caveolin-1$^{-/-}$ mice may result from the primary caveolin-1/caveolae deficiency and the consequent adipose tissue degradation and malfunction, and less so from the secondary insulin receptor deficiency. As noted (Razani *et al.*, 2002a; Cohen *et al.*, 2003), the phenotype and clinical manifestations of adipose tissue atrophy exhibited by the caveolin-1$^{-/-}$ mice are reminiscent of different forms of lipoatrophic syndromes, afflicted persons of which present with insulin resistance (Oral, 2003).

VI. INSULIN-STIMULATED GLUCOSE UPTAKE IN CAVEOLAE

A. Insulin Stimulates Glucose Uptake via the Insulin-Regulated Glucose Transporter GLUT4

The metabolic target organs of insulin—adipose tissue, skeletal, and heart muscles—very specifically express an insulin-regulated isoform of the glucose transporter proteins: GLUT4. Fibroblasts and myoblasts express this protein only when differentiated to adipocytes and myotubes, respectively. Hepatocytes and pancreatic islet β-cells express the low-affinity GLUT2 isoform. Under basal conditions, in the absence of insulin, GLUT4 is mostly found in intracellular membrane compartments (Cushman and Wardzala, 1980; Suzuki and Kono, 1980). Insulin receptor signaling releases these membrane vesicles from their intracellular tethering and translocates them to the cell surface, where they, presumably, fuse with the plasma membrane to productively insert the transporter protein in the membrane. Evidence cited below demonstrates that GLUT4 mainly appears in caveolae of the plasma membrane and that glucose uptake takes place in caveolae, at least in adipocytes. Adipose and muscle tissues also express GLUT1 that is largely constitutively found in the plasma membrane and appears to provide a basal uptake and supply of glucose for the cells. However, some GLUT1 is also found intracellularly, but segregated from GLUT4-containing vesicles, and to a small extent translocates to the cell surface in response to insulin. Because the effects on GLUT4 translocation and the consequent glucose uptake by insulin are much stronger, much attention has been focused on GLUT4 and the mechanisms underlying regulated translocation of this isoform in response to insulin.

B. Insulin-Stimulated Translocation of GLUT4 to Caveolae for Glucose Uptake

In contrast to the insulin receptor, GLUT4 has been obtained in a detergent-resistant fraction of whole 3T3-L1 adipocytes (Scherer *et al.*, 1994) or of the plasma membrane from primary rat adipocytes (Gustavsson *et al.*, 1996) after stimulation with insulin. Immunogold electron microscopy examination of plasma membranes of 3T3-L1 adipocytes has confirmed that GLUT4 is localized to caveolae (Karlsson *et al.*, 2001; Ros-Baro *et al.*, 2001). Interestingly, about 85% of GLUT4 in the plasma membrane of 3T3-L1 adipocytes is in caveolae (Karlsson *et al.*, 2001). Insulin increases the total amount of GLUT4 in the plasma membrane and in caveolae but does not affect the percentage of GLUT4 in caveolae. The localization of GLUT4 in caveolae has been verified in primary rat adipocytes by isolation of caveolae without using detergent extraction (Karlsson *et al.*, 2001).

The time course for insulin-induced translocation of GLUT4 to the plasma membrane and to caveolae of primary rat adipocytes (Gustavsson *et al.*, 1996) revealed that GLUT4 accumulated with the plasma membrane fraction with a $t_{1/2}$ of ca. 3 min (Fig. 4). GLUT4 appeared in caveolae, however, considerably later, with a $t_{1/2}$ of ca. 7 min. Significantly, glucose uptake by the cells closely paralleled the accumulation of GLUT4 in caveolae and not the total plasma membrane fraction. This suggests that GLUT4 rapidly translocates to the plasma membrane and in a slower process productively inserts in the caveolae membrane. This could be explained in two ways. GLUT4-containing vesicles dock and fuse with the plasma membrane outside caveolae, and then in a slower process GLUT4 moves laterally in the membrane into caveolae. Alternatively, GLUT4 vesicles dock with the caveolae membrane and then in a slower process fuse with the caveolae membrane. The fact that GLUT4 rapidly associates with the plasma membrane in response to insulin stimulation, but only becomes available for glucose transport and for labeling from the outside of intact cells several minutes later (Joost *et al.*, 1988; Smith *et al.*, 1991a; Satoh *et al.*, 1993; Gustavsson *et al.*, 1996), argues in favor of the latter explanation.

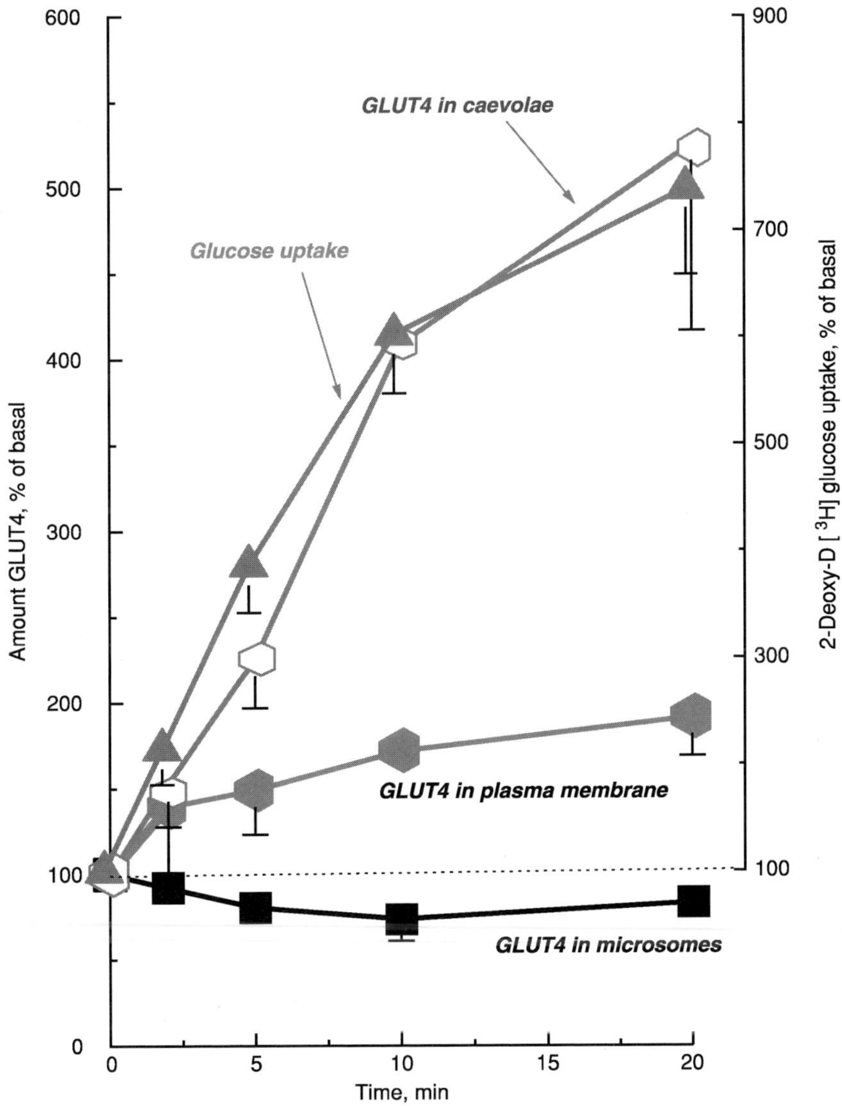

Figure 4. Translocation of GLUT4 to caveolae and glucose transport in caveolae. Rat adipocytes were incubated with insulin for the indicated time and subjected to subcellular fractionation. Caveolae were prepared from plasma membranes by treatment with non-ionic detergent and collecting the insoluble residue. The amount of GLUT4 protein in the different fractions was determined after SDS-PAGE and immunoblotting. Glucose uptake was measured from the same cell batches in parallel. From Gustavsson *et al.* (1996). (See Color Insert.)

C. Endocytosis of GLUT4 from Caveolae and the Plasma Membrane

On reversal of insulin stimulation in primary rat adipocytes, with isoproterenol and adenosine deaminase, GLUT4 makes its sortie from caveolae and the total plasma membrane in parallel, indicating that GLUT4 is internalized via caveolae (Gustavsson *et al.*, 1996). This interpretation is reinforced by experiments demonstrating that in 3T3-L1 adipocytes, the cholesterol-binding and caveolae-disrupting agent nystatin partially inhibits GLUT4 internalization after insulin removal (Ros-Baro *et al.*, 2001). Moreover, sphingomyelinase treatment of 3T3-L1 adipocytes, which reduces the plasma membrane content of sphingomyelin and cholesterol, causes an increased amount of GLUT4 in the plasma membrane on prolonged incubations (Liu *et al.*, 2004). Likewise, cholesterol depletion with methyl-β-cyclodextrin or a dominant-negative mutant (S80E) of caveolin-1 causes an accumulation of GLUT4 at the plasma membrane (Shigematsu *et al.*, 2003). This makes sense in light of the fact that cholesterol depletion of the plasma membrane compromises caveolae function and endocytosis, thus retaining GLUT4, which through default translocation accumulates at the plasma membrane.

D. Insulin Control of GLUT4 Insertion in the Caveolae Membrane

Further insight into GLUT4 trafficking to caveolae has lately been gained. Exo70, a component of the exocyst complex that targets vesicles to specific sites of the plasma membrane, was reported to translocate to the plasma membrane in response to insulin (Inoue *et al.*, 2003). Overexpression of an Exo70 mutant blocked the extracellular exposure of GLUT4 but not the translocation to the plasma membrane, indicating that Exo70 participates in insertion of GLUT4 in the plasma membrane. Insulin may recruit Exo70 to the plasma membrane to form the exocyst complex by way of TC10, a small G protein of the Rho family. Interestingly, TC10 appears to be localized in caveolae in 3T3-L1 adipocytes (Watson *et al.*, 2001), possibly by virtue of its double palmitoylation (Michaelson *et al.*, 2001). Exo70 was shown to interact with TC10 in a GTP-dependent way. TC10 may therefore direct GLUT4 to be inserted in the caveolae membrane by way of Exo70 and the exocyst complex.

Insulin was reported to activate TC10 via phosphorylation of the ubiquitin ligase protein Cbl (Baumann *et al.*, 2000) in the presence of the adapter protein APS (Ahmed *et al.*, 2000), which then, via the SoHo-domain-containing adapter protein CAP, can interact with flotillin in the caveolae membrane and recruit the TC10 activating protein complex C3G-CrkII to caveolae (Chiang *et al.*, 2001). C3G is a guanine nucleotide exchange factor acting on TC10. This signaling pathway for insertion of GLUT4 in caveolae complements the insulin receptor signaling via IRS1, PI3-kinase, and protein kinase B/protein kinase Cζ/λ discussed previously. In rat skeletal muscle, insulin does not induce phosphorylation of Cbl and association with CAP (Thirone *et al.*, 2003), which may mean that this pathway for insulin control of GLUT4 insertion in the plasma membrane is not operative in tissues other than adipose tissue. It will be interesting to see the central role of Cbl in transmitting this effect of insulin tested in Cbl knockout mice (Murphy *et al.*, 1998; Naramura *et al.*, 1998). If the model is correct, then adipocytes from $Cbl^{-/-}$ mice will not respond to insulin with enhanced glucose uptake. As a corollary, it will also be very interesting to see if these mice develop insulin resistance and diabetes.

GLUT4 translocation to caveolae in response to insulin obviously involves release from intracellular readily mobilizable stores, transport to the plasma membrane, and, finally, insertion in caveolae. Little further is known about how insulin accomplishes control of the upstream parts of this process. Caveolae or caveolin does not seem to be involved, though. It has been clearly demonstrated that in the intracellular membranes of 3T3-L1 adipocytes and skeletal muscle, GLUT4 is not present in detergent-insoluble membranes or in caveolin-containing membranes (Kandror *et al.*, 1995; Munoz *et al.*, 1996).

E. GLUT4 Translocation to and from Caveolae Is Disputed

Convincing though the above explanation may seem for GLUT4 trafficking to caveolae for glucose uptake, it has not gained universal acceptance, and there are experimental findings challenging this view. In particular, GLUT4 has been detected by electron microscopy in the following: in clathrin-coated pits, but caveolae were not examined (Slot *et al.*, 1991a,b); in clathrin-coated pits, but not in caveolae (Robinson *et al.*, 1992); and neither in caveolae nor in clathrin-coated pits (Voldstedlund *et al.*, 1993; Malide *et al.*, 2000). These discrepancies are likely explained by methodological differences (e.g., as discussed in Karlsson *et al.*, 2001; Ros-Baro *et al.*, 2001). Ros-Baro *et al.* directly demonstrated that the bleaching protocol used in Robinson *et al.* (1992) hindered detection of GLUT4 in caveolae (Ros-Baro *et al.*, 2001). It is also worth contemplating that in primary fat cells, which exhibit the most profound effect on GLUT4 translocation in response to insulin and insulin withdrawal, there are very few clathrin-coated pits (Sheldon *et al.*, 1962; Williamson, 1964; Smith and Jarett, 1983; Thorn *et al.*, 2003), making it unlikely that GLUT4 and glucose uptake take place there or that GLUT4 is endocytosed via these structures. A reported failure to detect GLUT4 in a detergent-insoluble fraction of the plasma membrane (Shigematsu *et al.*, 2003) could be explained by lack of attendance to the fact that the detergent insolubility of caveolae and rafts is not absolute. Addition of a relatively large volume of 1% Triton X100 to sheets of plasma membrane from a few cells at the bottom of a plate well is likely to solubilize most membranes. Indeed, in addition to GLUT4, much of the caveolin was also solubilized by the detergent (Shigematsu *et al.*, 2003).

VII. INSULIN SIGNALING AND CAVEOLAE IN OTHER TISSUES

The effects on adipose tissue and the insulin receptor in the caveolin-1$^{-/-}$ mouse were not mirrored in skeletal muscle or liver, tissues that normally do not express or, compared to adipose tissue, express only little caveolin-1, respectively (Cohen *et al.*, 2003). A function for caveolae in insulin signaling in liver or in primary hepatocytes has not been reported. Skeletal muscle expresses caveolin-3, though, and caveolae and caveolin have been demonstrated to be involved in T-tubule formation (Parton *et al.*, 1997). T-tubules and caveolae, which are prevalent in the sarcolemma, exhibit high levels of caveolin-3 (Ralston and Plough, 1999). The insulin receptor is obviously found in the sarcolemma of the muscle fiber, but nothing is known about the relation of the insulin receptor to caveolae and caveolin-3 in skeletal muscle. It has been found that GLUT4 translocates to both the sarcolemma and the T-tubules in response to insulin (Plough *et al.*, 1998). Considering the caveolar origin of the T-tubules, translocation of GLUT4 to T-tubules therefore is reminiscent of the translocation to caveolae in adipocytes. However, caveolin-$3^{-/-}$ mice did not show any signs of insulin resistance or diabetes (Galbiati *et al.*, 2001), although that was initially also the case with the caveolin-$1^{-/-}$ mouse (Drab *et al.*, 2001; Razani *et al.*, 2001).

VIII. CONCLUSION

Insulin receptor localization in caveolae and its ability to interact with caveolin-1 have been demonstrated, both *in vitro* and *in vivo*. Likewise, GLUT4 translocation to caveolae for glucose transport into the cells and the insulin signaling machinery for insertion of GLUT4 in the caveolae membrane have been demonstrated. Moreover, fatty acids are entering the fat cell via caveolae and are re-acylated to triacylglycerol in caveolae. Dysfunction in this system causes impaired insulin signaling and thus insulin resistance in isolated cells, animals, and human beings.

ACKNOWLEDGMENTS

I acknowledge the contributions of my colleagues Nabila Aboulaich, Anna Danielsson, Johanna Gustavsson, Margareta Karlsson, Fredrik Nyström, Santiago Parpal, Lilian Said, Karin Stenkula, Hans Thorn, Julia Vainonen, Unn Örtegren, and Anita Öst during different parts of the work performed in our laboratory. Financial support has been obtained from The Swedish Research Council, Swedish Diabetes Association, Swedish Foundation for Strategic Research, Lions Foundation, Swedish Society for Medical Research, Östergötland County Council.

REFERENCES

Aboulaich, N., Vainonen, J., Strålfors, P., and Vener, A. V. (2004). Vectorial proteomics reveal targeting, phosphorylation and specific fragmentation of polymerase I and transcript release factor (PTRF) at the surface of caveolae in human adipocytes. *Biochem. J.* **383,** 237–248.

Ahima, R. S. and Flier, J. S. (2000). Adipose tissue as an endocrine organ. *Trends Endocrinol. Metab.* **11,** 327–332.

Ahmed, Z., Smith, B. J., and Pillay, T. S. (2000). The APS adapter protein couples the insulin receptor to the phosphorylation of c-Cbl and facilitates ligand-stimulated ubiquitination of the insulin receptor. *FEBS Lett.* **475,** 31–34.

Anai, M., Ono, H., Funaki, M., Fukushima, Y., Inukai, K., Ogihara, T., Sakoda, H., Onishi, Y., Yazaki, Y., Kikuchi, M., Oka, Y., and Asano, T. (1998). Different subcellular distribution and regulation of expression of insulin receptor substrate (IRS)-3 from those of IRS-1 and IRS-2. *J. Biol. Chem.* **273,** 29686–29692.

Anderson, R. G. W. (1993). Caveolae: Where incoming and outgoing messengers meet. *Proc. Natl. Acad. Sci. USA* **90,** 10909–10913.

Baumann, C. A., Ribon, V., Kanzaki, M., Thurmond, D. C., Mora, S., Shigematsu, S., Bickel, P. E., Pessin, J. E., and Saltiel, A. R. (2000). CAP defines a second signalling pathway required for insulin-stimulated glucose transport. *Nature* **407,** 202–207.

Bayer, K. U. and Schulman, H. (2001). Regulation of signal transduction by protein targeting: The case for CaMKII. *Biochem. Biophys. Res. Commun.* **289,** 917–923.

Bluher, M., Michael, M. D., Peroni, O. D., Ueki, K., Carter, N., Kahn, B. B., and Kahn, C. R. (2002). Adipose tissue selective insulin receptor knockout protects against obesity and obesity-related glucose intolerance. *Dev. Cell* **3,** 25–38.

Calvo, M., Tebar, F., Lopez-Iglesias, C., and Enrich, C. (2001). Morphologic and functional characterization of caveolae in rat liver hepatocytes. *Hepatology* **33,** 1259–1269.

Cariou, B., Postic, C., Boudou, P., Burcelin, R., Kahn, C. R., Girard, J., Burnol, A.-F., and Mauvais-Jarvis, F. (2004). Cellular and molecular mechanisms of adipose tissue plasticity in muscle insulin receptor knockout mice. *Endocrinology* **145,** 1926–1932.

Carpentier, J. (1989). The cell biology of the insulin receptor. *Diabetologia* **32,** 627–635.

Carpentier, J., Obberghen, E. V., Gorden, P., and Orci, L. (1981). Surface redistribution of 125I-insulin in cultured human lymphocytes. *J. Cell Biol.* **91,** 17–25.

Carpentier, J. L., Fehlmann, M., Obberghen, E. V., Gorden, P., and Orci, L. (1985). Redistribution of 125I-insulin on the surface of rat hepatocytes as a function of dissociation time. *Diabetes* **34,** 1002–1007.

Charron, M. J., Katz, E. B., and Olson, A. L. (1999). *GLUT4* gene regulation and manipulation. *J. Biol. Chem.* **274,** 3253–3256.

Chiang, S.-H., Baumann, C. A., Kanzaki, M., Thurmond, D. C., Watson, R. T., Neudauer, C. L., Macara, I. G., Pessin, J. E., and Saltiel, A. R. (2001). Insulin-stimulated GLUT4 translocation requires the CAP-dependent activation of TC10. *Nature* **410,** 944–948.

Cohen, A. W., Razani, B., Wang, X. B., Combs, T. P., Williams, T. M., Scherer, P. E., and Lisanti, M. P. (2003). Caveolin-1 deficient mice show insulin resistance and defective insulin receptor protein expression in adipose tissue. *Am. J. Physiol.* **285,** C222–C235.

Cohen, P. T. W. (2002). Protein phosphatase 1 targeted in many directions. *J. Cell Sci.* **115,** 241–256.

Corley Mastick, C., Brady, M. J., and Saltiel, A. R. (1995). Insulin stimulates the tyrosine phosphorylation of caveolin. *J. Cell Biol.* **129,** 1523–1531.

Couet, J., Li, S., Okamoto, T., Ikezu, T., and Lisanti, M. P. (1997a). Identification of peptide and protein ligands for the caveolin-scaffolding domain. Implications for the interaction of caveolin with caveolae-associated proteins. *J. Biol. Chem.* **272,** 6525–6533.

Couet, J., Sargiacomo, M., and Lisanti, M. P. (1997b). Interaction of a receptor tyrosine kinase, EGF-R, with caveolins. Caveolin binding negatively regulates tyrosine and serine/threonine kinase activities. *J. Biol. Chem.* **272,** 30429–30438.

Cushman, S. W. and Wardzala, L. J. (1980). Potential mechanism of insulin action on glucose transport in the isolated rat adipose cell. *J. Biol. Chem.* **255,** 4758–4762.

Dorahy, D. J., Lincz, L. F., Meldrum, C. J., and Burns, G. F. (1996). Biochemical isolation of a membrane microdomain from resting platelets highly enriched in the plasma membrane glycoprotein CD36. *Biochem. J.* **319,** 67–72.

Drab, M., Verkade, P., Elger, M., Kasper, M., Lohn, M., Lauterbach, B., Menne, J., Lindschau, C., Mende, F., Luft, F. C., Schedl, A., Haller, H., and Kurzchalia, T. V. (2001). Loss of caveolae, vascular dysfunction, and pulmonary defects in caveolin-1 gene-disrupted mice. *Science* **293,** 2449–2452.

Fan, J. Y., Carpentier, J.-L., Gorden, P., Obberghen, E. v., Blackett, N. M., Grunfeld, C., and Orci, L. (1982). Receptor-mediated endocytosis of insulin: Role of microvilli, coated pits, and coated vesicles. *Proc. Natl. Acad. Sci. USA* **79,** 7788–7791.

Feliciello, A., Gottesman, M. E., and Avvedimento, E. V. (2001). The biological functions of A-kinase anchor proteins. *J. Mol. Biol.* **308,** 99–114.

Feron, O., Smith, T. W., Michel, T., and Kelly, R. A. (1997). Dynamic targeting of the agonist-stimulated m2 muscarinic acetylcholine receptor to caveolae in cardiac myocytes. *J. Biol. Chem.* **272,** 17744–17748.

Ferrannini, E. and Cobelli, C. (1987). The kinetics of insulin in man. I. General aspects. *Diabetes Metab. Rev.* **3,** 335–363.

Franckhauser, S., Munoz, S., Pujol, A., Casellas, A., Riu, E., Otaegui, P., Su, B., and Bosch, F. (2002). Increased fatty acid re-esterification by PEPCK overexpression in adipose tissue leads to obesity without insulin resistance. *Diabetes* **51,** 624–630.

Fruhbeck, G., Gómez-Ambrosi, J., Muruzábal, F. J., and Burrell, M. A. (2001). The adipocyte: A model for integration of endocrine and metabolic signaling in energy metabolism regulation. *Am. J. Physiol.* **280,** E827–E847.

Galbiati, F., Engelman, J. A., Volonte, D., Zang, X. L., Minetti, C., Li, M., H Hou, J., Kneitz, B., Edelmann, W., and Lisanti, M. P. (2001). Caveolin-3 null mice show a loss of caveolae, changes in the microdomain distribution of the dystrophin-glycoprotein complex, and T-tubule abnormalities. *J. Biol. Chem.* **276,** 21425–21433.

Goldrick, R. B. (1967). Effects of insulin on glucose metabolism in isolated human fat cells. *J. Lipid Res.* **8,** 581–588.

Gustavsson, J., Parpal, S., Karlsson, M., Ramsing, C., Thorn, H., Borg, M., Lindroth, M., Peterson, K. H., Magnusson, K.-E., and Strålfors, P. (1999). Localisation of the insulin receptor in caveolae of adipocyte plasma membrane. *FASEB J.* **13,** 1961–1971.

Gustavsson, J., Parpal, S., and Strålfors, P. (1996). Insulin-stimulated glucose uptake involves the transition of glucose transporters to a caveolae-rich fraction within the plasma membrane: Implications for type II diabetes. *Mol. Med.* **2,** 367–372.

Hedo, J. A., Collier, E., and Watkinson, A. (1987). Myristyl and palmityl acylation of the insulin receptor. *J. Biol. Chem.* **262,** 954–957.

Huo, H., Guo, X., Hong, S., Jiang, M., Liu, X., and Liao, K. (2003). Lipid rafts/caveolae are essential for insulin-like growth factor-1 receptor signaling during 3T3-L1 preadipocyte differentiation induction. *J. Biol. Chem.* **278,** 11561–11569.

Hurley, J. H. and Meyer, T. (2001). Subcellular targeting by membrane lipids. *Curr. Opin. Cell. Biol.* **13,** 146–152.

Ibrahimi, A. and Abumrad, N. A. (2002). Role of CD36 in membrane transport of long-chain fatty acids. *Curr. Opin. Clin. Nutr. Metab. Care* **5,** 139–145.

Imamura, T., Haruta, T., Takata, Y., Usui, I., Iwata, M., Ishihara, H., Ishiki, M., Ishibashi, O., Ueno, E., Sasaoka, T., and Kobayashi, M. (1998). Involvement of heat shock protein 90 in the degradation of mutant insulin receptors by the proteasome. *J. Biol. Chem.* **273,** 11183–11188.

Imamura, T., Takata, Y., Sasaoka, T., Takada, Y., Morioka, H., Haruta, T., Sawa, T., Iwanishi, M., Hu, Y. G., Suzuki, Y., Hamada, J., and Kobayashi, M. (1994). Two naturally occuring mutations in the kinase domain of the insulin receptor accelerate degradation of the insulin receptor and impair the kinase activity. *J. Biol. Chem.* **269,** 31019–31027.

Inoue, M., Chang, L., Hwang, J., Ciang, S.-H., and Saltiel, A. R. (2003). The exocyst complex is required for targeting of Glut4 to the plasma membrane by insulin. *Nature* **422,** 629–633.

Iwanishi, M., Haruta, T., Takata, Y., Ishibashi, O., Sasaoka, T., Egawa, K., Imamura, T., Naitou, K., Itazu, T., and Kobayashi, M. (1993). A mutation (Trp1193-Leu1193) in the tyrosine kinase domain of the insulin receptor associated with type A syndrome of insulin resistance. *Diabetologia* **36,** 414–422.

Joost, H. G., Weber, T. M., and Cushman, S. W. (1988). Qualitative and quantitative comparison of glucose transport activity and glucose transporter concentration in plasma membranes from basal and insulin-stimulated rat adipose cells. *Biochem. J.* **249,** 155–161.

Kandror, K. V., Stephens, J. M., and Pilch, P. F. (1995). Expression and compartmentalization of caveolin in adipose cells: Coordinate regulation with and structural segregation from GLUT4. *J. Cell Biol.* **129,** 999–1006.

Karlsson, M., Thorn, H., Danielsson, A., Stenkula, K. G., Öst, A., Gustavsson, J., Nystrom, F. H., and Strålfors, P. (2004). Colocalization of insulin receptor and insulin receptor substrate-1 to caveolae in primary human adipocytes. Cholesterol depletion blocks insulin signalling for metabolic and mitogenic control. *Eur. J. Biochem.* **271,** 2471–2479.

Karlsson, M., Thorn, H., Parpal, S., Strålfors, P., and Gustavsson, J. (2001). Insulin induces translocation of glucose transporter GLUT4 to plasma membrane caveolae in adipocytes. *FASEB J.* **16,** 249–251.

Khan, M. N., Savoie, S., Bergeron, J. J. M., and Possner, B. I. (1986). Characterization of rat liver endosomal fractions. *In vivo* activation of insulin-stimulatable receptor kinase in these structures. *J. Biol. Chem.* **261,** 8462–8472.

Kim, S. and Moustaid-Moussa, N. (2000). Secretory, endocrine and autocrine/paracrine function of the adipocyte. *J. Nutr.* **130,** 3110S–3115S.

Kimura, A., Mora, S., Shigematsu, S., Pessin, J. E., and Saltiel, A. R. (2002). The insulin receptor catalyzes the tyrosine phosphorylation of caveolin-1. *J. Biol. Chem.* **277,** 30153–30158.

Klein, H. H., Freidenberg, G. R., Matthaei, S., and Olefsky, J. M. (1987). Insulin receptor kinase following internalization in isolated rat adipocytes. *J. Biol. Chem.* **262,** 10557–10564.

Le, P. U., Guay, G., Altshuler, Y., and Nabi, I. R. (2002). Caveolin-1 is a negative regulator of caveolae-mediated endocytosis to the endoplasmic reticulum. *J. Biol. Chem.* **277,** 3371–3379.

Lee, H., Volonte, D., Galbiati, F., Iyengar, P., Lublin, D. M., Bregman, D. B., Wilson, M. T., Campos-Gonzalez, R., Bouzahzah, B., Pestell, R. G., Scherer, P. E., and Lisanti, M. P. (2000). Constitutive and growth factor-regulated phosphorylation of caveolin-1 occurs at the same site (Tyr-14) *in vivo*: Identification of c-Src/Cav-1/Grb7 signaling cassette. *Mol. Endocrinol.* **14,** 1750–1775.

Le Lay, S., Krief, S., Farnier, C., Lefrere, I., Le Liepvre, X., Bazin, R., Ferré, P., and Dugail, I. (2001). Cholesterol: A cell size dependent signal which regulates glucose metabolism and gene expression in adipocytes. *J. Biol. Chem.* **276,** 16904–16910.

Lisanti, M. P., Scherer, P. E., Tang, Z., and Sargiacomo, M. (1994a). Caveolae, caveolin and caveolin-rich membrane domains: A signalling hypothesis. *Trends Cell Biol.* **4,** 231–235.

Lisanti, M. P., Scherer, P. E., Vidugiriene, J., Tang, Z., Hermanowski-Vosatka, A., Tu, Y.-H., Cook, R. F., and Sargiacomo, M. (1994b). Characterization of caveolin-rich membrane domains isolated from an endothelial-rich source: Implications for human disease. *J. Cell Biol.* **126,** 111–126.

Liu, H., Kublaoui, B., Pilch, P. F., and Lee, J. (2000). Insulin activation of mitogen-activated protein (MAP) kinase and Akt is phosphatidylinositol 3-kinase-dependent in rat adipocytes. *Biochem. Biophys. Res. Commun.* **274,** 845–851.

Liu, P., Leffler, B. J., Weeks, L. K., Chen, G., Bouchard, C. M., Strawbridge, A. B., and Elmendorf, J. S. (2004). Sphingomyelinase activates GLUT4 translocation via a cholesterol-dependent mechanism. *Am. J. Physiol.* **286,** C317–C329.

Liu, P., Ying, Y., Ko, Y.-G., and Anderson, R. G. (1996). Localization of platelet-derived growth factor-stimulated phosphorylation cascade to caveolae. *J. Biol. Chem.* **271,** 10299–10303.

Luo, R. Z. T., Beniac, D. R., Fernades, A., Yip, C. C., and Ottensmeyer, F. P. (1999). Quaternary structure of the insulin-insulin receptor complex. *Science* **285,** 1077–1080.

Maggi, D., Biedi, C., Segat, D., Barbero, D., Panetta, D., and Cordera, R. (2002). IGF-1 induces caveolin 1 tyrosine phosphorylation and translocation in the lipid rafts. *Biochem. Biophys. Res. Commun.* **295,** 1085–1089.

Mahfoud, R., Garmy, N., Maresca, M., Yahi, N., Puigserver, A., and Fantini, J. (2002). Identification of a common sphingolipid-binding domain in Alzheimer, prion, and HIV-1 proteins. *J. Biol. Chem.* **277,** 11292–11296.

Malide, D., Ramm, G., Cushman, S. W., and Slot, J. W. (2000). Immunoelectron microscopic evidence that GLUT4 translocation explains the stimulation of glucose transport in isolated rat white adipose cells. *J. Cell Sci.* **113,** 4203–4210.

Marceau, F., Sabourin, T., Houle, S., Fortin, J. P., Petitclerc, E., Molinaro, G., and Adam, A. (2002). Kinin receptors: Functional aspects. *Int. Immunopharmacol.* **2,** 1729–1739.

Mastick, C. C. and Saltiel, A. R. (1997). Insulin-stimulated tyrosine phosphorylation of caveolin is specific for the differentiated adipocyte phenotype in 3T3-L1 cells. *J. Biol. Chem.* **272,** 20706–20714.

Melkonian, K. A., Ostermeyer, A. G., Chen, J. Z., Roth, M. G., and Brown, D. A. (1999). Role of lipid modifications in targeting proteins to detergent-resistant membrane rafts. Many raft proteins are acylated, while few are prenylated. *J. Biol. Chem.* **274,** 3910–3917.

Michaelson, D., Silletti, J., Murphy, G., D'Eustachio, P., Rush, M., and Philips, M. R. (2001). Differential localization of Rho GTPases in live cells: Regulation by hypervariable regions and RhoGDI binding. *J. Cell Biol.* **152,** 111–126.

Mineo, C., James, G. L., Smart, E. J., and Anderson, R. G. (1996). Localization of epidermal growth factor-stimulated ras/raf-1 interaction to caveolae membrane. *J. Biol. Chem.* **271,** 11930–11935.

Minokoshi, Y., Kahn, C. R., and Kahn, B. B. (2003). Tissue-specific ablation of the GLUT4 glucose transporter or the insulin receptor challenges assumptions about insulin action and glucose homeostasis. *J. Biol. Chem.* **278,** 33609–33612.

Moller, D. E., Benecke, H., and Flier, J. S. (1991). Biological activities of naturally occuring human insulin receptor mutations. Evidence that metabolic effects of insulin can be mediated by a kinase-deficient insulin receptor mutant. *J. Biol. Chem.* **266,** 10995–11001.

Moller, D. E., Yokota, A., Ginsberg-Fellner, F., and Flier, J. S. (1990). Functional properties of a naturally occuring Trp1200-Ser1200 mutation of the insulin receptor. *Mol. Endocrinol.* **4,** 1183–1191.

Muller, G., Jung, C., Wied, S., Welte, S., Jordan, H., and Frick, W. (2001). Redistribution of glycolipid raft domain components induces insulin-mimetic signaling in rat adipocytes. *Mol. Cell. Biol.* **21,** 4553–4567.

Munoz, P., Mora, S., Sevilla, L., Kaliman, P., Tomàs, E., Gumà, A., Testar, X., Palacín, M., and Zorzano, A. (1996). Expression and insulin-regulated distribution of caveolin in skeletal muscle. Caveolin does not colocalize with GLUT4 in intracellular membranes. *J. Biol. Chem.* **271,** 8133–8139.

Murphy, M. A., Schnall, R. G., Venter, D. J., Barnett, L., Bertoncello, I., Thien, C. B. F., Langdon, W. Y., and Bowtell, D. D. L. (1998). Tissue hyperplasia and enhanced T-cell signalling via ZAP-70 in c-Cbl-deficient mice. *Mol. Cell. Biol.* **18,** 4872–4882.

Naramura, M., Kole, H. K., Hu, R.-J., and Gu, H. (1998). Altered thymic positive selection and intracellular signals in Cbl-deficient mice. *Proc. Natl. Acad. Sci. USA* **95,** 15547–15552.

Nystrom, F. H., Chen, H., Cong, L.-N., Li, Y., and Quon, M. J. (1999). Caveolin-1 interacts with the insulin receptor and can differentially modulate insulin signaling in transfected Cos-7 cells and rat adipose cells. *Mol. Endocrinol.* **13,** 2013–2024.

Oral, E. A. (2003). Lipoatrophic diabetes and other related syndromes. *Rev. Endocr. Metab. Dis.* **4,** 61–77.

Örtegren, U., Karlsson, M., Blazic, N., Blomqvist, M., Nystrom, F. H., Gustavsson, J., Fredman, P., and Strålfors, P. (2004). Lipids and glycosphingolipids in caveolae and surrounding plasma membrane of primary rat adipocytes. *Eur. J. Biochem.* **271,** 2028–2036.

Öst, A., Örtegren, U., Gustavsson, J., Nystrom, F. H., and Strålfors, P. (2005). Triacylglycerol is synthesized in a specific subclass of caveolae in primary adipocytes. *J. Biol. Chem.* **280,** 5–8.

Ostrom, R. S., Gregorian, C., Drenan, R. M., Xiang, Y., Regan, J. W., and Insel, P. A. (2001). Receptor number and caveolar co-localization determined receptor coupling efficiency to adenylyl cyclase. *J. Biol. Chem.* **276,** 42063–42069.

Park, D. S., Cohen, A. W., Frank, P. G., Razani, B., Lee, H., Williams, T. M., Chandra, M., Shirani, J., Souza, A. P. D., Tang, B., Jelicks, L. A., Factor, S. M., Weiss, L. M., Tanowitz, H. B., and Lisanti, M. P. (2003). Caveolin-1 null (–/–) mice show dramatic reductions in life span. *Biochemistry* **42,** 15124–15131.

Parpal, S., Karlsson, M., Thorn, H., and Strålfors, P. (2001). Cholesterol depletion disrupts caveolae and insulin receptor signaling for metabolic control via IRS-1, but not for MAP-kinase control. *J. Biol. Chem.* **276,** 9670–9678.

Parton, R. G., Way, M., Zorzi, N., and Stang, E. (1997). Caveolin-3 associates with developing T-tubules during muscle differentiation. *J. Cell Biol.* **136,** 137–154.

Pelkmans, L., Puntener, D., and Helenius, A. (2002). Local actin polymerization and dynamin recruitment in SV40-induced internalization of caveolae. *Science* **296,** 535–539.

Plough, T., Deurs, B. V., Ai, H., Cushman, S. W., and Ralston, E. (1998). Analysis of GLUT4 distribution in whole skeletal muscle fibers: Identification of distinct storage compartments that are recruited by insulin and muscle contraction. *J. Cell Biol.* **142,** 1429–1446.

Rajala, M. W. and Scherer, P. E. (2003). Minireview: The adipocyte—at the crossroads of energy homeostasis, inflammation, and atherosclerosis. *Endocrinology* **144,** 3765–3773.

Ralston, E. and Plough, T. (1999). Caveolin-3 is associated with the T-tubules of mature skeletal muscle fibers. *Exp. Cell Res.* **246,** 510–515.

Razani, B., Combs, T. P., Wang, X. B., Frank, P. G., Park, D. S., Russel, R. G., Li, M., Tang, B., Jelicks, L. A., Scherer, P. E., and Lisant, M. P. (2002a). Caveolin-1 deficient mice are lean, resistant to diet-induced obesity, and show hyper-triglyceridemia with adipocyte abnormalities. *J. Biol. Chem.* **277,** 8635–8647.

Razani, B., Engelman, J. A., Wang, X. B., Schubert, W., Zhang, X. L., Marks, C. B., Macaluso, F., Russel, R. G., Li, M., Pestell, R. G., Vizio, D. D., Hou, H., Jr., Kneitz, B., Lagaud, G., Christ, G. J., Edelmann, W., and Lisanti, M. P. (2001). Caveolin-1 null mice are viable but show evidence of hyperproliferative and vascular abnormalities. *J. Biol. Chem.* **276,** 38121–38138.

Razani, B. and Lisanti, M. P. (2001). Caveolin-deficient mice: Insight into caveolar function and human disease. *J. Clin. Inv.* **108,** 1553–1561.

Razani, B., Rubin, C. S., and Lisanti, M. P. (1999). Regulation of cAMP-mediated signal transduction via interaction of caveolins with the catalytic subunit of protein kinase A. *J. Biol. Chem.* **274,** 32333–32341.

Razani, B., Wang, X. B., Engelman, J. A., Battista, M., Lagaud, G., Zhang, X. L., Kneitz, B., Hou, H. Jr., Christ, G. J., Edelmann, W., and Lisanti, M. P. (2002b). Caveolin-2-deficient mice show evidence of severe pulmonary dysfunction without disruption of caveolae. *Mol. Cell. Biol.* **22,** 2329–2344.

Razani, B., Woodman, S. E., and Lisanti, M. P. (2002c). Caveolae: From cell biology to animal physiology. *Pharmacol. Rev.* **54,** 431–467.

Robinson, L. J., Pang, S., Harris, D. S., Heuser, J., and James, D. E. (1992). Translocation of the glucose transporter (GLUT4) to the cell surface in permeabilized 3T3-L1 adipocytes: Effects of ATP, insulin, and GTPγS and localization of GLUT4 to clathrin lattices. *J. Cell Biol.* **117,** 1181–1196.

Ros-Baro, A., Lopez-Iglesias, C., Peiro, S., Bellido, D., Palacin, M., Zorzano, A., and Camps, M. (2001). Lipid rafts are required for GLUT4 internalization in adipose cells. *Proc. Natl. Acad. Sci. USA* **98,** 12050–12055.

Rybin, V. O., Xu, X., Lisanti, M. P., and Steinberg, S. F. (2000). Differential targeting of β-adrenergic receptor subtypes and adenylyl cyclase to cardiomyocyte caveolae. A mechanism to functionally regulate the cAMP signaling pathway. *J. Biol. Chem.* **275,** 41447–41457.

Sabourin, T., Bastien, L., Bachvarov, D. R., and Marceau, F. (2002). Agonist-induced translocation of the kinin B(1) receptor to caveolae-related rafts. *Mol. Pharmacol.* **61,** 546–553.

Saltiel, A. R. and Kahn, C. R. (2001). Insulin signalling and the regulation of glucose and lipid metabolism. *Nature* **414,** 799–806.

Satoh, S., Nishimura, H., Clark, A. E., Kozka, I. J., Vanucci, S. J., Simpson, I. A., Quon, M. J., Cushman, S. W., and Holman, G. D. (1993). Use of bismannose photolabel to elucidate insulin-regulated GLUT4 subcellular trafficking kinetics in rat adipose cells. *J. Biol. Chem.* **268,** 17820–17829.

Sawa, T., Imamura, T., Haruta, T., Sasaoka, T., Ishiki, M., Takata, Y., Takada, Y., Morioka, H., Ishihara, H., Usui, I., and Kobayashi, M. (1996). Hsp70 family molecular chaperones and mutant insulin receptor: Differential binding specificities of BiP and Hsp70/Hsc70 determines accumulation or degradation of insulin receptor. *Biochem. Biophys. Res. Commun.* **218,** 449–453.

Scherer, P. E., Lisanti, M. P., Baldini, G., Sargiacomo, M., Corley Mastick, C., and Lodish, H. F. (1994). Induction of caveolin during adipogenesis and association of GLUT4 with caveolin-rich vesicles. *J. Cell Biol.* **127,** 1233–1243.

Scherer, P. E., Williams, S., Fogliano, M., Baldini, G., and Lodish, H. F. (1995). A novel serum protein similar to C1q, produced exclusively in adipocytes. *J. Biol. Chem.* **270,** 26746–26749.

Shaul, P. W. and Anderson, R. G. W. (1998). Role of plasmalemmal caveolae in signal transduction. *Am. J. Physiol.* **275,** L843–L851.

Sheldon, H., Hollenberg, C. H., and Winegrad, A. I. (1962). Observations on the morphology of adipose tissue. I. The fine structure of cells from fasted and diabetic rats. *Diabetes* **11,** 378–387.

Shepherd, P. R., Gnudi, L., Tozzo, E., Yang, H., Leach, F., and Kahn, B. B. (1993). Adipose cell hyperplasia and enhanced glucose disposal in transgenic mice overexpressing GLUT4 selectively in adipose tissue. *J. Biol. Chem.* **268,** 22243–22246.

Shigematsu, S., Watson, R. T., Khan, A. H., and Pessin, J. E. (2003). The adipocyte plasma membrane caveolin functional/structural organization is necessary for the efficient endocytosis of GLUT4. *J. Biol. Chem.* **278,** 10683–10690.

Simons, K. and Toomre, D. (2000). Lipid rafts and signal transduction. *Nature Rev. Mol. Cell Biol.* **1,** 31–40.

Slot, J. W., Geuze, H. J., Gigengack, S., James, D. E., and Lienhard, G. E. (1991a). Translocation of the glucose transporter GLUT4 in cardiac myocytes of the rat. *Proc. Natl. Acad. Sci. USA* **88,** 7815–7819.

Slot, J. W., Geuze, H. J., Gigengack, S., Lienhard, G. E., and James, D. E. (1991b). Immuno-localization of the insulin regulatable glucose transporter in brown adipose tissue of the rat. *J. Cell Biol.* **113,** 123–135.

Smart, E. J., Graf, G. A., McNiven, M. A., Sessa, W. C., Engelman, J. A., Scherer, P. E., Okamoto, T., and Lisanti, M. P. (1999). Caveolins, liquid-ordered domains and signal transduction. *Mol. Cell. Biol.* **19,** 7289–7304.

Smith, R. M., Charron, M. J., Shah, N., Lodish, H. F., and Jarett, L. (1991a). Immunoelectron microscopic demonstration of insulin-stimulated translocation of glucose transporters to the plasma membrane of isolated rat adipocytes and masking of the carboxy-terminal epitope of intracellular GLUT4. *Proc. Natl. Acad. Sci. USA* **88,** 6893–6897.

Smith, R. M. and Jarett, L. (1983). Quantitative ultrastructural analysis of receptor-mediated insulin uptake into adipocytes. *J. Cell. Physiol.* **115,** 199–207.

Smith, R. M., Seely, B. L., Shah, N., Olefsky, J. M., and Jarett, L. (1991b). Tyrosine kinase-defective insulin receptors undergo insulin-induced microaggregation but do not concentrate in coated pits. *J. Biol. Chem.* **266,** 17522–17530.

Souto, R. P., Vallega, G., Wharton, J., Vinten, J., Tranum-Jensen, J., and Pilch, P. F. (2003). Immunopurification and characterization of rat adipocyte caveolae suggest their dissociation from insulin signaling. *J. Biol. Chem.* **278,** 18321–18329.

Strålfors, P. (1990). Autolysis of isolated adipocytes by endogenously produced fatty acids. *FEBS Lett.* **263,** 153–154.

Strålfors, P. (1997). Insulin second messengers. *BioEssays* **19,** 327–335.

Strålfors, P. and Belfrage, P. (1983). Phosphorylation of hormone-sensitive lipase by cyclic AMP-dependent protein kinase. *J. Biol. Chem.* **258,** 15146–15152.

Strålfors, P., Björgell, P., and Belfrage, P. (1984). Hormonal regulation of hormone-sensitive lipase in intact adipocytes: Identification of phosphorylated sites and effects on the phosphorylation by lipolytic hormones and insulin. *Proc. Natl. Acad. Sci. USA* **81,** 3317–3321.

Strålfors, P., Hiraga, A., and Cohen, P. (1985). The protein phosphatases involved in cellular regulation: Purification and characterisation of the glycogen-bound form of protein phosphatase-1 from rabbit skeletal muscle. *Eur. J. Biochem.* **149,** 295–303.

Strålfors, P. and Honnor, R. C. (1989). Insulin-induced dephosphorylation of hormone-sensitive lipase. Correlation with lipolysis and cAMP-dependent protein kinase activity. *Eur. J. Biochem.* **182,** 379–385.

Suzuki, K. and Kono, T. (1980). Evidence that insulin causes translocation of glucose transport activity to the plasma membrane from an intracellular storage site. *Proc. Natl. Acad. Sci. USA* **77,** 2542–2545.

Tansey, J. T., Huml, A. M., Vogt, R., Davis, K. E., Jones, J. M., Fraser, K. A., Brasaemle, D. L., Kimmel, A. R., and Londos, C. (2003). Functional studies on native and mutated forms of perilipins. A role in protein kinase A-mediated lipolysis of triacylglycerols. *J. Biol. Chem.* **278,** 8401–8406.

Thirone, A. C., Carvalheira, J. B., Hirata, A. E., Velloso, L. A., and Saad, M. J. (2003). Regulation of cbl-associated protein/cbl pathway in muscle and adipose tissue of two animal models of insulin resistance. *Endocrinology* **145,** 281–293.

Thomsen, P., Roepstorff, K., Stahlhut, M., and Deurs, B. V. (2002). Caveolae are highly immobile plasma membrane microdomains, which are not involved in constitutive endocytic trafficking. *Mol. Biol. Cell* **13,** 238–250.

Thorn, H., Stenkula, K. G., Karlsson, M., Örtegren, U., Nystrom, F. H., Gustavsson, J., and Strålfors, P. (2003). Cell surface orifices of caveolae and localization of caveolin to the necks of caveolae in adipocytes. *Mol. Biol. Cell* **14,** 3967–3976.

Trigatti, B. L., Anderson, R. G., and Gerber, G. E. (1999). Identification of caveolin-1 as a fatty acid binding protein. *Biochem. Biophys. Res. Commun.* **255,** 34–39.

Vainio, S., Heino, S., Månsson, J.-E., Fredman, P., Kuismanen, E., Vaarala, O., and Ikonen, E. (2002). Dynamic association of human insulin receptor with lipid rafts in cells lacking caveolae. *EMBO Rep.* **3,** 95–100.

Voldstedlund, M., Tranum-Jensen, J., and Vinten, J. (1993). Quantification of Na+/K+-ATPase and glucose transporter isoforms in rat adipocyte plasma membrane by immunogold labeling. *J. Membr. Biol.* **136,** 63–73.

Watson, R. T., Shigematsu, S., Chiang, S.-H., Mora, S., Kanzaki, M., Macara, I. G., Saltiel, A. R., and Pessin, J. E. (2001). Lipid raft microdomain compartmentalization of TC10 is required for insulin signaling and GLUT4 translocation. *J. Cell Biol.* **154,** 829–840.

White, M. F. (2002). IRS proteins and the common path to diabetes. *Am. J. Physiol.* **283,** E413–E422.

Williamson, J. R. (1964). Adipose tissue. Morphological changes associated with lipid mobilisation. *J. Cell Biol.* **20,** 57–74.

Wu, C., Butz, S., Ying, Y.-S., and Anderson, R. G. W. (1997). Tyrosine kinase receptors concentrated in caveolae-like domains from neuronal plasma membrane. *J. Biol. Chem.* **272,** 3554–3559.

Yamabhai, M. and Anderson, R. G. (2002). Second cysteine-rich region of epidermal growth factor receptor contains targeting information for caveolae/rafts. *J. Biol. Chem.* **277,** 24843–24846.

Yamamoto, M., Toya, Y., Schwencke, C., Lisanti, M. P., Myers, M. G., and Ishikawa, Y. (1998). Caveolin is an activator of insulin receptor signaling. *J. Biol. Chem.* **273,** 26962–26968.

Zhang, Y., Proenca, R., Maffei, M., Barone, M., Leopold, L., and Friedman, J. M. (1994). Positional cloning of the mouse obese gene and its human homologue. *Nature* **372,** 425–432.

Section IV

Caveolae and Pathologies

Chapter 9

Caveolin-3, Caveolae, and Hereditary Muscle Diseases

Matthias Vorgerd

Department of Neurology, Neuromuscular Center Ruhrgebiet, Ruhr-University Bochum, D-44789 Bochum, Germany

Caveolin-3 (CAV3) is the muscle-specific protein product of the caveolin gene family with a relative molecular weight of ~22 kDa that is expressed in skeletal, cardiac, and smooth muscle cells. About 14–16 CAV3 monomers homo-oligomerize to form high (~300–350 kDa) molecular multimers that are the integral membrane component of caveolae. Caveolae are small plasma membrane invaginations that participate in membrane trafficking, transport, and signal transduction (Kurzchalia and Parton, 1999; Razani and Lisanti, 2001). The expression of CAV3 is induced during the differentiation of skeletal myoblasts. CAV3 is localized in the plasma membrane, where it is associated with the dystrophin–glycoprotein complex. It has been recognized that mutations in the CAV3 gene (*CAV3*) give rise to very different phenotypes. Four clinical subgroups with various degrees of skeletal muscle

Advances in Molecular and Cell Biology, Vol. 36

ISBN: 0-444-51500-3
DOI: 10.1016/S0079-6603(05)36009-0

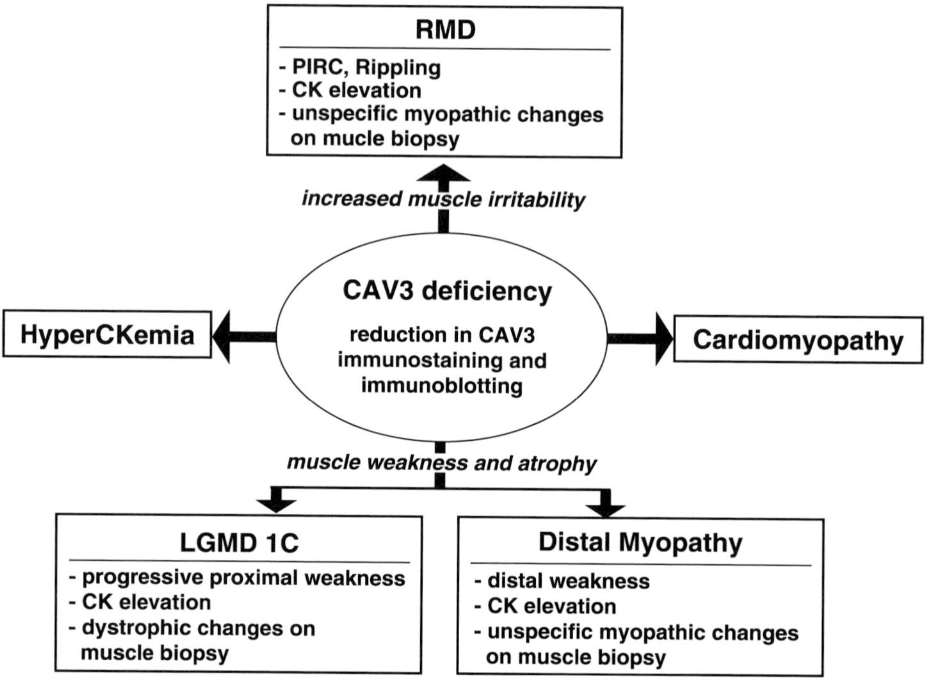

Figure 1. Schematic diagram of the different clinical phenotypes associated in CAV3 deficiency.

involvement have been identified. Patients at the severe end of the clinical spectrum have symptoms of a progressive limb-girdle muscular dystrophy (LGMD). Patients with a milder form of CAV3 deficiency either present with a distal myopathy or manifest only an isolated elevation of the muscle-specific creatine kinase enzyme (CK) activity in blood (hyperCKemia) without further neuromuscular symptoms. *CAV3* mutations can also cause nondystrophic rippling muscle disease (RMD); these patients show symptoms of muscle hyperexcitability such as percussion-induced rapid muscle contractions and stretch-induced rippling muscle contractions but typically no muscle weakness or atrophy. In addition, *CAV3* mutations can manifest with symptoms of a cardiomyopathy without evidence for skeletal muscle involvement. These distinct clinical subtypes can occur among families with the same *CAV3* mutation or can overlap in patients with CAV3 deficiency, indicating that other genes or factors in addition to CAV3 may be involved in the clinical expression. (See Fig. 1.)

I. CAVEOLIN-3 IN MUSCULAR DYSTROPHIES

A. Limb-Girdle Muscular Dystrophy (LGMD)

The term limb-girdle muscular dystrophy (LGMD) is used to describe a major group of non-congenital autosomal muscular dystrophies different from the X-linked Duchenne-Becker muscular dystrophy and also distinct from the autosomal dominant facioscapulohumeral forms. The LGMD group comprises clinical phenotypes with face-sparing, progressive weakness, onset in the limb-girdles, and histopathological evidence of a dystrophin-positive muscular dystrophy (muscle cell necrosis and regeneration, lipofibromatosis, variation of muscle fiber diameter, centrally located nuclei). Major advances in the genetic understanding of LGMD have led to a genetically based classification in which the autosomal dominant forms are designated as LGMD type 1 and the autosomal recessive forms as type 2. There are now six genes or gene loci that correspond to LGMD 1 (LGMD 1A–1F) and 10 associated with LGMD 2 (LGMD 2A–2J). The LGMD 1 forms are relatively rare and represent less than 10% of all LGMD.

Minetti *et al.* (1998) reported for the first time a muscle disorder associated with *CAV3* mutations. They described two Italian families with clinical features of a mild–to-moderate LGMD that started in early childhood, mild dystrophic features on muscle biopsies, and an almost complete loss of CAV3 immunostaining at the sarcolemma. Mutation analysis of *CAV3* identified distinct heterozygous mutations in these two autosomal dominant LGMD families: a missense mutation P104L within the membrane-spanning domain, and a nine base pair microdeletion that removes three amino acids ΔTFT 63–65 within the scaffolding domain. Therefore, LGMD caused by *CAV3* mutations is designated as subtype LGMD 1C. Subsequent ultrastructural studies in these LGMD 1C patients revealed an almost complete loss of caveolae in the sarcolemma (Minetti *et al.*, 2002). Moreover, changes in the T-tubule system were evident in these LGMD 1C patients and in *CAV3* knockout mice, suggesting that dysregulation of muscle calcium homeostasis may be associated with loss of CAV3 (Galbiati *et al.*, 2001; Minetti *et al.*, 2002).

Herrmann *et al.* (2000) reported a 4-year-old girl suffering from myalgia and muscle cramps. Her muscle biopsy showed dystrophic features and absence of CAV3 immunostaining at the sarcolemma. Mutational studies revealed a missense mutation (A45T) in the N-terminal signature sequence of CAV3. Interestingly, a secondary profound reduction of nNOS and of α-dystroglycan immunolabeling at the sarcolemma was found.

Matsuda *et al.* (2001) described an early onset LGMD 1C patient from Japan with a T63P missense mutation in *CAV3*. This patient had progressive proximal muscle weakness, exercise-induced myalgia, mildly elevated CK activities, and dystrophic changes on muscle biopsy (fiber size variation, necrotic and regenerating fibers, centrally located nuclei, fibromatosis). Immunohistochemistry revealed absence of CAV3 and reduction of dysferlin immunolabeling at the sarcolemma. Western blotting confirmed the reduction of CAV3, but the expression of dysferlin was normal.

In contrast to the LGMD 1C patients with an early onset of symptoms in childhood, a late beginning LGMD 1C was reported in a 71-year old French woman caused by a R26Q mutation in *CAV3*. She had mild dystrophic changes in her muscle biopsy, a drastically reduced CAV3 and decreased dysferlin immunolabeling at the sarcolemma. CAV3 was not detectable, but dysferlin expression was unaltered in Western blotting. Electron microscopy demonstrated a lack of caveolae (Figarella-Branger *et al.*, 2003).

It has been suggested that *CAV3* mutations (G55S, C71W, R125H) might also cause autosomal recessive LGMD (McNally *et al.*, 1998). However, the presence of these alterations in healthy controls and the normal CAV3 immunostaining in the sarcolemma in patients with the G55S alteration indicate that these alterations may be benign polymorphisms (DePaula *et al.*, 2001).

B. Secondary Alteration of Caveolin-3 in Duchenne Muscular Dystrophy (DMD)

DMD is an X-linked recessive disorder caused by mutations in the dystrophin gene. The primary deficiency of dystrophin leads to a concomitant reduction in the expression of dystrophin-associated glycoproteins (e.g., sarcoglycans and dystroglycans) and of nNOS. In contrast to the markedly reduced dystrophin–glycoprotein complex, there are several observations that suggest a compensatory upregulation of CAV3 in DMD. It has been shown that the amount of CAV3 protein is elevated and—consistent with this—the ultrastructural studies revealed an increased number of caveolae in DMD patients (Bonilla *et al.*, 1981; Vaghy *et al.*, 1998). Similar findings were also demonstrated in the *mdx* mice, an animal model of DMD (Repetto *et al.*, 1999).

II. RIPPLING MUSCLE DISEASE (RMD)

A. Heterozygous RMD

Hereditary rippling muscle disease (RMD) is a rare autosomal dominant disorder characterized by signs of increased muscle irritability such as percussion/pressure-induced rapid muscle contractions (PIRC), painful localized mounding of muscle induced by tapping, as well as rolling movements across a muscle group ("rippling") triggered by tapping or passive muscle stretch. Muscle weakness is not a prominent feature in RMD. The main complaints of RMD patients include muscle stiffness, exercise-induced muscle pain, and cramp-like sensations. In most patients these symptoms begin before the age of 20 years. RMD is a rather benign myopathy, usually not progressive, is not accompanied by dystrophic changes, and has a normal life expectancy. Some patients may need symptomatic therapy for muscle cramps and myalgia that includes Ca^{2+} antagonists and/or benzodiazepines for muscle relaxation. Elderly patients with a longer duration of the disease may develop some proximal muscle weakness in addition to the RMD phenotype.

Torbergsen described RMD in a Norwegian family in 1975. Ricker *et al.* (1989) reported on two German RMD families and coined the name "rippling muscle disease." A large American RMD family was investigated by Stephan *et al.* (1994). Vorgerd *et al.* (1999) described two additional German RMD families, outlined diagnostic criteria for RMD, and introduced "percussion-induced rapid muscle contractions" (PIRC) as a clinical feature of RMD that is even more reliable than "rippling." So *et al.* (2001) reported on clinical observations in another American family and confirmed the diagnostic value of PIRC.

The first RMD locus was mapped to the long arm of chromosome 1 in a single four-generation Oregon pedigree, but this locus has been excluded in all other informative RMD families described so far. In 2000, Betz *et al.* showed that RMD is linked to chromosome 3p25 and associated with mutations in *CAV3*. They described five families with autosomal-dominant RMD from two European countries (Germany and Norway) caused by different heterozygous *CAV3* mutations (R26Q, A45T, A45V, P104L), including the original Norwegian RMD pedigree. These *CAV3* mutations resulted in a reduced plasma membrane expression of CAV3 in stably transfected C2C12 cells. Moreover, a significant increase in nitric oxidic production was shown in mutant C2C12 myotubes, and it was speculated that increased inducibility of neuronal nitric oxide synthase (nNOS) contributes to the mechanically induced muscle contractions in RMD. This study demonstrated the allelism of dystrophic LGMD 1C and nondystrophic RMD diseases. Thereafter, two additional autosomal dominant RMD families from Japan and a sporadic German RMD patient were reported, caused by the same R26Q mutation in *CAV3* (Vorgerd *et al.*, 2001; Yabe *et al.*, 2003). Clinical diagnosis was based on PIRC and rippling muscle along with elevations of CK activity. Muscle biopsy studies showed unspecific myopathic changes (centrally located nuclei, muscle fiber hypertrophy, but no dystrophic features such as necrosis, regeneration, or lipofibromatosis) and an absent CAV3 immunostaining. A preserved nNOS staining at the muscle fiber membrane was reported in the sporadic German RMD patient.

Schara *et al.* (2002) presented a detailed study on seven children with RMD and proven *CAV3* mutations. They pointed out that initial symptoms of RMD include inability to walk on heels, tiptoe walking with myalgia, and a warm-up phenomenon, calf hypertrophy and elevated serum CK. All children had PIRC and CK elevation. Painful muscle mounding on percussion and mechanically induced rippling muscle contractions were less frequent, thus confirming that PIRC was a sensitive and reliable sign for RMD even at the early stages of the disease. (See Fig. 2.)

B. Homozygous RMD

Recently, two unrelated patients with novel homozygous mutations (L86P and A92T) in *CAV3* were reported (Kubisch *et al.*, 2003). Both presented with a more severe clinical phenotype than usually seen in RMD. Immunohistochemical and immunoblot analyses of muscle biopsies revealed a strong reduction of CAV3 in both homozygous RMD patients, similar to the findings in heterozygous RMD. Electron microscopy studies showed a nearly complete absence of caveolae in the sarcolemma. Additional plasma membrane irregularities (small plasmalemmal discontinuities, subsarcolemmal vacuoles, abnormal papillary projections) were more pronounced in homozygous than in heterozygous RMD patients. Interestingly, these latter ultrastructural changes looked similar to the earliest pathological alterations reported in

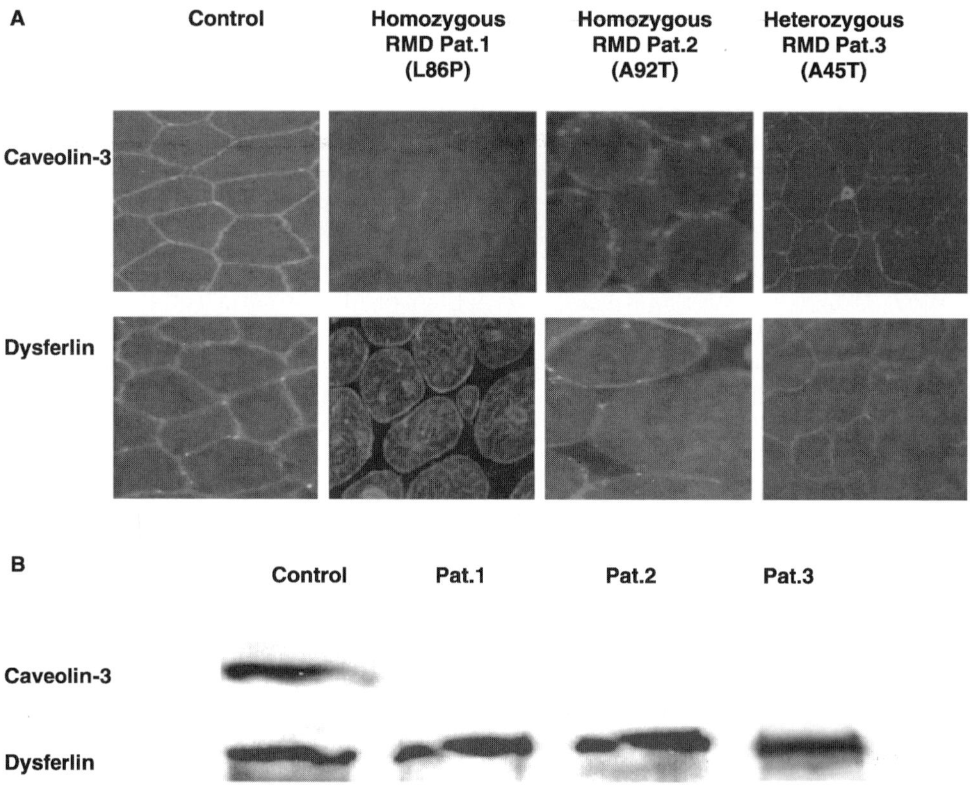

Figure 2. (A) Immunofluorescence and (B) immunoblot analysis of CAV3 and dysferlin on muscle biopsies of two patients with homozgyous (L86P, A92T) and one heterozygous (A45T) RMD. In the RMD patients there is a drastic reduction in the intensity of CAV3 immunostaining at the sarcolemma and in immunoblotting. Dysferlin is reduced in RMD by immunohistochemistry, but normal in immunoblotting (Kubisch *et al.*, 2003). (See Color Insert.)

dysferlinopathy (Selcen *et al.*, 2001). Like in LGMD1C, dysferlin immunoreactivity was reduced in RMD and more pronounced in the homozygous RMD patients. This study further extended the phenotypic variability of muscle caveolinopathies by identification of a severe form of RMD associated with homozygous *CAV3* mutations.

C. RMD of Unknown Etiology

Evidence of autosomal recessive inheritance of RMD has been reported in two families from Oman who—in addition to PIRC and rippling muscle waves—showed cardiac involvement, short stature, and a delayed bone age (Koul *et al.*, 2001). This phenotype clearly differs from the hitherto reported RMD patients with *CAV3* mutations, in whom symptoms were restricted to the skeletal muscle. This family remains to be further studied on the genetical level.

III. DISTAL MYOPATHY

Tateyama *et al.* (2002) reported a sporadic patient with distal myopathy restricted to the intrinsic hand and feet muscle that was associated with the R26Q *CAV3* mutation. Histopathological examination revealed unspecific myopathic features (variation in fiber size, centrally located nuclei, type 1 fiber predominance), a drastically reduced but not absent CAV3 and dysferlin immunostaining at the plasma membrane, and absent CAV3, but normal dysferlin expression on Western blotting. This report expanded the clinical spectrum of CAV3 deficiency by adding the distal myopathy phenotype. It was emphasized that reduced dysferlin immunostaining on muscle biopsy should be analyzed for a possible *CAV3* mutations.

IV. HyperCKemia

Elevated serum activity of CK without additional symptoms of a neuromuscular disease is defined as hyperCKemia. Carbone *et al.* (2000) reported two unrelated children with hyperCKemia and a greatly reduced CAV3 staining in the sarcolemma without any other major alterations on muscle biopsies. They identified the R26Q *CAV3* mutation that occurred *de novo* in both patients. Merlini *et al.* (2002) described a family with hyperCKemia and CAV3 deficiency on muscle biopsy associated with a missense mutation (P28L). Therefore, CAV3 immunostaining should be part of the routine diagnostic evaluation of muscle biopsies in patients with hyperCKemia, although *CAV3* mutations seem to be very rare in this situation (Capasso *et al.*, 2003).

V. CARDIOMYOPATHY

Recently, Hayashi *et al.* (2004) described two Japanese siblings with juvenile onset hypertrophic cardiomyopathy associated with a T63S *CAV3* mutation who had neither symptoms of skeletal muscle involvement nor CK elevation. Functional studies of the T63S mutation in cultured cells showed decreased but detectable immunolabeling of mutant CAV3 at the cell surface. CAV3 retention in the cytoplasm was observed, and a mild reduction in the CAV3 expression was found by Western blotting. Interestingly, these changes were milder than those found in T63P and ΔTFT 63–65 mutants. This indicated that the functional differences in the CAV3 distribution and possibly the cardiac phenotype depend on the mutations themselves. This report came to the conclusion that cardiomyopathy may be another phenotypic manifestation of CAV3 deficiency and further expands the already known clinical spectrum of caveolinopathies.

VI. INTRAFAMILIAL CLINICAL HETEROGENEITY IN *CAV3* FAMILIES

A remarkable intrafamilial phenotypic heterogeneity was observed in two large *CAV3* families. Fischer *et al.* (2003) reported a German multigenerational family associated with a D27E mutation in *CAV3*. Immunohistochemical and Western blot analysis showed a profound reduction of CAV3 expression in skeletal muscle that was attributed to a dominant-negative effect of the *CAV3* mutation. Ultrastructural studies revealed an abnormal folding of the plasma membrane and multiple vacuoles in the subsarcolemmal region. Interestingly, signs of RMD were present either alone in some affected individuals or coexisting with distal myopathy or LGMD-like findings in the other patients within this family. This study pointed out that the clinical phenotypes in CAV3 deficiency may overlap very well and that the clinical spectrum may be continuous rather than distinct. In one Italian family, a recently described ΔF96 mutation in the transmembrane domain of *CAV3* resulted in RMD, LGMD, or hyperCKemia in all affected family members. Analysis of skeletal muscle biopsies revealed a greatly reduced expression of CAV3, normal expression of α-dystroglycan and nNOS, and absent caveolae. In contrast, a myocardial biopsy from one patient with RMD and pelvic-girdle weakness showed only a mild reduction in CAV3 immunostaining and Western blot. In addition, cardiac caveolae were preserved on electron microscopy analysis. It was speculated that differential factors are possibly regulating the expression of CAV3 in skeletal muscle and myocardium (Cagliani *et al.*, 2003). These observations of intrafamilial phenotypic heterogeneity clearly confirm that other hitherto unidentified modifying factors/genes are involved in the clinical outcome of *CAV3* mutations.

VII. PATHOGENESIS OF CAV3 DEFICIENCY

At present it is difficult to see a common pathogenetic pathway of action in primary CAV3 deficiency that leads to the distinct clinical phenotypes. Likewise, in other muscle disorders substantially different phenotypes even within the same family and identical mutations are known. This is true for LGMD 2I caused by a defect in the fukutin-related protein (FKRP) gene with a wide phenotypic spectrum of congenital muscular dystrophy at the severe end, varying degrees of limb-girdle muscular dystrophy (LGMD 2I) with onset between early childhood and late adulthood, and even asymptomatic *FKRP* homozygotes at the other end of the spectrum. Mutations in the nuclear membrane protein lamin A/C also display striking variability causing skeletal muscle disease (LGMD 1B, Emery-Dreifuss muscular dystrophy), cardiomyopathy with AV-block, familial partial lipodystrophy (Dunnigan-type), axonal peripheral neuropathy (Charcot-Marie-Tooth disease 2A), mandibulo-acral dysplasia, and multisystem Hutchinson–Gilford progeria syndrome (postnatal growth retardation, midface hypoplasia, micrognathia, premature atherosclerosis, absence of subcutaneous fat, alopecia, osterodysplasia). (See Fig. 3.)

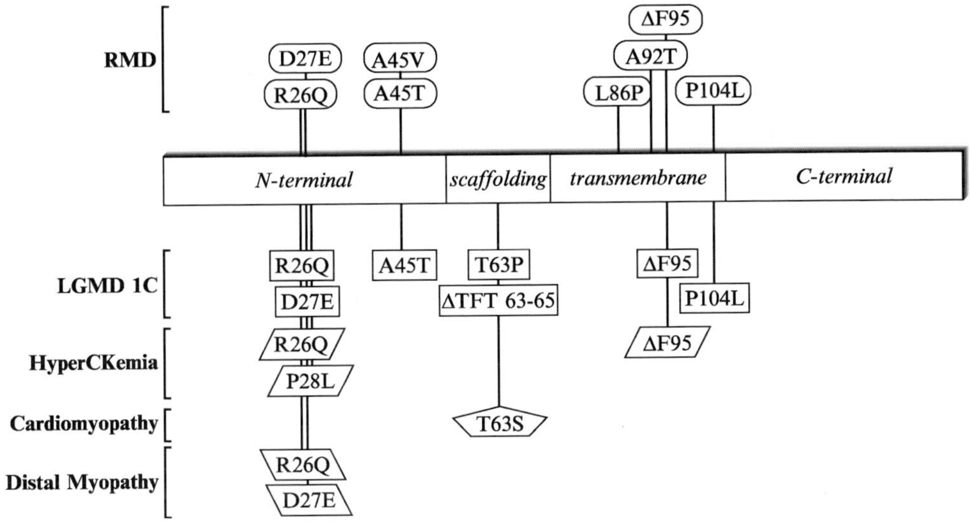

Figure 3. Schematic diagram of the CAV3 protein with the hitherto reported mutations associated with RMD, LGMD 1C, distal myopathy, hyperCKemia, and cardiomyopathy.

All *CAV3* mutations reported so far are associated with a drastic reduction of CAV3 immunostaining at the sarcolemma. The molecular mechanisms underlying this sarcolemmal CAV3 deficiency have not yet been fully identified. One possible mechanism is that heterozygous CAV3 mutations exert a dominant-negative effect that induces rapid degradation of both wild-type and mutant CAV3 proteins. This has been shown in elegant heterologous expression studies in NIH 3T3 cells. P104L and ΔTFT 63–65 mutants of CAV3 form unstable high-molecular-mass CAV3 aggregates that are retained in the Golgi apparatus and are not targeted to the cell surface (Galbiati *et al.*, 1999). In addition, the P104L and ΔTFT 63–-65 CAV3 mutants underwent ubiquitination and proteasomal degradation, further reducing the cell surface expression of CAV3 (Galbiati *et al.*, 2000). Another molecular explanation was recently proposed for the R26Q mutation and may involve haplo-insufficiency instead of the dominant-negative pathway. It has been shown that transfected cells with R26Q mutant CAV3 form oligomers of a much larger size than wild-type CAV3 that are retained within the Golgi complex. When co-expressed with mutant CAV3, however, the reduced level of wild-type CAV3 was insufficient for an appropriate CAV3 localization at the plasma membrane, although it was not disturbed by mutant CAV3 protein (Sotgia *et al.*, 2003).

In addition to the models of how different *CAV3* mutations may directly act on the CAV3 protein turnover, alterations of the complex CAV3 network including the nNOS pathway, the CAV3 interacting partner dysferlin, and the T-system may also be relevant in the pathogenesis.

A. nNOS

CAV3 directly interacts with nNOS, the nitric oxide synthase isoform expressed in skeletal muscle (Kusner and Kaminski, 1996). Interaction with CAV3 results in the inhibition of nNOS catalytic activity (Michel and Feron, 1997; Stamler and Meissner, 2001).

It was shown that cytokine-stimulated NOS activity is increased in C2C12 myotubes transfected with mutant *CAV3* (Betz *et al.*, 2001). Consistent with this observation, transgenic mice expressing the P104L *CAV3* mutant in skeletal muscle revealed an increase of nNOS activity in skeletal muscle (Sunada *et al.*, 2001). Recently, nicotinamide dinucleotide phosphate diaphorase (NDP) assays in RMD patients with homozygous *CAV3* mutations showed an increased sarcolemmal staining consistent with increased nNOS activity (Kubisch *et al.*, 2003). In contrast, a severe reduction of nNOS expression was shown in LGMD1C (Herrmann *et al.*, 2000) and in Duchenne muscular dystrophy (Brenman *et al.*, 1995). It has been speculated that increased nNOS activity may contribute to a benign RMD phenotype without dystrophic changes and signs of muscle hyperirritability, whereas the combined loss of CAV3 and nNOS may lead to a dystrophic LGMD phenotype (Betz *et al.*, 2001; Kubisch *et al.*, 2003). Additional studies on the nNOS pathway are necessary to further characterize its role and functional consequence in CAV3 deficiency.

B. Dysferlin

Dysferlin is a large sarcolemmal protein distinct from the dystrophin–glycoprotein complex. It has a large intracellular domain with six C2 domains, contains multiple CAV3-binding domains, and presumably is involved in calcium-dependent membrane fusion and repair by regulating vesicle fusion with the plasma membrane (Davis *et al.*, 2002). Mutations in the human dysferlin gene cause a spectrum of different clinical presentations that include a limb-girdle muscular dystrophy phenotype (LGMD 2B), a distal myopathy, or mixed presentations, even with identical mutations (Weiler *et al.*, 1999). Matsuda *et al.* (2001) showed by immunoprecipitation that dysferlin and CAV3 interact in skeletal muscle. They described a reduced dysferlin immunostaining but normal dysferlin expression on Western blotting in muscle biopsies of *CAV3* mutant (R26Q, T63P) LGMD 1C patients. This was confirmed in another LGMD 1C patient (R26Q), in distal myopathy with *CAV3* mutation (R26Q), and also in patients with RMD (Figarella-Branger *et al.*, 2003; Kubisch *et al.*, 2003) and can be interpreted as secondary changes in the subcellular distribution of dysferlin in response to the almost complete absence of CAV3. Vice versa, a secondary reduction of CAV3 but normal organization of caveolae was detected in a subset of patients with autosomal recessive dysferlin deficiency (LGMD 2B) (Walter *et al.*, 2003). Altogether, the disturbed interaction between CAV3 and dysferlin may contribute to the pathogenesis in caveolinopathies, but the functional relevance of these findings awaits further elucidation.

C. T-Tubule System

A transient association of CAV3 with transverse tubules (T-tubules) during differentiation of skeletal fibers was demonstrated, indicating a possible role of muscle cell caveolae in the formation or organization of the T-tubule system (Parton *et al.*, 1997). Ultrastructural studies of skeletal muscle from patients with LGMD 1C and RMD with proven *CAV3* mutations showed vacuolar changes in the subsarcolemmal region (Minetti *et al.*, 2002; Fischer *et al.*, 2003; Kubisch *et al.*, 2003) that probably are caused by a striking disorganization of the T-tubular system openings at the subsarcolemmal region (Minetti *et al.*, 2002). These findings are in agreement with T-system alterations found in *CAV3*-null mice. Immunostaining with the T-tubule marker dihydropyridine receptor-1α and ryanodine receptor showed diffuse and disorganized labeling. Moreover, ultrastructural studies revealed dilated and irregularly oriented T-tubules in *CAV3*-null skeletal muscle (Galbiati *et al.*, 2001). It is tempting to speculate that these alterations in the T-tubular system trigger an inappropriate calcium release from the sarcoplasmic reticulum that are related to the signs of muscular hyperirritability in RMD.

VIII. SUMMARY

CAV3 mutations are associated with a wide spectrum of different clinical phenotypes. Most patients with *CAV3* mutations reported so far present with RMD; some patients develop LGMD 1C. More rarely, they have distal myopathy, hyperCKemia, or cardiomyopathy. However, systematic genetical studies evaluating *CAV3* mutations among larger series of biopsy-proven LGMD patients or patients with only unspecific changes on muscle biopsy are necessary to assess the exact prevalence of caveolinopathies and the related phenotypes.

All patients have a profound reduction in CAV3 and caveolar disorganization at the sarcolemma in common, irrespective of the clinical phenotype and type of *CAV3* mutation. So far, there seems to be no clear correlation between the residual CAV3 protein expression, *CAV3* mutations, and the clinical phenotype. This suggests that alterations in the expression of other muscle proteins (e.g., CAV3-related proteins) and changes in the CAV3 network contribute to the pathogenesis and clinical presentation of CAV3 deficiency.

ACKNOWLEDGMENT

M.V. thanks Dr. C. Kubisch, Bonn, and Dr. B. Schoser, Munich, for cooperation, and Dr. B. Voss and Dr. R. Kley for critical reading of the manuscript. M.V. is a member of the German network on muscular dystrophies (*MD-NET*, 01GM0302) funded by the German Ministry of Education and Research (BMBF, Bonn, Germany).

REFERENCES

Betz, R. C., Schoser, B. G. H., Kasper, D., Ricker, K., Ramirez, A., Stein, V., Torbergsen, T., Lee, Y. A., Nothen, M. M., Wienker, T. F., Malin, J. P., Propping, P., Reis, A., Mortier, W., Jentsch, T. J., Vorgerd, M., and Kubisch, C. (2001). Mutations in CAV3 cause mechanical hyperirritability of skeletal muscle in rippling muscle disease. *Nat. Genet.* **28,** 218–219.

Bonilla, E., Fischbeck, K., and Schotland, D. L. (1981). Freeze-fracture studies of muscle caveolae in human muscular dystrophy. *Am. J. Pathol.* **104,** 167–173.

Brenman, J. E., Chao, D. S., Xia, H., Aldape, K., and Bredt, D. S. (1995). Nitric oxide synthase complexed with dystrophin and absent from skeletal muscle sarcolemma in Duchenne muscular dystrophy. *Cell* **82,** 743–752.

Cagliani, Bresolin,N., Prelle, N., Gallanti, A., Fortunato, F., Sironi, M., Ciscato, P., Fagiolari, G., Bonato, S., Galbiati, S., Corti, S., Lamperti, C., Moggio, M., and Comi, G. P. (2003). A CAV3 microdeletion differentially affects skeletal muscle and myocardium. *Neurology* **61,** 1513–1519.

Capasso, M., Lupo, S., Di Muzio, A., Uncini, A., and Servidei, S. (2003). Immunohistochemical study of caveolin-3 in idiopathic hyperCKaemia. *J. Neurol. Neurosurg. Psychiatr.* **74,** 547–548.

Carbone, I., Bruno, C., Sotgia, F., Bado, M., Broda, P., Masetti, E., Panella, A., Zara, F., Bricarelli, F. D., Codone, G., Lisanti, M. P., and Minetti, C. (2000). Mutation in the CAV3 gene causes partial caveolin-3 deficiency and hyper-CKemia. *Neurology* **54,** 1373–1376.

Davis, D. B., Doherty, K. R., Delmonte, A. J., and McNally, E. M. (2002). Calcium-sensitive phospholipid binding properties of normal and mutant ferlin C2 domains. *J. Biol. Chem.* **277,** 22883–22888.

De Paula, F., Vainzof, M., Bernardino, A. L. F., McNally, E., Kunkel, L. M., and Zatz, M. (2001). Mutations in the caveolin-3 gene: When are they pathogenic? *Am. J. Med. Genet.* **99,** 303–307.

Figarella-Branger, D., Ponget, J., Bernard, R., Krahn, M., Fernandez, C., Levy, N., and Pellissier, J. F. (2003). Limb-girdle muscular dystrophy in a 71-year-old woman with an R27Q mutation in the CAV3 gene. *Neurology* **61,** 562–564.

Fischer, D., Schroers, A., Blümcke, I., Urbach, H., Zerres, K., Mortier, W., Vorgerd, M., and Schroder, R. (2003). Consequences of a novel caveolin-3 mutation in a large German family. *Ann. Neurol.* **53,** 233–241.

Galbiati, F., Engelman, J. A., Volonte, D., Zhang, X. L., Minetti, C., Li, M., Hou, H., Jr., Kneitz, B., Edelmann, W., and Lisanti, M. P. (2001). Caveolin-3 null mice show a loss of caveolae, changes in the microdomain distribution of the dystrophin-glycoprotein complex, and T-tubule abnormalities. *J. Biol. Chem.* **276,** 21425–21433.

Galbiati, F., Volonte, D., Minetti, C., Bregman, D. B., and Lisanti, M. P. (2000). Limb-girdle muscular dystrophy (LGMD-1C) mutants of caveolin-3 undergo ubiquitination and proteasomal degradation. Treatment with proteasomal inhibitors blocks the dominant negative effect of LGMD-1C mutants and rescues wild-type caveolin-3. *J. Biol. Chem.* **275,** 37702–37711.

Galbiati, F., Volonte, D., Minetti, C., Chu, J. B., and Lisanti, M. P. (1999). Phenotypic behavior of caveolin-3 mutations that cause autosomal dominant limb girdle muscular dystrophy (LGMD-1C). Retention of LGMD-1C caveolin-3 mutants within the golgi complex. *J. Biol. Chem.* **274,** 25632–25641.

Hayashi, T., Arimura, T., Ueda, K., Shibata, H., Hohda, S., Takahashi, M., Hori, H., Koga, Y., Oka, N., Imaizumi, T., Yasunami, M., and Kimura, A. (2004). Identification and functional analysis of a caveolin-3 mutation associated with familial hypertrophic cardiomyopathy. *Biochem. Biophys. Res. Commun.* **313,** 178–184.

Herrmann, R., Straub, V., Blank, M., Kutzick, C., Franke, N., Jacob, E. N., Lenard, H. G., Kroger, S., and Voit, T. (2000). Dissociation of the dystroglycan complex in caveolin-3 deficient limb girdle muscular dystrophy. *Hum. Mol. Genet.* **9,** 2335–2340.

Koul, R. L., Chand, R. P., Chacko, A., Ali, M., Brown, K. M., Bushnarmuth, S. R., Escolar, D. M., and Stephan, D. A. (2001). Severe autosomal recessive rippling muscle disease. *Muscle Nerve* **24,** 1542–1547.

Kubisch, C., Schoser, B. G., von Düring, M., Betz, R. C., Goebel, H. H., Zahn, S., Ehrbrecht, A., Aasly, J., Schroers, A., Popovic, N., Lochmuller, H., Schroder, J. M., Bruning, T., Malin, J. P., Fricke, B.,

Recently, there has been a rapidly growing interest as to the function of caveolae e vascular system, including their role in angiogenesis, as well as vascular cell entiation and proliferation (Anderson, 1998; Razani *et al.*, 2002). Caveolae are o 80-nm flask-shaped invaginations of the plasma membrane, enriched in choles- and sphingolipids. Their basic structural protein units are the caveolin gene ly, which encodes three isoforms. Caveolin-1 (Cav-1) and caveolin-2 (Cav-2) are nt in most cells, whereas caveolin-3 (Cav-3) is muscle specific (most strongly essed in striated muscle cells).

The caveolin molecules form oligomers in the plasma membrane and interact tly with cholesterol, giving rise to the characteristic omega-like shape of caveolae. y classes of signaling molecules accumulate in caveolae and interact with caveolin ts scaffolding domain. Notably, association with caveolin is inhibitory (in most) and maintains various signaling molecules in an inactive state (Razani *et al.*,). Once the proper ligand is available, a given signal transducer dissociates from olin, becomes activated, and leaves the caveolae microenvironment—thereby ting a signaling cascade.

CAVEOLAE AND CAVEOLINS IN ENDOTHELIAL CELLS

The endothelium is a highly dynamic cell layer that is involved in a multitude of iological functions, including the control of vasomotor tone, trafficking of cells nutrients, maintenance of blood fluids, and the growth of new blood vessels. Over past decade, caveolae and the caveolin proteins have been recognized as major ers in various endothelial functions. The EC is one of the cell types that expresses ighest levels of Cav-1 (Lisanti *et al.*, 1994b; Frank *et al.*, 2003). As a consequence, cells contain large numbers of caveolae. The generation of Cav-1-deficient mice $-1^{-/-}$) using standard homologous recombination techniques has allowed many stigators to define the role of Cav-1 in caveolae formation, as well as other aspects veolae function (Drab *et al.*, 2001; Razani *et al.*, 2001b). Caveolae are especially dant in wild-type aortic ECs. Interestingly, Cav-1-deficient ECs clearly lack any gnizable caveolar structures (Fig. 1) (Frank *et al.*, 2003).

Several functional roles have been attributed to caveolae in ECs. Caveolae and structural proteins, the caveolins, are implicated in many of their normal iological functions, as well as pathological states.

Trans-Endothelial Transport

A widely accepted role for caveolae is their ability to promote the trans-endothelial sport (transcytosis) of solutes, such as albumin, across vascular endothelia ionescu *et al.*, 2002). During transcytosis, caveolae bud-off and detach from the na membrane at one side of the cell and move into the cytoplasm as free vesicles

Meinck, H. M., Torbergsen, T., Engels, H., Voss, B., and Vorgerd, M. (2003). Homozygous mutations in caveolin-3 cause a severe form of rippling muscle disease. *Ann. Neurol.* **53,** 512–520.

Kurzchalia, T. V. and Parton, R. G. (1999). Membrane microdomains and caveolae. *Curr. Opin. Cell Biol.* **11,** 424–431.

Kusner, L. L. and Kaminski, H. J. (1996). Nitric oxide synthase is concentrated at the skeletal muscle endplate. *Brain Res.* **730,** 238–242.

Matsuda, C., Hayashi, Y. K., Ogawa, M., Aoki, M., Murayama, K., Nishino, I., Nonaka, I., Arahata, K., and Brown, R. H., Jr. (2001). The sarcolemmal proteins dysferlin and caveolin-3 interact in skeletal muscle. *Hum. Mol. Genet.* **10,** 1761–1766.

McNally, E. M., de Sa Moreira, E., Duggan, D. J., Bonnemann, C. G., Lisanti, M. P., Lidov, H. G., Vainzof, M., Passos-Bueno, M. R., Hoffman, E. P., Zatz, M., and Kunkel, L. M. (1998). Caveolin-3 in muscular dystrophy. *Hum. Mol. Genet.* **7,** 871–877.

Merlini, L., Carbone, I., Capanni, C., Sabatelli, P., Tortorelli, S., Sotgia, F., Lisanti, M. P., Bruno, C., and Minetti, C. (2002). Familial isolated hyperCKaemia associated with a new mutation in the caveolin-3 (CAV-3) gene. *J. Neurol. Neurosurg. Psychiatr.* **73,** 65–67.

Michel, T. and Feron, O. (1997). Nitric oxide synthases: Which, where, how, and why? *J. Clin. Invest.* **100,** 2146–2152.

Minetti, C., Bado, M., Broda, P., Sotgia, F., Bruno, C., Galbiati, F., Volonte, D., Lucania, G., Pavan, A., Bonilla, E., Lisanti, M. P., and Cordone, G. (2002). Impairment of caveolae formation and T-system disorganization in human muscular dystrophy with caveolin-3 deficiency. *Am. J. Pathol.* **160,** 265–270.

Minetti, C., Sotgia, F., Bruno, C., Scartezzini, P., Broda, P., Bado, M., Masetti, E., Mazzocco, M., Egeo, A., Donati, M. A., Volonte, D., Galbiati, F., Cordone, G., Bricarelli, F. D., Lisanti, M. P., and Zara, F. (1998). Mutations in the caveolin-3 gene cause autosomal dominant limb-girdle muscular dystrophy. *Nat. Genet.* **18,** 365–368.

Parton, R. G., Way, M., Zorzi, N., and Stang, E. (1997). Caveolin-3 associated with developing T-tubules during muscle differentiation. *J. Cell Biol.* **136,** 137–154.

Razani, B. and Lisanti, M. P. (2001). Caveolin-deficient mice: Insights into caveolar function and human disease. *J. Clin. Invest.* **108,** 1553–1561.

Repetto, S., Bado, M., Broda, P., Lucania, G., Masetti, E., Sotgia, F., Carbone, I., Pavan, A., Bonilla, E., Cordone, G., Lisanti, M. P., and Minetti, C. (1999). Increased number of caveolae and caveolin-3 overexpression in Duchenne muscular dystrophy. *Biochem. Biophys. Res. Commun.* **261,** 547–550.

Ricker, K., Moxley, R. T., and Rohkamm, R. (1989). Rippling muscle disease. *Arch. Neurol.* **48,** 405–408.

Schara, U., Vorgerd, M., Popovic, N., Schoser, B. G., Ricker, K., and Mortier, W. (2002). Rippling muscle disease in childhood. *J. Child Neurol.* **17,** 483–490.

Selcen, D., Stilling, G., and Engel, A. G. (2001). The earliest pathologic alterations in dysferlinopathy. *Neurology* **56,** 1472–1481.

So, Y. T., Zu, L., Barraza, C., Figueroa, K. P., and Pulst, S. M. (2001). Rippling muscle disease: Evidence for phenotypic and genetic heterogeneity. *Muscle Nerve* **24,** 340–344.

Sotgia, F., Woodman, S. E., Bonuccelli, G., Capozza, F., Minetti, C., Scherer, P. E., and Lisanti, M. P. (2003). Phenotypic behavior of caveolin-3 R26Q, a mutant associated with hyperCKemia, distal myopathy, and rippling muscle disease. *Am. J. Physiol. Cell Physiol.* **285,** C1150–C1160.

Stamler, J. S. and Meissner, G. (2001). Physiology of nitric oxide in skeletal muscle. *Physiol. Rev.* **81,** 209–237.

Stephan, D. A., Buist, N. R. M., Chittenden, A. B., Ricker, K., Zhou, J., and Hoffman, E. P. (1994). A rippling muscle disease gene is localized to 1q41: Evidence for multiple genes. *Neurology* **44,** 1915–1920.

Sunada, Y., Ohi, H., Hase, A., Ohi, H., Hosono, T., Arata, S., Higuchi, S., Matsumura, K., and Shimizu, T. (2001). Transgenic mice expressing mutant caveolin-3 show severe myopathy associated with increased nNOS activity. *Hum. Mol. Genet.* **10,** 173–178.

Tateyama, M., Aoki, M., Nishino, I., Hayashi, Y. K., Sekiguchi, S., Shiga, Y., Takahashi, T., Onodera, Y., Haginoya, K., Kobayashi, K., Iinuma, K., Nonaka, I., Arahata, K., and Itoyama, Y. (2002). Mutation in the caveolin-3 gene causes a peculiar form of distal myopathy. *Neurology* **58,** 323–325.

Torbergsen, T. (1975). A family with dominant hereditary myotonia, muscular hypertrophy, and increased muscular irritability, distinct from myotonia congenita *Thomsen. Acta Neurol. Scand.* **1,** 225–232.
Vaghy, P. L., Fang, J., Wu, W., and Vaghy, L. P. (1998). Increased caveolin-3 levels in mdx mouse muscles. *FEBS Lett.* **431,** 125–127.
Vorgerd, M., Bolz, H., Patzold, T., Kubisch, C., Malin, J. P., and Mortier, W. (1999). Phenotypic variability in rippling muscle disease. *Neurology* **52,** 1453–1459.
Vorgerd, M., Ricker, K., Ziemssen, F., Kress, W., Goebel, H. H., Nix, W. A., Kubisch, C., Schoser, B. G., and Mortier, W. (2001). A sporadic case of rippling muscle disease caused by a de novo caveolin-3 mutation. *Neurology* **57,** 2273–2277.
Walter, M., Braun, C., Vorgerd, M., Poppe, M., Thirion, C., Schmidt, C., Schreiber, H., Knirsch, U. I., Brummer, D., Muller-Felber, W., Pongratz, D., Muller-Hocker, J., Huebner, A., and Lochmuller, H. (2003). Variable reduction of caveolin-3 in patients with LGMD2B/MM. *J. Neurol.* **250,** 1431–1438.
Weiler, T., Bashir, R., Anderson, L. V., Davison, K., Moss, J. A., Britton, S., Nylen, E., Keers, S., Vafiadaki, E., Greenberg, C. R., Bushby, C. R., and Wrogemann, K. (1999). Identical mutations in patients with limb girdle muscular dystrophy type 2B or Miysoshi myopathy suggests a role for modifier gene(s). *Hum. Mol. Genet.* **8,** 871–877.
Yabe, I., Kawashima, A., Kikuchi, S., Higashi, T., Fukazawa, T., Hamada, T., Sasaki, H., and Tashiro, K. (2003). Caveolin-3 gene mutation in Japanese with rippling muscle disease. *Acta Neurol. Scand.* **108,** 47–51.

Chapter 10

Caveolae and Caveolins in the Vascular Syste Functional Roles in Endothelia, Macrophages, Smooth Muscle Cells

Ghada S. Hassan,[1] Michael P. Lisanti,[1] and Philippe G. Frank

[1]*Department of Molecular Pharmacology, Albert Einstein College of Medicine, Bronx, Ne*
[2]*Department of Urology, Albert Einstein College of Medicine, Bronx, New York*

The vascular system transports blood through the body in hollow t tures called blood vessels. The blood vessel is a dynamic structure t necessary "irrigation" to all the organs and tissues. The vessel wall is three inter-connected layers: the intima, the media, and the adventitia. contains a monolayer of endothelial cells (ECs) that is attached to a bas brane formed by extracellular matrix proteins. In contrast, the medial l posed mainly of smooth muscle cells surrounded by extracellular matrix. adventitial layer consists of fibroblasts, collagen, and elastic fibers and is the *vasa vasora* in the case of larger vessels (Griendling *et al.*, 2001).

This chapter focuses on the two main cell types in the vasculature: en smooth muscle cells. These cell types are implicated in most homeostati pathological, conditions of the vascular system. The chapter also tou importance of macrophages in this system.

Advances in Molecular and Cell Biology, Vol. 36

DOI: 10.1016/S0

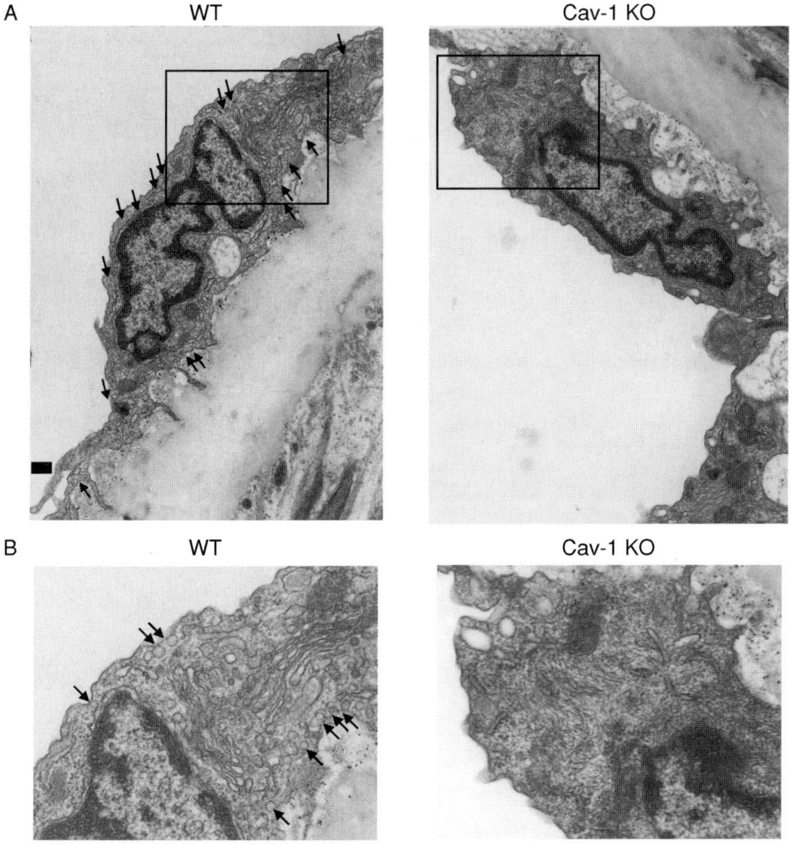

Figure 1. Caveolae in endothelial cells from aorta. Electron micrographs of wild-type (WT) and Cav-1$^{-/-}$ aortas are shown. Wild-type endothelial cells exhibit numerous caveolae at the cell surface (arrows) in addition to detached plasmalemmal vesicles. These structures are clearly absent from caveolin-1-deficient endothelial cells (A). Scale bar = 500 nm. (B) shows a 1.5×-magnified view of the boxed section presented in (A). Modified from Frank *et al.* (2003).

that directionally fuse with the plasma membrane at the other side of the cell (van Deurs *et al.*, 2003). Caveolar channels have also been proposed as a transport mechanism in arterial ECs, without the need for free caveolae vesicles (Ogawa *et al.*, 2001). This process is plausible since ECs are very flat; therefore, caveolae could form short-lived vesicles or even transient channels across the cell (Simionescu *et al.*, 2002; van Deurs *et al.*, 2003). Importantly, it was shown that endothelial cytosol and plasmalemmal vesicles (caveolae) contain the necessary proteins and lipids for fission, fusion, and targeting, including molecules such as N-ethylmaleimide-sensitive factor (NSF), NSF-attachment proteins and their soluble receptors, dynamin, and Rab 5 (Schnitzer *et al.*, 1995b; Oh *et al.*, 1998; Predescu *et al.*, 2001).

Albumin is one of the most studied proteins that is transferred across ECs (Simionescu *et al.*, 2002). Studies have provided clear molecular genetic evidence to support a role for caveolae in the transcytosis of albumin (Schubert *et al.*, 2001). Using Cav-1$^{-/-}$ mice and perfusion with gold-labeled albumin, it was demonstrated that Cav-1 expression is required for the trans-endothelial transport of albumin. Results showed that ECs from Cav-1$^{-/-}$ mice failed to transport gold-labeled albumin, which remained in the blood vessel lumen. On the other hand, ECs from wild-type mice efficiently transported albumin to the sub-endothelial space via caveolae (Schubert *et al.*, 2001).

Many plasma proteins and molecules use the caveolar transcytotic pathway for their transport to the sub-endothelial tissue space (Simionescu *et al.*, 2002). These molecules include albumin (Ghitescu *et al.*, 1986; Schnitzer *et al.*, 1994b,1995a; Tiruppathi *et al.*, 1997), insulin (Schnitzer *et al.*, 1994a; Bendayan and Rasio, 1996), ferritin (Schnitzer *et al.*, 1995a), and native and modified low-density lipoproteins (LDL) (Vasile *et al.*, 1983; Kim *et al.*, 1994; Dehouck *et al.*, 1997). Transcytosis can be constitutive (e.g., fluid-phase transport) and/or regulated (e.g., receptor mediated) (Fig. 2). For instance, the trans-endothelial transport of albumin requires the interaction of the albumin-docking protein, gp60, with caveolin and the subsequent activation of c-Src tyrosine kinase signaling (Tiruppathi *et al.*, 1997; Minshall *et al.*, 2000; Minshall *et al.*, 2002).

B. Vascular Permeability

The regulation of endothelial microvascular permeability is another important and major role for caveolae/caveolins. The selective permeability of ECs to a specific molecule may be controlled not only by the size of caveolar vesicles, but also by the presence of specific receptors within caveolae. The presence of other caveolae-related structures and their regulation might also be critical. For example, the clustering of caveolae into structures termed vesico-vacuolar vesicles (VVOs) requires a specific combination of surface matrix exposure and vascular endothelial growth factor (VEGF) (Vasile *et al.*, 1999). In addition, VEGF was found to induce caveolae expression and their fission and fusion. It also stimulated the formation of elongated caveolin-decorated structures and fenestrae, both responsible for the increase in vascular permeability (Chen *et al.*, 2002).

The overall vascular permeability appears to be controlled by two processes: (1) the endothelial transcytosis pathway, in which molecules are taken up by caveolae vesicles and transported across the cell, and (2) the paracellular pathway, whereby transported molecules are transferred to the sub-endothelial space by passing between ECs, across tight junctions (Majno and Palade, 1961; Predescu and Palade, 1993; Schubert *et al.*, 2002) (Fig. 2).

Schubert *et al.* have shown that even though Cav-1 deficiency inhibited gold-labeled albumin transcytosis across the endothelium, it dramatically increased microvascular permeability via increases in the para-cellular pathway (Schubert *et al.*, 2002).

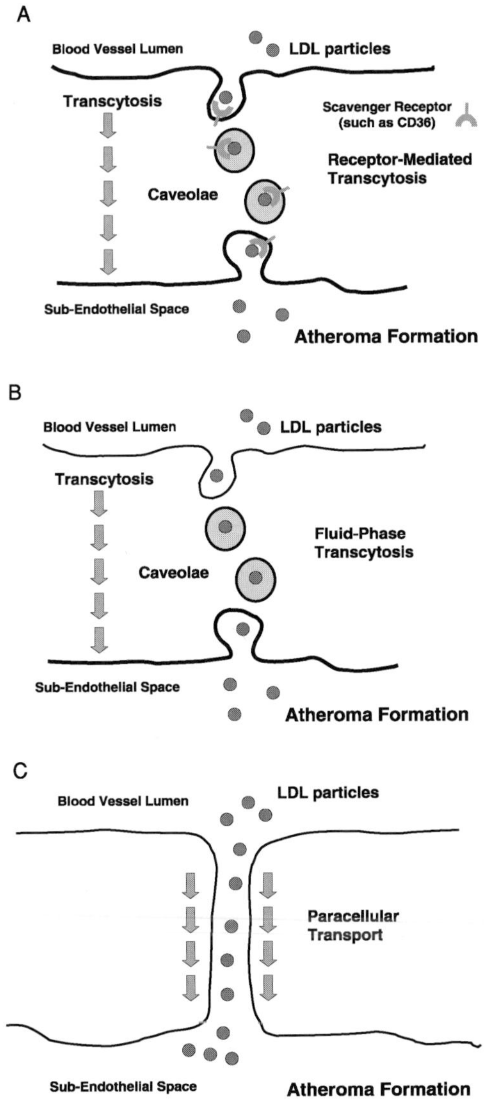

Figure 2. Models of trans-endothelial transport. (A and B) Caveolae-mediated transcytosis. In this model, LDL particles are transferred from the blood vessel lumen to the intima. Two possible pathways exist for this endothelial transcellular transport. (A) An LDL particle may bind to a specific receptor found in endothelial cell caveolae and induce the endocytosis/transcytosis of caveolae vesicle, and the particle is then transferred to the subendothelial space (receptor-mediated transcytosis). (B) Alternatively, an LDL particle can be transported via a non-specific, or fluid-phase, form of caveolar transcytosis. (C) Para-cellular transport of LDL particles across the endothelial cell barrier. Under certain conditions, inter-endothelial tight junctions may be altered, increasing endothelium permeability. Modified from Frank and Lisanti (2004). (See Color Insert.)

In this study, Cav-$1^{-/-}$ mice cleared radiolabeled albumin from the circulation at a faster rate than control wild-type mice, due to the induction of para-cellular transport in these animals. In support of these findings, Schubert *et al.*, using electron microscopy, also reported defects in tight junction morphology and abnormalities in lung capillaries' EC attachment to the basement membrane. This increase in microvascular permeability was found to be nitric oxide (NO)-dependent, as it was reversed by the nitric oxide synthase inhibitor L-NAME. In fact, the enzyme endothelial nitric oxide synthase (eNOS) is constitutively activated in Cav-$1^{-/-}$ mice (Drab *et al.*, 2001; Razani *et al.*, 2001b).

Therefore, in addition to playing a role in endothelial transcytosis, Cav-1 can influence vascular permeability by affecting the para-cellular transport pathway, through its action on eNOS as well as on the organization of protein interactions between cells and the extracellular matrix (Frank *et al.*, 2003). Endothelial permeability plays a major role in lesion formation and is increased in blood vessels with atherosclerotic lesions. As a matter of fact, areas of the aorta that present with increased permeability are more susceptible to atherosclerosis (McGill *et al.*, 1957; Packham *et al.*, 1967; Somer and Schwartz, 1971; Fry *et al.*, 1993). Therefore, the regulation of vascular permeability is one mechanism through which caveolae and caveolins may play a major role in influencing atherosclerotic lesion formation.

C. Caveolae/Caveolin and eNOS: Roles in Atherosclerosis

Atherosclerosis is the underlying cause of various types of arterial disease and predisposes vessels toward vasospasm and/or thrombus formation (Fuster *et al.*, 1992). Atherosclerosis is known to be associated with EC injury or dysfunction, resulting in lipid and leukocyte deposition in the arterial wall (Ross, 1993; Harrison and Ohara, 1995). ECs, as well as macrophages in the intimal layer, release growth factors, which induce the migration of vascular smooth muscle cells (VSMCs) from the media to the intima. These activated VSMCs then undergo proliferation. The accumulation of lipids, foam cells, proliferating cells, and extracellular matrix proteins in the subendothelial layer leads to the eventual formation of a fibrous cap lesion (Ross, 1999).

Endothelium-derived NO has a profound effect on vessel tonicity and permeability. Within the plasma membrane, the eNOS protein is highly enriched in caveolae, where caveolin proteins are also localized. In addition, eNOS activity is found to be nine- to tenfold more prevalent in the caveolae fraction as compared to whole plasma membrane and undetectable in non-caveolar membranes (Shaul *et al.*, 1996). Moreover, within an EC, eNOS was shown to target to lipid rafts/caveolae of the plasma membrane and the Golgi apparatus, where its activity is inhibited by binding to Cav-1 (Garcia-Cardena *et al.*, 1996; Michel *et al.*, 1997). The interaction between Cav-1 and eNOS has thoroughly been discussed in Chapter 5 of this volume. In brief, optimal eNOS activity occurs when the interaction between eNOS and Cav-1 is competitively disrupted by calcium/calmodulin binding to eNOS (Ju *et al.*, 1997; Michel *et al.*, 1997).

Therefore, Cav-1 acts as a negative modulator of eNOS activity and, thus, is a regulator of vascular tone and permeability.

The first direct *in vivo* evidence of a negative regulatory role for Cav-1 in tonically inhibiting eNOS was provided by studies involving Cav-1$^{-/-}$ mice. Razani *et al.* showed that Cav-1-deficient mice exhibited impaired phenylephrine-induced vasoconstriction as well as a potentiated acetylcholine-induced relaxation (both NO-dependent phenomena) due to increased eNOS activity in these mice (Razani *et al.*, 2001b). In addition, Drab *et al.* (2001) also showed similar abnormalities in the NO-dependent pathways in their independently generated Cav-1$^{-/-}$ mice (Drab *et al.*, 2001).

The reduced bioavailability of the endothelium-derived signaling molecule NO is a key process in the early pathogenesis of hypercholesterolemia-induced vascular disease and atherosclerosis (Vane *et al.*, 1990). During the early stages of atherosclerosis, eNOS was found to influence leukocyte adhesion to ECs, a process mediated by adhesion molecules expressed at the surface of activated ECs. For instance, eNOS and NO were shown to downregulate the level of the vascular cell adhesion molecule-1 (VCAM-1) expressed in ECs (Takahashi *et al.*, 1996; Kawashima *et al.*, 2001). In accordance with this finding, VCAM-1 expression was shown to be decreased in Cav-1$^{-/-}$ ApoE$^{-/-}$ mice, in which elevated NO levels and activity are observed (Frank *et al.*, 2004). In addition, the anti-atherogenic role of eNOS can also be explained by its inhibitory action on SMC migration and proliferation (Garg and Hassid, 1989; Kariya *et al.*, 1989). In accordance with these findings, atheroma formation was remarkably less severe in Cav-1$^{-/-}$ ApoE$^{-/-}$ mice compared to ApoE$^{-/-}$ mice treated identically. This effect may be due, at least in part, to the increased eNOS activity observed in these animals (Frank *et al.*, 2004). The negative regulation of eNOS by Cav-1 provides further evidence to support a role for the caveolin gene family in the pathogenesis of atherosclerosis.

D. Cholesterol Transport and Homeostasis

Since caveolae have a high cholesterol content, they have also been functionally implicated in cholesterol transport and homeostasis. The role of caveolae in cholesterol trafficking has been investigated in many cell types. Fielding and Fielding have shown that while clathrin-coated pits mediate the transfer of free cholesterol from LDL into the cell (Fielding and Fielding, 1996), caveolar microdomains may be the sites of efflux of newly synthesized cholesterol as well as LDL-derived free cholesterol (Fielding and Fielding, 1995). Moreover, Cav-1 was found to mediate the transport of cholesterol from the endoplasmic reticulum (ER) to plasma membrane caveolar sites (Smart *et al.*, 1996). The interaction between cholesterol and caveolin is reciprocal. An increase in cellular free cholesterol content upregulates Cav-1 mRNA via the SREBP pathway (Bist *et al.*, 1997). Moreover, Cav-1 translocation to the plasma membrane is also enhanced by incubation of ECs with LDL. In this process, Ras, an important signaling mediator, is activated in the presence of LDL and translocates to caveolae, where it associates with Cav-1 (Zhu *et al.*, 2000). These findings suggest that, in addition to its role in cellular cholesterol transport, Cav-1 is also involved in the regulation of signaling pathways activated by LDL in ECs.

As a major player in the metabolism of lipoproteins, Cav-1 could shift the distribution of lipoproteins toward a more atherogenic profile, as suggested by preliminary studies in Cav-1-deficient mice (Razani *et al.*, 2001a). Interestingly, Cav-1$^{-/-}$ mice have normal plasma cholesterol levels but develop hypertriglyceridemia (with elevated VLDL/chylomicrons). Triglyceride clearance is delayed in these mice, independently of lipoprotein lipase activity, which was found to be unaltered from wild-type animals. Therefore, Cav-1 deficiency may disturb the transport and/or storage of fatty acids and triglycerides in adipocytes, leading to defects in lipid homeostasis (Razani *et al.*, 2001a).

Cav-1-mediated cholesterol trafficking may be one of the mechanisms by which caveolae play a role in the initiation and/or progression of atherosclerosis. As a consequence, the relationship between caveolins and the good cholesterol, transported by high-density lipoprotein (HDL), has been thoroughly investigated. The importance of HDL in cholesterol elimination has been suggested by several studies that showed an inverse correlation between the development of atherosclerosis and HDL cholesterol levels (Gordon *et al.*, 1977; Miller *et al.*, 1977; Gordon and Rifkind, 1989; Wilson *et al.*, 1994; Castelli, 1996). HDL particles function as extracellular acceptors for cellular cholesterol. After cholesterol is transferred to HDL, it is esterified by the enzyme lecithin:cholesterol acyltransferase (LCAT) and then transferred to apoB-containing lipoproteins by cholesterol ester transfer protein (CETP) for elimination by the liver and removal from the body.

While we have shown that transformed NIH/3T3 cells in which Cav-1 expression is downregulated can transfer cholesterol to HDL more rapidly than the untransformed parental cell line (Frank *et al.*, 2001a), it appears that, in most cases, Cav-1 has no or minimal effect on HDL-mediated cellular cholesterol efflux from untransformed cells (Matveev *et al.*, 2001; Frank *et al.*, 2002; Wang *et al.*, 2003). However, further studies are required to clarify the matter. HDL can, in turn, negatively regulate Cav-1, at both the transcriptional and protein levels. We have shown that downregulation of Cav-1 induced by chronic exposure to HDL can prevent the uptake of oxLDL by ECs (Frank *et al.*, 2001a). These results provide further evidence linking caveolae and Cav-1 to oxLDL uptake by ECs and atherosclerosis. These findings also suggest a new mechanism to explain the anti-atherogenic effect of HDL: HDL could reduce the uptake of oxLDLs by ECs and prevent their accumulation in the subendothelial space—by downregulating the expression of Cav-1 and, thus, blocking caveolae formation and LDL transcytosis (Frank *et al.*, 2001a).

Further evidence for the relationship between HDL and Cav-1 was provided by studies involving the HDL receptor [scavenger receptor class B type I (SR-BI)]. Interestingly, SR-BI is associated with caveolae (Babitt *et al.*, 1997). While the uptake of LDL cholesterol occurs by internalization and degradation of the entire particle in the cell, the uptake of esterified cholesterol from HDL is more selective and involves SR-BI (Graf *et al.*, 1999). Recently, we and others have shown that SR-BI-mediated selective HDL CE uptake into cells is inhibited by Cav-1. However, SR-BI-mediated cellular cholesterol efflux was not affected by Cav-1 expression (Matveev *et al.*, 2001; Frank *et al.*, 2002). These results identify SR-BI as a key receptor in the cellular uptake of cholesterol via caveolae. We also showed that SR-BI stabilized Cav-1 and induced significant clustering of the protein at the plasma membrane level. Similarly, the cellular free cholesterol content was found to stabilize Cav-1 (Frank *et al.*, 2002), a finding that is in accordance with our previous work in which a reciprocal relationship was reported between Cav-1 expression and cellular cholesterol levels (Frank *et al.*, 2001a). These findings add to the importance of Cav-1 in regulating cellular cholesterol homeostasis.

E. Relationships between Lipoproteins, eNOS, and Caveolins

Several lines of evidence have indicated that the specialized lipid environment within caveolae influences the targeting and regulation of eNOS within this membrane microdomain. Blair *et al.* have shown that Cav-1 and eNOS remained highly enriched in caveolae membranes when cells are exposed to lipoprotein-deficient serum, HDL, or native LDL (Blair *et al.*, 1999). However, ECs exposed to oxLDL show the translocation of eNOS and Cav-1 from plasmalemmal caveolae to an internal membrane fraction containing ER, the Golgi apparatus, mitochondria, and other intracellular organelles (Blair *et al.*, 1999). In addition to inducing the redistribution of eNOS in ECs, oxLDL attenuated the activity of eNOS, as evaluated by acetylcholine-induced eNOS activation (Blair *et al.*, 1999).

OxLDL may cause a marked depletion of caveolae cholesterol by acting as an acceptor of cholesterol, as determined by measuring the sterol content of caveolae isolated from [^{3}H]acetate-labeled ECs. This process was found to mediate eNOS displacement, since the use of a cholesterol-chelating agent, namely, cyclodextrin, produced similar results to oxLDL and caused the redistribution of eNOS and Cav-1 from caveolae to the intracellular membrane fraction. Thus, oxLDL causes a depletion of cholesterol from caveolae, resulting in the redistribution of eNOS and an inhibition of acetylcholine-induced activation of the enzyme (Blair *et al.*, 1999). This process may be an important mechanism early in the pathogenesis of atherosclerosis.

In agreement with these findings, Feron *et al.* have shown that exposure of ECs to serum from hypercholesterolemic patients impairs basal NO release. This effect was paralleled by the enhanced formation of the inhibitory Cav-1–eNOS complex (Feron *et al.*, 1999).

Since there is a strong negative correlation between HDL levels and the risk for atherosclerosis (Gordon *et al.*, 1977; Gordon and Rifkind, 1989; Assmann *et al.*, 1993), several studies have focused on the interaction between HDL, eNOS, and caveolae. HDL was found to activate eNOS in cultured ECs and to enhance NO-dependent relaxation in mouse aorta via SR-BI. Moreover, HDL was shown to counteract the effects of oxLDL on eNOS-Cav-1 redistribution and on eNOS activation (Uittenbogaard *et al.*, 2000). The addition of HDL to medium containing oxLDL prevented eNOS and Cav-1 displacement from caveolae and additionally restored

Ach-induced stimulation of eNOS. Mechanistically, the effects of HDL on eNOS targeting were mediated by the function of HDL as a sterol donor to restore the cholesterol content of caveolae membranes, which is normally removed by oxLDL (Uittenbogaard *et al.*, 2000; Shaul, 2003). CD36 was identified as the receptor responsible for mediating oxLDL-induced depletion of cholesterol from caveolae. Reversion of this process by HDL was mediated by SR-BI (Uittenbogaard *et al.*, 2000; Shaul, 2003).

Therefore, the interactions between caveolae, eNOS, and lipoproteins play a key role in vascular disease, particularly atherosclerosis. OxLDL, a major risk factor for vascular disease, was shown to alter eNOS activation and, thus, NO production by disrupting the caveolin activation complex. HDL, a cardioprotective lipoprotein, prevents the disruption of the caveolin activation complex by oxLDL, by donating cholesterol to cell surface caveolae or, alternatively, may protect caveolae cholesterol from oxidation (Everson and Smart, 2001).

Direct *in vivo* evidence for the involvement of Cav-1 in the development of atherosclerosis was recently provided using knockout mice. In brief, Cav-1$^{-/-}$ ApoE$^{-/-}$ double-knockout (dKO) mice were obtained by interbreeding Cav-1$^{-/-}$ mice with ApoE-null mice [ApoE$^{-/-}$], an atherosclerosis-prone mouse strain. Interestingly, genetic ablation of Cav-1 provides dramatic protection against atherosclerosis (Fig. 3). We showed that these findings are likely due to the reduced expression of certain pro-atherogenic molecules, such as CD36 and VCAM-1, in dKO mice. The increased eNOS activity observed in the ApoE/Cav-1 dKO model could also account for the reduced pathological appearance of these lesions (Frank *et al.*, 2004). In addition, the elimination of Cav-1 may also reduce the transcytosis of LDL across ECs and their accumulation in the sub-endothelial space.

F. Caveolin-1 and Angiogenesis

During both tumor development and vascular diseases, the demand for an increased blood supply drives the formation of new vessels. This process, termed angiogenesis, involves the formation of new vasculature that is derived from pre-existing blood vessels (Hanahan and Folkman, 1996). Angiogenesis is characterized by the initial proliferation of ECs, their migration, and differentiation in their new residence.

Since Cav-1 plays a role in cell multiplication and differentiation, it was therefore not surprising to observe that Cav-1 expression was downregulated during EC migration but was upregulated during capillary tube formation (Liu *et al.*, 2002). Studies have now shown that Cav-1 downregulation significantly reduces capillary tube formation (Griffoni *et al.*, 2000; Brouet *et al.*, 2001; Liu *et al.*, 2002).

The exact role of Cav-1 has not yet been completely established, but it is definitely a major regulator of angiogenesis. This function is also complicated by the fact that caveolae contain several important molecules, which play a major role in this process (Frank *et al.*, 2003), including the VEGF receptor (VEGFR) (Stahl and Mueller, 1995;

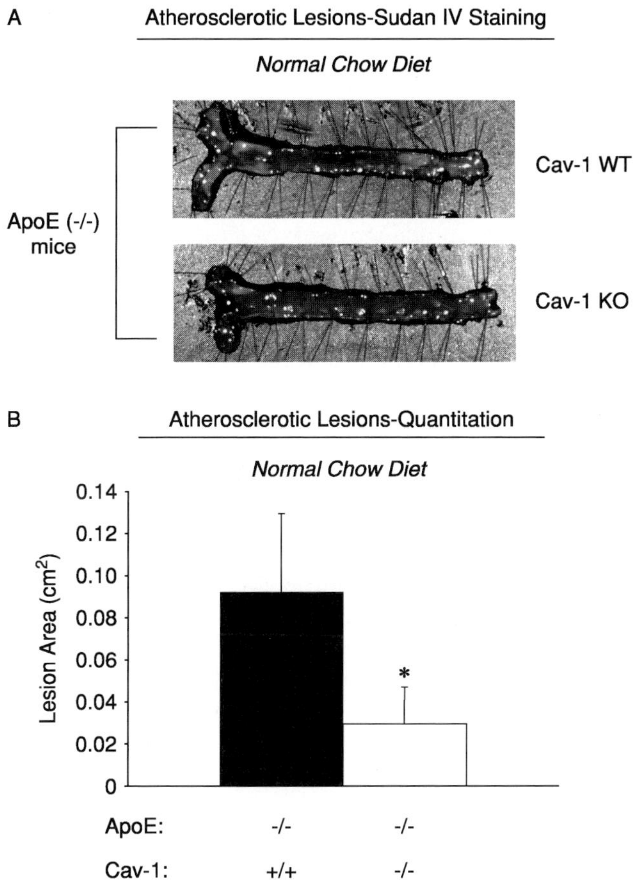

Figure 3. Caveolin-1 deficiency reduces the extent of atheromatous lesions in ApoE$^{-/-}$ mice. (A) En face visualization of aortas, which were harvested from ApoE$^{-/-}$/Cav-1$^{+/+}$ and ApoE$^{-/-}$ /Cav-1$^{-/-}$ mice, fed a normal chow diet. (B) Quantification of the total area occupied by atherosclerotic lesions per aorta. Note that in animals fed a normal chow diet, a Cav-1 deficiency leads to dramatic reductions (~70%) in the total area occupied by these atherosclerotic lesions. An asterisk (*) indicates statistical significance ($p < 0.05$); $n = 5$ mice for each experimental group. Modified from Frank *et al.* (2004). (See Color Insert.)

Labrecque *et al.*, 2003) and the urokinase receptor (uPAR) (Stahl and Mueller, 1995). Recently, metalloproteinases (MMPs), enzymes implicated in degradation of the extracellular matrix and major factors involved in the initiation of angiogenesis, were also found to be associated with caveolae. These microdomains were identified as the endocytic pathway used by the membrane type 1 matrix metalloproteinase (MT1-MMP) in ECs. MT1-MMP co-localized with Cav-1 and required caveolar trafficking for its proper localization and function (Galvez *et al.*, 2004).

On the other hand, caveolae and caveolins are associated with factors that inhibit angiogenesis, such as eNOS (Brouet *et al.*, 2001) and prostacyclin synthase

(Spisni *et al.*, 2001). Sprouty-1 and -2, two negative regulators of angiogenesis, have been localized to caveolae (Impagnatiello *et al.*, 2001). Also, the naturally occurring inhibitor of angiogenesis, endostatin, was found to localize to lipid rafts/caveolae together with its integrin receptor and Cav-1. The lipid raft localization of endostatin was shown to be essential for its downstream signaling, i.e., the Src-dependent disassembly of the actin cytoskeleton (Wickstrom *et al.*, 2002, 2003). An important aspect concerning the role of caveolae/caveolin-1 was established using Cav-1$^{-/-}$ mice. *In vivo* studies using Cav-1$^{-/-}$ mice implanted with Matrigel plugs [supplemented with the basic fibroblast growth factor, bFGF] clearly show a positive role for Cav-1 in the regulation of angiogenesis (Woodman *et al.*, 2003). It was also shown that smaller tumors were formed in these mice with the melanoma cell line B16-F10. Taken together, these data suggest that Cav-1 deficiency reduces angiogenesis.

Overall, these findings suggest a complex role for Cav-1 in the regulation of angiogenesis and demonstrate the need for future investigations into the role of caveolins in angiogenesis-related disease states.

In conclusion, endothelial caveolae have been implicated in the following diverse functions: (1) the transcytosis of molecules across ECs, (2) inactivation of eNOS, (3) the control of cholesterol trafficking, and (4) cell differentiation and proliferation (related to angiogenesis). As previously proposed, "Cav-1 and caveolae are required for the proper organization of signaling pathways within ECs, allowing for a rapid, efficient, and regulated response to specific extracellular stimuli" (Lisanti *et al.*, 1994a; Frank *et al.*, 2003). Disruption of this organization may lead to a variety of vascular abnormalities, such as those observed in atherosclerosis, restenosis after angioplasty, and hypertension.

II. CAVEOLIN-1 IN MACROPHAGES

The important role of macrophages in vascular disease, particularly atherosclerosis, cannot be overlooked. It has now been well established that macrophages indeed express Cav-1 and Cav-2 (Baorto *et al.*, 1997; Matveev *et al.*, 1999; Gargalovic and Dory, 2001), and studies have indicated that Cav-1 may play a role in regulating cellular cholesterol metabolism in these cells. For example, Cav-1 is believed to be involved in the selective uptake of cholesteryl esters from HDL. In J774 and RAW macrophages, transfection of Cav-1 inhibited HDL-mediated cholesterol uptake without affecting the cholesterol efflux (Matveev *et al.*, 2001). On the other hand, the same group showed that Cav-1 expression, which was upregulated upon macrophage differentiation, correlated with increased selective uptake of HDL CE in human THP-1 macrophages (Matveev *et al.*, 1999). This discrepancy as to the role of Cav-1 in cholesterol uptake in macrophages might be due to different levels of CD36 and SR-BI, two members of the class B scavenger receptors known to selectively bind HDL and transfer core cholesteryl esters to cells (Huh *et al.*, 1996; Matveev *et al.*, 2001).

In the plasma membrane, cholesterol localizes either within lipid rafts domains (where it tightly binds sphingolipids) or within non-raft areas (where it associates with phospholipids). However, non-raft domains may represent a more favorable site for cholesterol efflux, because of their loosely packed structure as compared to lipid rafts. In fact, ABCA1, a major protein implicated in the regulation of cellular cholesterol efflux, is not associated with lipid raft domains (Mendez *et al.*, 2001). However, Cav-1 may still play a role in cholesterol efflux by facilitating the transfer of newly synthesized or exogenous cholesterol from the ER to the plasma membrane, where cholesterol first associates with lipid rafts, and then diffuses to non-rafts areas—finally to be effluxed to cholesterol acceptors, such as HDL (Gargalovic and Dory, 2003a). In addition, it was shown that SR-BI and increased cholesterol levels stabilize Cav-1 expression in HEK-293 cells (Frank *et al.*, 2002). This finding further highlights the importance of the Cav-1–cholesterol interaction, which might be of major significance in cells at risk for cholesterol overload, such as macrophages.

Increased levels of Cav-1 are also associated with apoptosis in macrophages. In one study, Cav-1 was found to co-localize with phosphatidylserine and to play a role in the externalization of this phospholipid on the surface of apoptotic cells (Gargalovic and Dory, 2003b). The role of caveolins in the apoptotic behavior of macrophages could have important implications for the development of atherosclerosis and, in particular, during plaque rupture.

In addition, Cav-1 expression may also play a role in the regulation of lipoprotein uptake, especially modified LDL. This process may be regulated in a manner similar to that observed in ECs (Frank *et al.*, 2001b). Caveolae and Cav-1 may therefore have a negative role in the regulation of foam cell formation. Finally, Cav-1 has been shown to regulate several signaling proteins involved in atherosclerosis. This pathway includes protein kinase A and protein kinase C (Smart *et al.*, 1999), which have been shown to play an important role in macrophage metabolism (Bernard *et al.*, 1991; Li *et al.*, 1997).

The above-mentioned findings suggest an important regulatory role for Cav-1 in cholesterol metabolism in macrophages, despite the discrepancies observed in some of the reported results. Nevertheless, further examination of the role of caveolins in the maintenance of cholesterol homeostasis and macrophages-foam cell formation should be undertaken. Availability of Cav-1$^{-/-}$ mice provides a powerful tool for these future studies.

III. CAVEOLAE AND CAVEOLINS IN VASCULAR SMOOTH MUSCLE CELLS

Smooth muscle cells (SMCs) are the building blocks of the medial layer of the arterial wall. Normally, SMCs are highly differentiated contractile cells that are involved in the control of blood pressure and flow. Upon vascular injury, they undergo a transition from a contractile phenotype (characterized by a high myofilament content) to a synthetic phenotype (a less differentiated state). In the synthetic state, they gain the ability to migrate from the media to the intima, to undergo intimal proliferation, and to secrete extracellular matrix components. These phenotypic changes highlight the participation of the SMCs in the pathogenesis of intimal thickening, e.g., during atherosclerosis and restenosis after angioplasty (Ross, 1993; Lusis, 2000; Thyberg, 2003).

Arterial SMCs display abundant caveolae (Somlyo, 1980). Interestingly, they are one of the only cell type that co-expresses all three caveolin family members (Cav-1, Cav-2, and Cav-3) (Doyle *et al.*, 2003; Woodman *et al.*, 2004). However, prolonged culture of primary arterial SMCs leads to the selective downregulation of Cav-3 expression (without affecting the expression of Cav-1 or Cav-2). This is likely due to dedifferentiation, which may occur in culture (Doyle *et al.*, 2003). Several studies have investigated the role of caveolae and the caveolin proteins in various signaling pathways in VSMCs—by following the Ca^{2+} response of SMCs to vasoactive substances and by assessing contractile function, cell differentiation, and, last but not least, cell proliferation.

A. Vascular Contractility

Caveolae play an important regulatory role in maintaining vascular tone and regulating smooth muscle contraction. Dreja *et al.* assessed arterial contractility in vessels treated with a cholesterol-chelating agent, namely, methyl-β-cyclodextrin. Cholesterol depletion disrupted caveolae and reduced the force responses to 5-hydroxytryptamine (5-HT), vasopressin, and endothelin. The increase in intracellular free Ca^{2+} concentration and Ca^{2+} waves activity in response to 5-HT were also attenuated. These responses were reversed by addition of exogenous cholesterol, restoring caveolae formation. These findings suggest that caveolae mediate part of the signaling pathways implicated in smooth muscle contraction (Dreja *et al.*, 2002).

Moreover, caveolar microdomains are involved in the initiation of Ca^{2+} sparks in SMCs, regulating excitation-contraction coupling. Ca^{2+} sparks arise from the openings of ryanodine receptor (RyR) channels and are generated by the precise delivery of Ca^{2+} ions through Ca^{2+} channels to RyRs of the sarcoplasmic reticulum. A novel signaling model defined by Lohn *et al.* suggests that transient elevations in Ca^{2+} at the inner mouth of a single caveolar Ca^{2+} channel, usually an L-type Ca^{2+} channel, induces simultaneous activation and the opening of several RyRs to generate a local Ca^{2+} release event, that is, a Ca^{2+} spark (Lohn *et al.*, 2000). These results are in agreement with those of Isshiki *et al.* (Isshiki *et al.*, 2002a,b).

In addition, Cav-1$^{-/-}$ mice show abnormalities in vascular tone and SMC contractility (Drab *et al.*, 2001; Razani *et al.*, 2001b). Drab *et al.* described reduced myogenic tone, a partially contracted state that is an important determinant of blood pressure, as well as a weaker calcium-dependent contractile response to various known vasoconstrictors in arteries isolated from Cav-1$^{-/-}$ mice (Drab *et al.*, 2001).

Taken together, these findings suggest an important role for caveolae and caveolins in regulating VSMC contractility and, therefore, in influencing the status of vascular tone in both health and disease states.

B. Mitogenic Signaling in VSMCs

In addition to regulating Ca^{2+} responses in vascular smooth muscle, several studies have demonstrated that caveolae and caveolins play a prominent role in the control of mitogenic signaling in SMCs. Evidence of such a role for caveolae was provided by Thyberg *et al.* who demonstrated that the number of caveolae decreased upon phenotypic alteration of smooth muscle cells from a contractile (differentiated) to a synthetic (proliferative) state. This was accompanied by an internalization and redistribution of caveolin to intracellular Golgi-associated vesicles (Thyberg *et al.*, 1997).

Serum-induced cell proliferation was found to be associated with the downregulation of Cav-1 and Cav-2 protein levels in VSMCs in culture. This process was reversed and the expression of caveolins was restored by heparin treatment, a known inhibitor of VSMC proliferation (Peterson *et al.*, 1999). These findings represent early observations implicating caveolins in the control of VSMC proliferation.

Further studies highlighted the significance of caveolin and caveolae in the control of mitogenic signaling in vascular SMCs. Peterson *et al.* investigated the interactions between Cav-1 and the potent SMC mitogen platelet-derived growth factor (PDGF). They reported that PDGF exposure of cultured VSMCs resulted in the downregulation of Cav-1 protein expression, despite a marked increase in the Cav-1 mRNA levels, owing to the lysosomal degradation of Cav-1 in these cells. An *in vivo* confirmation of these findings was obtained in a rabbit arterial injury model. Neointimal SMCs revealed reduced Cav-1 immunostaining, as compared with SMCs of the media (Peterson *et al.*, 2003).

These authors further evaluated how Cav-1 overexpression affects SMCs using an adenoviral vector approach. The proliferation response to PDGF and to normal growth medium was inhibited in cells transduced with the Ad-Cav-1 vector. In addition, Cav-1 was shown to block entrance into the G_1 phase of the cell cycle by preventing the upregulation of cyclin D1, a process normally observed during cell cycle progression. Moreover, cells overexpressing Cav-1 showed increased apoptosis, as $\sim$20% of the SMCs became apoptotic, and exhibited cleavage of caspase-9 and poly-ADP ribose polymerase (Peterson *et al.*, 2003).

Another study investigating the mitogenic signaling in VSMCs examined the role of cholesterol-rich plasma membrane domains, such as caveolae, in the growth of vascular SMCs in response to stretch. Vessel strips exposed to methyl-β-cyclodextrin, the cholesterol-chelating agent, exhibited the expected decrease in the number of SMC caveolae. Moreover, when caveolae were depleted of cholesterol, the stretch-induced stimulation of protein and DNA synthesis was reduced in cultured strips, as was the stretch-induced phosphorylation of the p42/44 MAP kinases (ERK-1/2). It was also reported that cyclodextrin abolished ERK-1/2 phosphorylation in response to endothelin-1 (ET-1), a possible endogenous mediator of the stretch effect. These results suggest that cholesterol-rich microdomains, such as caveolae, mediate stretch-induced growth in VSMCs by a mechanism involving ET-1 signaling (Zeidan *et al.*, 2003).

On the other hand, Kawabe *et al.* have shown that treatment of VSMCs with cyclodextrin enhances ERK activation by decreasing the total amount of caveolae-localized caveolin. In this study, the authors described a new model for ERK activation, involving caveolins at non-caveolar sites. They showed that stretch-induced stress induces the translocation of the caveolin proteins to non-caveolar sites and their association with a β_1-integrin–Fyn–Shc complex, components that are required for ERK activation (Kawabe *et al.*, 2004). It was suggested that while caveolins residing in caveolar microdomains work on inhibiting ERK activation, caveolins that have translocated to non-caveolar sites could allow ERK activation and, thus, activate the appropriate downstream pathways.

Increased mechanical stress is one of the major factors contributing to pathological conditions of the vascular system such as hypertension, atherosclerosis, and restenosis. The aforementioned findings identified caveolae and caveolins as important mediators of the stretch-induced responses of VSMCs observed in vascular diseases.

Recently, we showed that Cav-1 is implicated in SMCs proliferation using Cav-1$^{-/-}$ mice. We used a model developed by Kumar and Lindner (1997) whereby interruption of blood flow by ligating the common carotid artery induces neointima formation. Our results demonstrated more pronounced neointimal hyperplasia consisting mainly of SMCs in Cav-1$^{-/-}$ mice compared to wild-type (Fig. 4). Interestingly, elevated expression levels of phospho-p42/44 MAP kinase (ERK-1/2), phospho-Rb, and cyclin D1 were observed in Cav-1$^{-/-}$ neointimal regions. This *in vivo* study is the first to show that a genetic deficiency of Cav-1 in mice stimulates smooth muscle cell proliferation (neointimal hyperplasia) (Hassan *et al.*, 2004).

Neointimal hyperplasia is a signature feature of early atherosclerosis, restenosis after angioplasty, and bypass graft failure. Therefore, we can conclude that caveolae, and Cav-1 in particular, play an important role in diseases of the vascular system, not only by influencing EC function but also by regulating the VSMC phenotype.

Caveolae and caveolin proteins control many cellular processes in VSMCs such as the Ca^{2+} response, apoptosis, and proliferation. Nevertheless, the analysis of caveolin and caveolae functions in SMC physiology and pathophysiology is still in its infancy. Future work will facilitate the development of caveolin-based therapeutics for the treatment of vascular disease.

IV. CONCLUSIONS

The vascular system is very dynamic—simple in its structure, yet complex in its functions. Caveolar microdomains and their basic protein units, the caveolins, are abundant in cells of the vasculature. ECs express exceptionally high levels of both Cav-1 and Cav-2, while smooth muscle cells express all three caveolin isoforms (Cav-1,-2, and-3). Caveolins (mainly Cav-1) play key regulatory roles in the normal physiology of vascular cells. In this regard, Cav-1 acts as a negative regulator of proliferative signaling pathways in the vascular system as well as a modulator of trans-endothelial transport, microvascular permeability, angiogenesis, vascular relaxation/contraction,

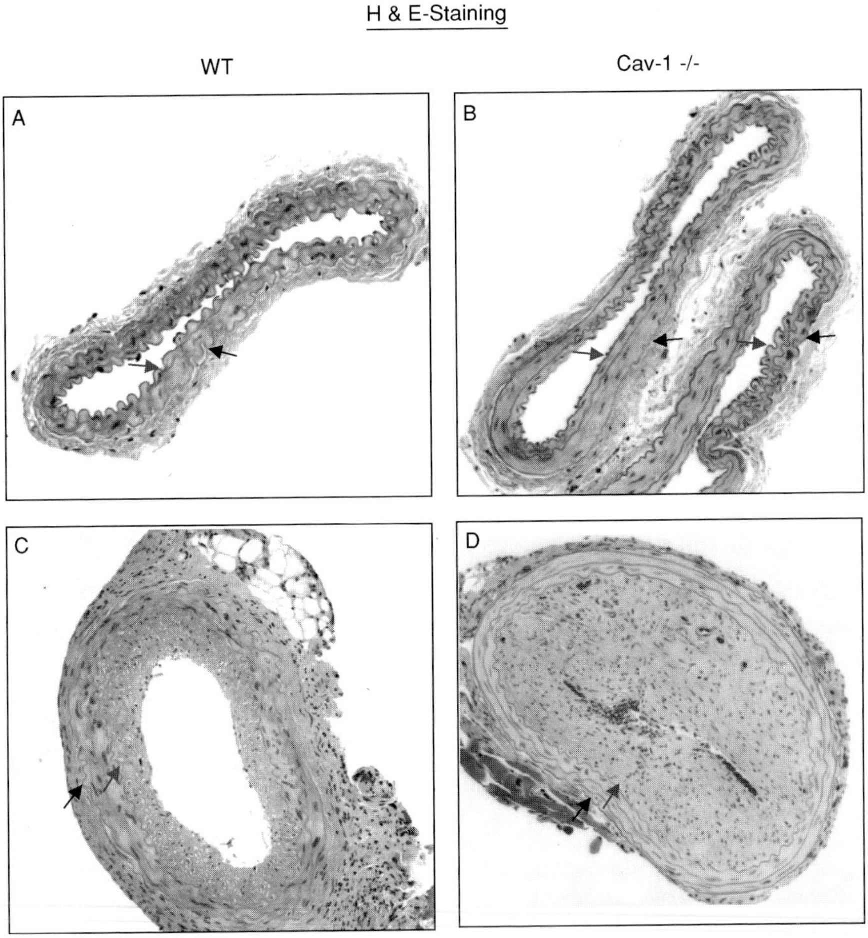

Figure 4. Histological analysis of wild-type (WT) and Cav-1$^{-/-}$ carotid arteries, retrieved 4 weeks post-ligation. (A and B) H&E staining of control wild-type (A) and control Cav-1$^{-/-}$ (B) non-ligated arteries. Note that both WT and Cav-1$^{-/-}$ carotid arteries appear normal, without any evidence of baseline SMC hyperproliferation or migration. (C and D) H&E staining of ligated wild-type (C) and ligated Cav-1$^{-/-}$ (D) arteries. Note that the carotids of Cav-1$^{-/-}$ mice show significantly more intimal hyperplasia, with subtotal luminal obstruction, as compared to WT mice. In (A)–(D), red arrows denote the internal elastic lamina and black arrows denote the external elastic lamina. Modified from Hassan *et al.* (2004). (See Color Insert.)

and cell differentiation. Thus, the study of both caveolae and caveolins has important implications for understanding the pathogenesis of numerous human diseases, including atherosclerosis, restenosis, and hypertension. Importantly, the findings discussed in this chapter clearly identify caveolins and caveolae organelles as targets for future drug development, toward the pharmacological prevention of atherosclerosis and other related vascular diseases in humans.

ACKNOWLEDGMENTS

This work was supported by grants from National Institutes of Health (NIH), the Susan G. Komen Breast Cancer Foundation, and the American Heart Association, as well as Hirschl/Weil-Caulier Career Scientist Award (all to M. P. L.). G.S.H. is the recipient of a post-doctoral fellowship from the Foundation of Health Research (FRSQ), Quebec, Canada. P. G. F. was the recipient of a Scientist Development Grant from the American Heart Association.

REFERENCES

Anderson, R. G. W. (1998). The caveolae membrane system. *Ann. Rev. Biochem.* **67,** 199–225.

Assmann, G., Von Eckardstein, A., and Funke, H. (1993). High density lipoproteins, reverse transport of cholesterol, and coronary artery disease: Insights from mutations. *Circulation* **87**(Suppl.), III28–III34.

Babitt, J., Trigatti, B., Rigotti, A., Smart, E. J., Anderson, R. G., Xu, S., and Krieger, M. (1997). Murine SR-BI, a high density lipoprotein receptor that mediates selective lipid uptake, is N-glycosylated and fatty acylated and colocalizes with plasma membrane caveolae. *J. Biol. Chem.* **272,** 13242–13249.

Baorto, D. M., Gao, Z., Malaviya, R., Dustin, M. L., van der Merwe, A., Lublin, D. M., and Abraham, S. N. (1997). Survival of FimH-expressing enterobacteria in macrophages relies on glycolipid traffic. *Nature* **389,** 636–639.

Bendayan, M. and Rasio, E. A. (1996). Transport of insulin and albumin by the microvascular endothelium of the rete mirabile. *J. Cell Sci.* **109**(Pt. 7), 1857–1864.

Bernard, D. W., Rodriguez, A., Rothblat, G. H., and Glick, J. M. (1991). cAMP stimulates cholesteryl ester clearance to high density lipoproteins in J774 macrophages. *J. Biol. Chem.* **266,** 710–716.

Bist, A., Fielding, P. E., and Fielding, C. J. (1997). Two sterol regulatory element-like sequences mediate up-regulation of caveolin gene transcription in response to low density lipoprotein free cholesterol. *Proc. Natl. Acad. Sci. USA* **94,** 10693–10698.

Blair, A., Shaul, P. W., Yuhanna, I. S., Conrad, P. A., and Smart, E. J. (1999). Oxidized low density lipoprotein displaces endothelial nitric-oxide synthase (eNOS) from plasmalemmal caveolae and impairs eNOS activation. *J. Biol. Chem.* **274,** 32512–32519.

Brouet, A., Sonveaux, P., Dessy, C., Moniotte, S., Balligand, J. L., and Feron, O. (2001). Hsp90 and caveolin are key targets for the proangiogenic nitric oxide-mediated effects of statins. *Circ. Res.* **89,** 866–873.

Castelli, W. P. (1996). Lipids, risk factors and ischaemic heart disease. *Atherosclerosis* **124**(Suppl.), S1–S9.

Chen, J., Braet, F., Brodsky, S., Weinstein, T., Romanov, V., Noiri, E., and Goligorsky, M. S. (2002). VEGF-induced mobilization of caveolae and increase in permeability of endothelial cells. *Am. J. Physiol. Cell Physiol.* **282,** C1053–C1063.

Dehouck, B., Fenart, L., Dehouck, M. P., Pierce, A., Torpier, G., and Cecchelli, R. (1997). A new function for the LDL receptor: Transcytosis of LDL across the blood-brain barrier. *J. Cell Biol.* **138,** 877–889.

Doyle, D. D., Upshaw-Earley, J., Bell, E., and Palfrey, H. C. (2003). Expression of caveolin-3 in rat aortic vascular smooth muscle cells is determined by developmental state. *Biochem. Biophys. Res. Commun.* **304,** 22–25.

Drab, M., Verkade, P., Elger, M., Kasper, M., Lohn, M., Lauterbach, B., Menne, J., Lindschau, C., Mende, F., Luft, F. C., Schedl, A., Haller, H., and Kurzchalia, T. V. (2001). Loss of caveolae, vascular dysfunction, and pulmonary defects in caveolin-1 gene-disrupted mice. *Science* **293,** 2449–2452.

Dreja, K., Voldstedlund, M., Vinten, J., Tranum-Jensen, J., Hellstrand, P., and Sward, K. (2002). Cholesterol depletion disrupts caveolae and differentially impairs agonist-induced arterial contraction. *Arterioscler. Thromb. Vasc. Biol.* **22,** 1267–1272.

Everson, W. V. and Smart, E. J. (2001). Influence of caveolin, cholesterol, and lipoproteins on nitric oxide synthase. implications for vascular disease. *Trends Cardiovasc. Med.* **11,** 246–250.

Feron, O., Dessy, C., Moniotte, S., Desager, J. P., and Balligand, J. L. (1999). Hypercholesterolemia decreases nitric oxide production by promoting the interaction of caveolin and endothelial nitric oxide synthase. *J. Clin. Invest.* **103,** 897–905.

Fielding, P. E. and Fielding, C. J. (1995). Plasma membrane caveolae mediate the efflux of cellular free cholesterol. *Biochemistry* **34,** 14288–14292.

Fielding, P. E. and Fielding, C. J. (1996). Intracellular transport of low density lipoprotein derived free cholesterol begins at clathrin-coated pits and terminates at cell surface caveolae. *Biochemistry* **35,** 14932–14938.

Frank, P. G., Galbiati, F., Volonte, D., Razani, B., Cohen, D. E., Marcel, Y. L., and Lisanti, M. P. (2001a). Influence of caveolin-1 on cellular cholesterol efflux mediated by high-density lipoproteins. *Am. J. Physiol.* **280,** 1204–1214.

Frank, P. G., Lee, H., Park, D. S., Tandon, N. N., Scherer, P. E., and Lisanti, M. P. (2004). Genetic ablation of caveolin-1 confers protection against atherosclerosis. *Arterioscler. Thromb. Vasc. Biol.* **24,** 98–105.

Frank, P. G., Marcel, Y. L., Connelly, M. A., Lublin, D. M., Franklin, V., Williams, D. L., and Lisanti, M. P. (2002). Stabilization of caveolin-1 by cellular cholesterol and scavenger receptor class B type I. *Biochemistry* **41,** 11931–11940.

Frank, P. G., Pedraza, A., Cohen, D. E., and Lisanti, M. P. (2001b). Adenovirus-mediated expression of caveolin-1 in mouse liver increases plasma high-density lipoprotein levels. *Biochemistry* **40,** 10892–10900.

Frank, P. G., Woodman, S. E., Park, D. S., and Lisanti, M. P. (2003). Caveolin, caveolae, and endothelial cell function. *Arterioscler. Thromb. Vasc. Biol.* **23,** 1161–1168.

Fry, D. L., Herderick, E. E., and Johnson, D. K. (1993). Local intimal-medial uptakes of 125I-albumin, 125I-LDL, and parenteral Evans blue dye protein complex along the aortas of normocholesterolemic minipigs as predictors of subsequent hypercholesterolemic atherogenesis. *Arterioscler. Thromb.* **13,** 1193–1204.

Fuster, V., Badimon, L., Badimon, J. J., and Chesebro, J. H. (1992). The pathogenesis of coronary artery disease and the acute coronary syndromes (2). *N. Engl. J. Med.* **326,** 310–318.

Galvez, B. G., Matias-Roman, S., Yanez-Mo, M., Vicente-Manzanares, M., Sanchez-Madrid, F., and Arroyo, A. G. (2004). Caveolae are a novel pathway for membrane-type 1 matrix metalloproteinase traffic in human endothelial cells. *Mol. Biol. Cell* **15,** 678–687.

Garcia-Cardena, G., Fan, R., Stern, D., Liu, J., and Sessa, W. C. (1996). Endothelial nitric oxide synthase is regulated by tyrosine phosphorylation and interacts with caveolin-1. *J. Biol. Chem.* **271,** 27237–27240.

Garg, U. C. and Hassid, A. (1989). Nitric oxide-generating vasodilators and 8-bromo-cyclic guanosine monophosphate inhibit mitogenesis and proliferation of cultured rat vascular smooth muscle cells. *J. Clin. Invest.* **83,** 1774–1777.

Gargalovic, P. and Dory, L. (2001). Caveolin-1 and caveolin-2 expression in mouse macrophages. High density lipoprotein 3-stimulated secretion and a lack of significant subcellular co-localization. *J. Biol. Chem.* **276,** 26164–26170.

Gargalovic, P. and Dory, L. (2003a). Caveolins and macrophage lipid metabolism. *J. Lipid Res.* **44,** 11–21.

Gargalovic, P. and Dory, L. (2003b). Cellular apoptosis is associated with increased caveolin-1 expression in macrophages. *J. Lipid Res.* **44,** 1622–1632.

Ghitescu, L., Fixman, A., Simonescu, M., and Simonescu, N. (1986). Specific binding sites for albumin restricted to plasmalemmal vesicles of continuous capillary endothelium: Receptor-mediated transcytosis. *J. Cell Biol.* **102,** 1304–1311.

Gordon, D. J. and Rifkind, B. M. (1989). High-density lipoprotein—The clinical implications of recent studies. *N. Engl. J. Med.* **321,** 1311–1316.

Gordon, T., Castelli, W. P., Hjortland, M. C., Kannel, W. B., and Dawber, T. (1977). High density lipoprotein as a protective factor against coronary heart disease. The Framingham Study. *Am. J. Med.* **62,** 707–714.

Graf, G. A., Connell, P. M., Van der Westhuyzen, D. R., and Smart, E. J. (1999). The class B, type I scavenger receptor promotes the selective uptake of high density lipoprotein cholesterol ethers into caveolae. *J Biol. Chem.* **274,** 12043–12048.

Griendling, K. K., Harrison, D. G., and Alexander, R. W. (2001). Molecular and cellular biology of blood vessels. In: *Hurst's The Heart* (Fuster, V., Alexander, R. W., and R. A., O. R., Eds.), Vol. 1, pp. 127–145. McGraw-Hill, New York.

Griffoni, C., Spisni, E., Santi, S., Riccio, M., Guarnieri, T., and Tomasi, V. (2000). Knockdown of caveolin-1 by antisense oligonucleotides impairs angiogenesis *in vitro* and *in vivo*. *Biochem. Biophys. Res. Commun.* **276,** 756–761.

Hanahan, D. and Folkman, J. (1996). Patterns and emerging mechanisms of the angiogenic switch during tumorigenesis. *Cell* **86,** 353–364.

Harrison, D. G. and Ohara, Y. (1995). Physiologic consequences of increased vascular oxidant stresses in hypercholesterolemia and atherosclerosis: Implications for impaired vasomotion. *Am. J. Cardiol.* **75,** 75B–81B.

Hassan, G. S., Jasmin, J.-F., Schubert, W., Frank, P. G., and Lisanti, M. P. (2004). Caveolin-1 deficiency stimulates neointima formation during vascular injury. *Biochemistry* **43,** 8312–8321.

Huh, H. Y., Pearce, S. F., Yesner, L. M., Schindler, J. L., and Silverstein, R. L. (1996). Regulated expression of CD36 during monocyte-to-macrophage differentiation: Potential role of CD36 in foam cell formation. *Blood* **87,** 2020–2028.

Impagnatiello, M. A., Weitzer, S., Gannon, G., Compagni, A., Cotten, M., and Christofori, G. (2001). Mammalian sprouty-1 and -2 are membrane-anchored phosphoprotein inhibitors of growth factor signaling in endothelial cells. *J. Cell Biol.* **152,** 1087–1098.

Isshiki, M., Ando, J., Yamamoto, K., Fujita, T., Ying, Y., and Anderson, R. G. (2002a). Sites of Ca(2+) wave initiation move with caveolae to the trailing edge of migrating cells. *J. Cell Sci.* **115,** 475–484.

Isshiki, M., Ying, Y. S., Fujita, T., and Anderson, R. G. (2002b). A molecular sensor detects signal transduction from caveolae in living cells. *J. Biol. Chem.* **277,** 43389–43398.

Ju, H., Zou, R., Venema, V. J., and Venema, R. C. (1997). Direct interaction of endothelial nitric-oxide synthase and caveolin-1 inhibits synthase activity. *J. Biol. Chem.* **272,** 18522–18525.

Kariya, K., Kawahara, Y., Araki, S., Fukuzaki, H., and Takai, Y. (1989). Antiproliferative action of cyclic GMP-elevating vasodilators in cultured rabbit aortic smooth muscle cells. *Atherosclerosis* **80,** 143–147.

Kawabe, J., Okumura, S., Lee, M. C., Sadoshima, J., and Ishikawa, Y. (2004). Translocation of caveolin regulates stretch-induced ERK activity in vascular smooth muscle cells. *Am. J. Physiol. Heart Circ. Physiol.* **286,** H1845–H1852.

Kawashima, S., Yamashita, T., Ozaki, M., Ohashi, Y., Azumi, H., Inoue, N., Hirata, K., Hayashi, Y., Itoh, H., and Yokoyama, M. (2001). Endothelial NO synthase overexpression inhibits lesion formation in mouse model of vascular remodeling. *Arterioscler. Thromb. Vasc. Biol.* **21,** 201–207.

Kim, M.-J., Dawes, J., and Jessup, W. (1994). Transendothelial transport of modified low-density lipoproteins. *Atherosclerosis* **108,** 5–17.

Kumar, A., and Lindner, V. (1997). Remodeling with neointima formation in the mouse carotid artery after cessation of blood flow. *Arterioscler. Thromb. Vasc. Biol.* **17,** 2238–2244.

Labrecque, L., Royal, I., Surprenant, D. S., Patterson, C., Gingras, D., and Beliveau, R. (2003). Regulation of vascular endothelial growth factor receptor-2 activity by caveolin-1 and plasma membrane cholesterol. *Mol. Biol. Cell* **14,** 334–347.

Li, Q. Q., Tsujita, M., and Yokoyama, S. (1997). Selective down-regulation by protein kinase C inhibitors of apolipoprotein-mediated cellular cholesterol efflux in macrophages. *Biochemistry* **36,** 12045–12052.

Lisanti, M. P., Scherer, P., Tang, Z.-L., and Sargiacomo, M. (1994a). Caveolae, caveolin and caveolin-rich membrane domains: A signalling hypothesis. *Trends Cell Biol.* **4,** 231–235.

Lisanti, M. P., Scherer, P. E., Vidugiriene, J., Tang, Z., Hermanowski-Vosatka, A., Tu, Y. H., Cook, R. F., and Sargiacomo, M. (1994b). Characterization of caveolin-rich membrane domains isolated from an endothelial-rich source: Implications for human disease. *J. Cell Biol.* **126,** 111–126.

Liu, J., Wang, X. B., Park, D. S., and Lisanti, M. P. (2002). Caveolin-1 expression enhances endothelial capillary tubule formation. *J. Biol. Chem.* **277,** 10661–10668.

Lohn, M., Furstenau, M., Sagach, V., Elger, M., Schulze, W., Luft, F. C., Haller, H., and Gollasch, M. (2000). Ignition of calcium sparks in arterial and cardiac muscle through caveolae. *Circ. Res.* **87,** 1034–1039.

Lusis, A. J. (2000). Atherosclerosis. *Nature* **407,** 233–241.

Majno, G. and Palade, G. E. (1961). Studies on inflammation. 1. The effect of histamine and serotonin on vascular permeability: An electron microscopic study. *J. Biophys. Biochem. Cytol.* **11,** 571–605.

Matveev, S., Uittenbogaard, A., van Der Westhuyzen, D., and Smart, E. J. (2001). Caveolin-1 negatively regulates SR-BI mediated selective uptake of high-density lipoprotein-derived cholesteryl ester. *Eur. J. Biochem.* **268,** 5609–5616.

Matveev, S., van der Westhuyzen, D. R., and Smart, E. J. (1999). Co-expression of scavenger receptor-BI and caveolin-1 is associated with enhanced selective cholesteryl ester uptake in THP-1 macrophages. *J. Lipid Res.* **40,** 1647–1654.

McGill, H. C., Jr., Geer, J. C., and Holman, R. L. (1957). Sites of vascular vulnerability in dogs demonstrated by Evans blue. *AMA Arch. Pathol.* **64,** 303–311.

Mendez, A. J., Lin, G., Wade, D. P., Lawn, R. M., and Oram, J. F. (2001). Membrane lipid domains distinct from cholesterol/sphingomyelin-rich rafts are involved in the ABCA1-mediated lipid secretory pathway. *J. Biol. Chem.* **276,** 3158–3166.

Michel, J. B., Feron, O., Sacks, D., and Michel, T. (1997). Reciprocal regulation of endothelial nitric-oxide synthase by Ca^{2+}-calmodulin and caveolin. *J. Biol. Chem.* **272,** 15583–15586.

Miller, N. E., Thelle, D. S., Forde, O. H., and Mjos, O. D. (1977). The Tromso heart-study. High-density lipoprotein and coronary heart disease: A prospective case-control study. *Lancet* **1,** 965–968.

Minshall, R. D., Tiruppathi, C., Vogel, S. M., and Malik, A. B. (2002). Vesicle formation and trafficking in endothelial cells and regulation of endothelial barrier function. *Histochem. Cell Biol.* **117,** 105–112.

Minshall, R. D., Tiruppathi, C., Vogel, S. M., Niles, W. D., Gilchrist, A., Hamm, H. E., and Malik, A. B. (2000). Endothelial cell-surface gp60 activates vesicle formation and trafficking via G(i)-coupled Src kinase signaling pathway. *J. Cell Biol.* **150,** 1057–1070.

Ogawa, K., Imai, M., Ogawa, T., Tsukamoto, Y., and Sasaki, F. (2001). Caveolar and intercellular channels provide major transport pathways of macromolecules across vascular endothelial cells. *Anat. Rec.* **264,** 32–42.

Oh, P., McIntosh, D. P., and Schnitzer, J. E. (1998). Dynamin at the neck of caveolae mediates their budding to form transport vesicles by GTP-driven fission from the plasma membrane of endothelium. *J. Cell Biol.* **141,** 101–114.

Packham, M. A., Rowsell, H. C., Jorgensen, L., and Mustard, J. F. (1967). Localized protein accumulation in the wall of the aorta. *Exp. Mol. Pathol.* **7,** 214–232.

Peterson, T. E., Guicciardi, M. E., Gulati, R., Kleppe, L. S., Mueske, C. S., Mookadam, M., Sowa, G., Gores, G. J., Sessa, W. C., and Simari, R. D. (2003). Caveolin-1 can regulate vascular smooth muscle cell fate by switching platelet-derived growth factor signaling from a proliferative to an apoptotic pathway. *Arterioscler. Thromb. Vasc. Biol.* **23,** 1521–1527.

Peterson, T. E., Kleppe, L. S., Caplice, N. M., Pan, S., Mueske, C. S., and Simari, R. D. (1999). The regulation of caveolin expression and localization by serum and heparin in vascular smooth muscle cells. *Biochem. Biophys. Res. Commun.* **265,** 722–727.

Predescu, D. and Palade, G. E. (1993). Plasmalemmal vesicles represent the large pore system of continuous microvascular endothelium. *Am. J. Physiol.* **265,** H725–H733.

Predescu, S. A., Predescu, D. N., and Palade, G. E. (2001). Endothelial transcytotic machinery involves supramolecular protein-lipid complexes. *Mol. Biol. Cell* **12,** 1019–1033.

Razani, B., Combs, T. P., Wang, X. B., Frank, P. G., Park, D. S., Russell, R. G., Li, M., Tang, B., Jelicks, L. A., Scherer, P. E., and Lisanti, M. P. (2001a). Caveolin-1 deficient mice are lean, resistant to diet-induced obesity, and show hyper-triglyceridemia with adipocyte abnormalities. *J. Biol. Chem.* **277,** 8635–8647.

Razani, B., Engelman, J. A., Wang, X. B., Schubert, W., Zhang, X. L., Marks, C. B., Macaluso, F., Russell, R. G., Li, M., Pestell, R. G., Di Vizio, D., Hou, H., Jr., Kneitz, B., Lagaud, G., Christ, G. J., Edelmann, W., and Lisanti, M. P. (2001b). Caveolin-1 null mice are viable but show evidence of hyperproliferative and vascular abnormalities. *J. Biol. Chem.* **276,** 38121–38138.

Razani, B., Park, D. S., Miyanaga, Y., Ghatpande, A., Cohen, J., Wang, X. B., Scherer, P. E., Evans, T., and Lisanti, M. P. (2002). Molecular cloning and developmental expression of the caveolin gene family in the amphibian Xenopus laevis. *Biochemistry* **41,** 7914–7924.

Ross, R. (1993). The pathogenesis of atherosclerosis: A perspective for the 1990s. *Nature* **362,** 801–809.

Ross, R. (1999). Atherosclerosis is an inflammatory disease. *Am. Heart J.* **138,** S419–S420.

Schnitzer, J. E., Allard, J., and Oh, P. (1995a). NEM inhibits transcytosis, endocytosis, and capillary permeability: Implication of caveolae fusion in endothelia. *Am. J. Physiol.* **268,** H48–H55.

Schnitzer, J. E., Liu, J., and Oh, P. (1995b). Endothelial caveolae have the molecular transport machinery for vesicle budding, docking, and fusion including VAMP, NSF, SNAP, annexins, and GTPases. *J. Biol. Chem.* **270,** 14399–14404.

Schnitzer, J. E., Oh, P., Pinney, E., and Allard, J. (1994). Filipin-sensitive caveolae-mediated transport in endothelium: Reduced transcytosis, scavenger endocytosis, and capillary permeability of select macromolecules. *J. Cell Biol.* **127,** 1217–1232.

Schubert, W., Frank, P. G., Razani, B., Park, D. S., Chow, C. W., and Lisanti, M. P. (2001). Caveolae-deficient endothelial cells show defects in the uptake and transport of albumin *in vivo*. *J. Biol. Chem.* **276,** 48619–48622.

Schubert, W., Frank, P. G., Woodman, S. E., Hyogo, H., Cohen, D. E., Chow, C. W., and Lisanti, M. P. (2002). Microvascular hyperpermeability in caveolin-1 (–/–) knock-out mice. Treatment with a specific nitric-oxide synthase inhibitor, L-NAME, restores normal microvascular permeability in Cav-1 null mice. *J. Biol. Chem.* **277,** 40091–40098.

Shaul, P. W. (2003). Endothelial nitric oxide synthase, caveolae and the development of atherosclerosis. *J. Physiol.* **547,** 21–33.

Shaul, P. W., Smart, E. J., Robinson, L. J., German, Z., Yuhanna, I. S., Ying, Y. S., Anderson, R. G. W., and Michel, T. (1996). Acylation targets endothelial nitric-oxide synthase to plasmalemmal caveolae. *J. Biol. Chem.* **271,** 6518–6522.

Simionescu, M., Gafencu, A., and Antohe, F. (2002). Transcytosis of plasma macromolecules in endothelial cells: a cell biological survey. *Microsc. Res. Tech.* **57,** 269–288.

Smart, E. J., Graf, G. A., McNiven, M. A., Sessa, W. C., Engelmen, J. A., Scherer, P. E., Okamoto, T., and Lisanti, M. P. (1999). Caveolins, liquid-ordered domains, and signal transduction. *Mol. Cell. Biol.* **19,** 7289–7304.

Smart, E. J., Ying, Y.-s., Donzell, W. C., and Anderson, R. G. W. (1996). A role for caveolin in transport of cholesterol from endoplasmic reticulum to plasma membrane. *J. Biol. Chem.* **271,** 29427–29435.

Somer, J. B. and Schwartz, C. J. (1971). Focal 3 H-cholesterol uptake in the pig aorta. *Atherosclerosis* **13,** 293–304.

Somlyo, A. (1980). Ultrastructure of the microvascular wall: Functional correlations. In: *Handbook of Physiology, the Cardiovascular System, Vascular Smooth Muscle*, Vol. II, pp. 33–67. American Physiological Society, Bethesda, MD. Oxford University Press.

Spisni, E., Griffoni, C., Santi, S., Riccio, M., Marulli, R., Bartolini, G., Toni, M., Ullrich, V., and Tomasi, V. (2001). Colocalization prostacyclin (PGI2) synthase—caveolin-1 in endothelial cells and new roles for PGI2 in angiogenesis. *Exp. Cell Res.* **266,** 31–43.

Stahl, A. and Mueller, B. M. (1995). The urokinase-type plasminogen activator receptor, a GPI-linked protein, is localized in caveolae. *J. Cell Biol.* **129,** 335–344.

Takahashi, M., Ikeda, U., Masuyama, J., Funayama, H., Kano, S., and Shimada, K. (1996). Nitric oxide attenuates adhesion molecule expression in human endothelial cells. *Cytokine* **8,** 817–821.

Thyberg, J. (2003). Caveolin-1 and caveolae act as regulators of mitogenic signaling in vascular smooth muscle cells. *Arterioscler. Thromb. Vasc. Biol.* **23,** 1481–1483.

Thyberg, J., Roy, J., Tran, P. K., Blomgren, K., Dumitrescu, A., and Hedin, U. (1997). Expression of caveolae on the surface of rat arterial smooth muscle cells is dependent on the phenotypic state of the cells. *Lab. Invest.* **77,** 93–101.

Tiruppathi, C., Song, W., Bergenfeldt, M., Sass, P., and Malik, A. B. (1997). Gp60 activation mediates albumin transcytosis in endothelial cells by tyrosine kinase-dependent pathway. *J. Biol. Chem.* **272,** 25968–25975.

Uittenbogaard, A., Shaul, P. W., Yuhanna, I. S., Blair, A., and Smart, E. J. (2000). High density lipoprotein prevents oxidized low density lipoprotein-induced inhibition of endothelial nitric-oxide synthase localization and activation in caveolae. *J. Biol. Chem.* **275,** 11278–11283.

van Deurs, B., Roepstorff, K., Hommelgaard, A. M., and Sandvig, K. (2003). Caveolae: Anchored, multifunctional platforms in the lipid ocean. *Trends Cell Biol.* **13,** 92–100.

Vane, J. R., Anggard, E. E., and Botting, R. M. (1990). Regulatory functions of the vascular endothelium. *N. Engl. J. Med.* **323,** 27–36.

Vasile, E., Qu, H., Dvorak, H. F., and Dvorak, A. M. (1999). Caveolae and vesiculo-vacuolar organelles in bovine capillary endothelial cells cultured with VPF/VEGF on floating Matrigel-collagen gels. *J. Histochem. Cytochem.* **47,** 159–167.

Vasile, E., Simionescu, M., and Simionescu, N. (1983). Visualization of the binding, endocytosis, and transcytosis of low density lipoprotein in the arterial endothelium in situ. *J. Cell Biol.* **96,** 1677–1689.

Wang, L., Connelly, M. A., Ostermeyer, A. G., Chen, H. H., Williams, D. L., and Brown, D. A. (2003). Caveolin-1 does not affect SR-BI-mediated cholesterol efflux or selective uptake of cholesteryl ester in two cell lines. *J. Lipid Res.* **44,** 807–815.

Wickstrom, S. A., Alitalo, K., and Keski-Oja, J. (2002). Endostatin associates with integrin alpha5beta1 and caveolin-1, and activates Src via a tyrosyl phosphatase-dependent pathway in human endothelial cells. *Cancer Res.* **62,** 5580–5589.

Wickstrom, S. A., Alitalo, K., and Keski-Oja, J. (2003). Endostatin associates with lipid rafts and induces reorganization of the actin cytoskeleton via down-regulation of RhoA activity. *J. Biol. Chem.* **278,** 37895–37901.

Wilson, P. W., Anderson, K. M., Harris, T., Kannel, W. B., and Castelli, W. P. (1994). Determinants of change in total cholesterol and HDL-C with age: The Framingham Study. *J. Gerontol.* **49,** M252–M257.

Woodman, S. E., Ashton, A. W., Schubert, W., Lee, H., Williams, T. M., Medina, F. A., Wyckoff, J. B., Combs, T. P., and Lisanti, M. P. (2003). Caveolin-1 knockout mice show an impaired angiogenic response to exogenous stimuli. *Am. J. Pathol.* **162,** 2059–2068.

Woodman, S. E., Cheung, M. W., Tarr, M., North, A. C., Schubert, W., Lagaud, G., Marks, C. B., Russell, R. G., Hassan, G. S., Factor, S. M., Christ, G. J., and Lisanti, M. P. (2004). Urogenital alterations in aged male caveolin-1 knockout mice. *J. Urol.* **171,** 950–957.

Zeidan, A., Broman, J., Hellstrand, P., and Sward, K. (2003). Cholesterol dependence of vascular ERK1/2 activation and growth in response to stretch: Role of endothelin-1. *Arterioscler. Thromb. Vasc. Biol.* **23,** 1528–1534.

Zhu, Y., Liao, H. L., Wang, N., Yuan, Y., Ma, K. S., Verna, L., and Stemerman, M. B. (2000). Lipoprotein promotes caveolin-1 and Ras translocation to caveolae: Role of cholesterol in endothelial signaling. *Arterioscler. Thromb. Vasc. Biol.* **20,** 2465–2470.

Chapter 11

Caveolin Proteins in Cardiopulmonary Disease and Lung Cancers

Jean-François Jasmin,[1] Philippe G. Frank,[1,2] and Michael P. Lisanti[1]

[1]*Department of Molecular Pharmacology, Albert Einstein College of Medicine, Bronx, New York 10461;*
[2]*Department of Urology, Albert Einstein College of Medicine, Bronx, New York 10461*

I. CAVEOLIN PROTEINS AND THE CARDIOPULMONARY SYSTEM

Three caveolin (Cav) isoforms have been identified in the heart. Cav-3 is found at the myocyte sarcolemmal membrane and within intercalated disks, while Cav-1 and Cav-2 are expressed in the vascular endothelium and cardiac fibroblasts (Feron *et al.*, 1996; Woodman *et al.*, 2002; Cohen *et al.*, 2003).

Caveolin proteins are also highly expressed in the lungs (Lisanti *et al.*, 1994). Although all three caveolin isoforms have been identified in the lungs (Kawabe *et al.*, 2001), Cav-1 still remains the most extensively studied isoform. Cav-1 is found in many cell types of the lungs, such as endothelial cells, smooth muscle cells, fibroblasts, bronchial epithelial cells, and alveolar type I pneumocytes (Lisanti *et al.*, 1994; Newman *et al.*, 1999; Racine *et al.*, 1999). However, Cav-1 is poorly expressed in the cuboidal alveolar type II cells (Newman *et al.*, 1999). Interestingly, freshly isolated alveolar type II cells lack Cav-1 expression but gradually express Cav-1 as they acquire an alveolar type I-like cell phenotype (Campbell *et al.*, 1999).

Advances in Molecular and Cell Biology, Vol. 36

ISBN: 0-444-51500-3
DOI: 10.1016/S0079-6603(05)36011-9

A. Caveolins and Cardiopulmonary Development

Cav-1 might play an important role in the development of the lungs, as its alpha isoform can be detected in primitive pulmonary blood vessels as early as the 10th embryonic day (Ramirez *et al.*, 2002). Two Cav-1 mRNAs, a full-length and a 5′ variant, which translate into the alpha and beta isoforms, respectively, can also be detected at embryonic day 12 (Ramirez *et al.*, 2002). Cav-1 alpha has been proposed as an early marker for lung vasculogenesis, as it is expressed in lung fetal endothelial cells but not in lung fetal alveolar type I cells (Ramirez *et al.*, 2002). Cell-type specific expression of the two isoforms has recently been suggested, with Cav-1 alpha being expressed in endothelial cells and Cav-1 beta being expressed in alveolar type I cells (Kogo *et al.*, 2004). However, the presence of Cav-1 alpha in adult alveolar type I cells has previously been shown by other groups (Ramirez *et al.*, 2002). These differences in expression of Cav-1 alpha have recently been ascribed to differential transcriptional regulation of Cav-1 in lung epithelial and endothelial cells (Kathuria *et al.*, 2004). Indeed, Cav-1 expression was shown to be regulated by an ETS *cis*-element in a lung epithelial cell line, but not in a lung endothelial cell line (Kathuria *et al.*, 2004).

Similarly, caveolin proteins have been implicated in the development of the heart, as the Cav-3 mRNA was shown to be detectable in the wall of the *bulbus cordis* as early as embryonic day 11 and in the entire myocardium at embryonic day 15 (Biederer *et al.*, 2000).

B. Cardiopulmonary Phenotypes of Caveolin-Deficient Mice

The generation of caveolin-deficient mice has provided dramatic support for the hypothesis that caveolin proteins play an important role in the heart. For example, Cav-3-deficient (Cav-$3^{-/-}$) mice lack cardiomyocyte caveolae and develop a progressive form of cardiomyopathy (Woodman *et al.*, 2002). The hearts of Cav-3 null mice showed eccentric hypertrophy of the left ventricle, a 20% decrease in fractional shortening, myocyte hypertrophy, interstitial perivascular fibrosis, as well as hyperactivation of the p42/44 MAP kinase cascade (Fig. 1) (Woodman *et al.*, 2002).

Interestingly, Cav-$1^{-/-}$ mice also develop cardiac hypertrophy (Fig. 2) (Cohen *et al.*, 2003). Indeed, the hearts of Cav-$1^{-/-}$ mice showed right and left atrium dilation, right ventricular dilation, concentric left ventricular hypertrophy, decreased left ventricular fractional shortening, myocyte hypertrophy, and interstitial/perivascular fibrosis, with hyperactivation of the p42/44 MAP kinase cascade (Cohen *et al.*, 2003).

As expected, mice deficient in both Cav-1 and Cav-3 develop a more severe cardiomyopathic phenotype (Fig. 3) (Park *et al.*, 2002). Indeed, hearts of Cav-1/3 double knockout mice lack all morphological identifiable caveolae and develop right ventricular hypertrophy, concentric left ventricular hypertrophy, myocyte hypertrophy, and interstitial fibrosis, as early as 2 months of age (Park *et al.*, 2002). The left

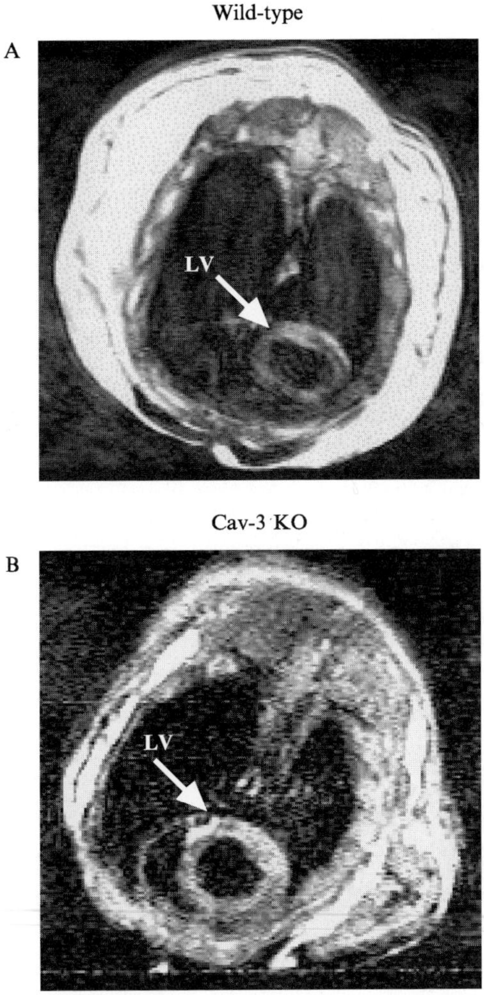

Figure 1. Cav-3$^{-/-}$ mice exhibit left ventricular hypertrophy (lower panel), as compared to wild type mice (upper panel). Modified from Woodman *et al.* (2002).

ventricular hypertrophy observed in the Cav-1/3 double knockout mice was clearly more pronounced than the one reported in age-matched Cav-1$^{-/-}$ or Cav-3$^{-/-}$ mice, thus suggesting that the dual ablation of both Cav-1 and Cav-3 exerts synergistic effects (Park *et al.*, 2002).

In contrast, the hearts of Cav-2$^{-/-}$ mice did not show any ventricular hypertrophy, interstitial fibrosis, or hyperactivation of the p42/44 MAP kinase pathway (Cohen *et al.*, 2003).

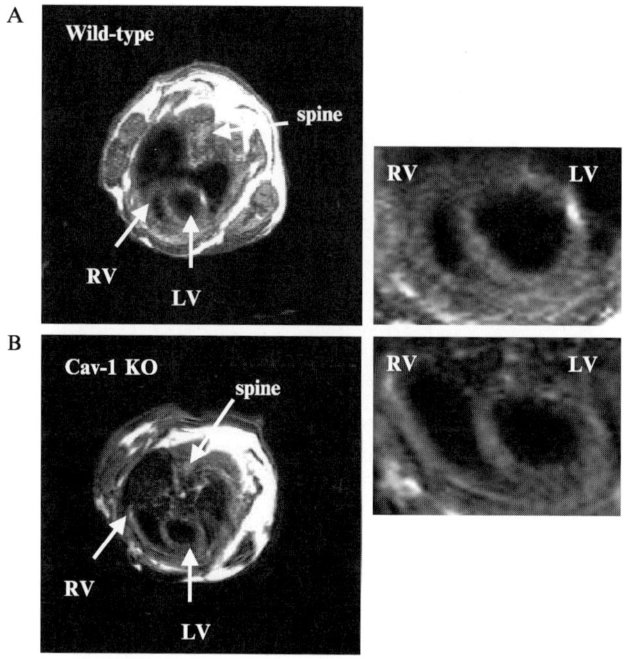

Figure 2. Cav-$1^{-/-}$ mice (B) exhibit right ventricular dilation and concentric left ventricular hypertrophy, as compared to wild-type mice (A). Modified from Cohen *et al.* (2003).

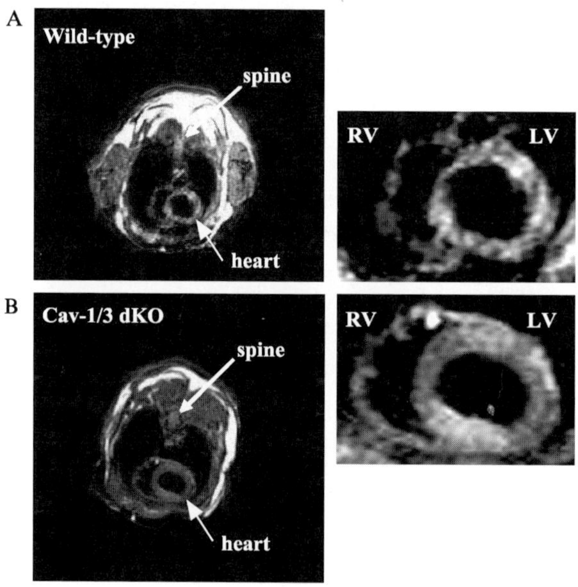

Figure 3. Cav-1/3 dKO mice (B) exhibit right ventricular hypertrophy, as well as concentric left ventricular hypertrophy, as compared to wild-type mice (A). Modified from Park *et al.* (2002).

Similarly, the generation of caveolin-deficient mice strongly supports the crucial roles of caveolin proteins in the lungs. Examination of the lung parenchyma of Cav-1$^{-/-}$ mice showed abnormalities in pulmonary structure and function, as demonstrated by thickening of the alveolar septa, hypercellularity, increased cellular staining for Flk-1, a marker of non-differentiated endothelial cells and hematopoietic progenitors, as well as interstitial fibrosis (Fig. 4) (Drab *et al.*, 2001; Razani *et al.*, 2001). These pulmonary abnormalities consequently resulted in markedly reduced exercise tolerance (Drab *et al.*, 2001; Razani *et al.*, 2001). However, since Cav-1$^{-/-}$ mice also have a near-complete ablation of Cav-2 expression, the functional implications of Cav-1 in those mice remained unclear.

Interestingly, Cav-2$^{-/-}$ mice showed the same thickening of the alveolar septa, hypercellularity, endothelial cell proliferation, and exercise intolerance as Cav-1$^{-/-}$ mice (Fig. 4) (Razani *et al.*, 2002). However, Cav-2$^{-/-}$ mice did not show any significant decreases in Cav-1 expression, defects in caveolar formation, abnormal vascular responses, or altered lipid homeostasis, suggesting a selective role for Cav-2 in lung function (Razani *et al.*, 2002).

Interestingly, it was recently shown that Cav-1$^{-/-}$ mice also develop pulmonary hypertension (PH), with marked increases in pulmonary artery pressure and lung weight, as well as the development of right ventricular hypertrophy (Zhao *et al.*, 2002). The role of Cav-2 in the development of PH is still unknown. Similarly, the pulmonary phenotype of the Cav-3$^{-/-}$ mice still remains to be explored.

C. Caveolins and Cardiopulmonary Endocytosis, Vesicular Trafficking, and Signal Transduction

The roles of caveolins in endocytosis, vesicular trafficking, and signal transduction are described in detail elsewhere in this volume and, therefore, only information relevant to the heart and lungs is discussed in the present chapter.

The isolation of caveolin-rich membrane domains from mouse lung tissue first identified several known proteins as caveolar components, such as CD36 (a glycoprotein receptor for oxidized-LDL), RAGE (receptor for advanced glycosylation end-products), glycosyl-phosphatidylinositol (GPI)-linked proteins, protein kinase C (PKC)-alpha, extracellular signal-related kinase-2 (ERK-2), heterotrimeric G-protein subunits, albumin, surfactant protein A, annexin II, myosin II, gelsolin, and actin (Lisanti *et al.*, 1994). These results suggest that lung caveolae may function in (1) the transcytosis of macromolecules (albumin, low-density lipoprotein [LDL], oxidized-LDL, advanced glycated end products [AGEs]), (2) the uptake of small molecules, (3) the compartmentalization of several signaling molecules (G-protein subunits, PKC-alpha, ERK-2), and (4) interactions with the actin-based cytoskeleton (annexin II, gelsolin, myosin II) (Lisanti *et al.*, 1994).

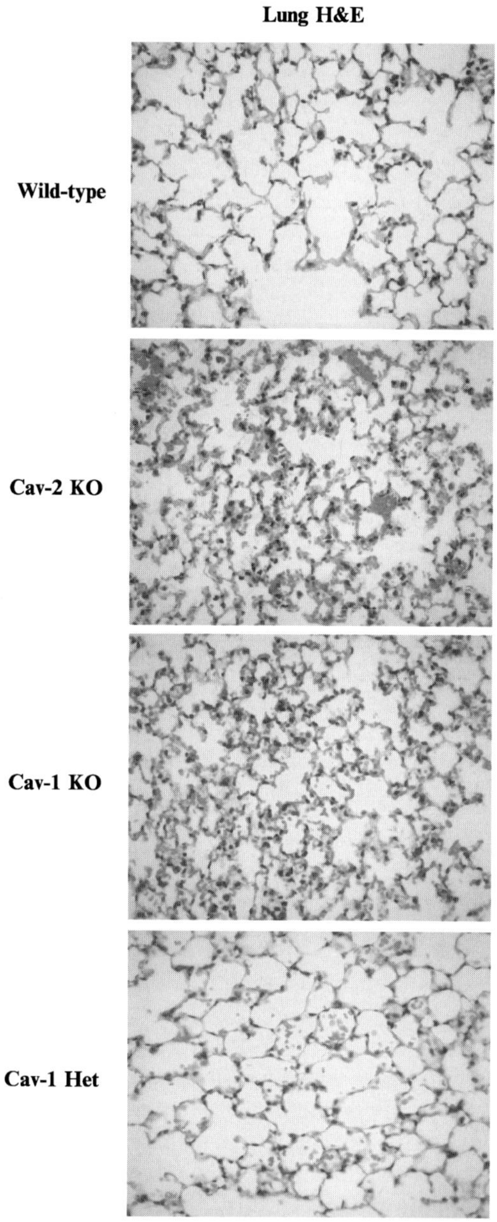

Figure 4. Cav-1$^{-/-}$ and Cav-2$^{-/-}$ mice show lung abnormalities characterized by hypercellularity and thickening of the alveolar septa, as compared to wild-type and Cav-1$^{-/+}$ heterozygous mice. Modified from Razani *et al.* (2002). (See Color Insert.)

1. Caveolins in Endocytosis and Vesicular Trafficking

The identification of several proteins implicated in vesicle formation, docking, and fusion in rat lung caveolae provides further concrete evidence of a role for caveolae in endocytosis and transcytosis (Schnitzer *et al.*, 1995a). These caveolae-associated proteins included vesicle-associated membrane protein-2 (VAMP-2), the N-ethylmaleimide (NEM)-sensitive fusion protein (NSF), and the soluble NSF attachment protein (SNAP).

Early studies showed that gold-conjugated albumin endocytosis was restricted to uncoated pits and plasmalemmal vesicles (caveolae) in mouse lung (Ghitescu *et al.*, 1986). A 60-kDa albumin-binding protein (gp60) was later found to localize within the caveolae of bovine pulmonary microvessel endothelial cells (Tiruppathi *et al.*, 1997). Activation of gp60 stimulated the phosphorylation of the Src-tyrosine kinase and stimulated the transcytosis of albumin across endothelial cells (Tiruppathi *et al.*, 1997). The generation of Cav-1$^{-/-}$ mice directly demonstrated the importance of lung caveolae in albumin endocytosis (Schubert *et al.*, 2001). Indeed, lung endothelial cells of Cav-1$^{-/-}$ mice showed dramatic defects in the uptake and transport of gold-conjugated albumin (Schubert *et al.*, 2001).

Cardiopulmonary caveolae have also been implicated in the transcapillary transport of water (Schnitzer and Oh, 1996). Indeed, lung endothelial caveolae have been shown to concentrate the mercury-sensitive transmembrane water channel aquaporin-1 (Schnitzer and Oh, 1996). Similarly, the presence of aquaporin-1 within caveolae isolated from atrial and ventricular cardiomyocytes has subsequently been shown (Page *et al.*, 1998).

2. Caveolins in Signal Transduction

As mentioned above, caveolar domains can also play a role in signal transduction via the compartmentalization of several signaling molecules (Table I). Different heterotrimeric G-protein subunits were shown to target discrete cell surface microdomains of mouse and rat lungs (Lisanti *et al.*, 1994; Oh and Schnitzer, 2001).

As mentioned above, PKC was also reported to be localized within caveolin-rich membrane domains isolated from mouse lung tissue (Lisanti *et al.*, 1994). Moreover, various PKC isoforms (alpha, delta, and epsilon) have been reported to translocate into cardiomyocyte caveolae upon agonist stimulation (Rybin *et al.*, 1999).

An association between the signal transducer and activator of transcription-3 (STAT3) and Cav-1 was shown in alveolar type II-like cells, human lung fibroblasts, and bovine pulmonary endothelial cells (Shah *et al.*, 2002). Interestingly, hyperactivation of the STAT3 signaling cascade in Cav-1$^{-/-}$ and Cav-2$^{-/-}$ mouse lungs recently suggested that the caveolin proteins could act as negative regulators of STAT3 activity (Jasmin *et al.*, 2004).

The pulmonary lipid phosphate phosphohydrolase (LPP), which converts phosphatidic acid into diacylglycerol, was also shown to localize within the caveolae/lipid

Table I
Receptors and Other Signaling Proteins Enriched in Cardiac and Pulmonary Caveolar Domains

Signaling proteins	Tissue	Reference
Adenosine receptors	Heart	Lasley *et al.*, 2000
Atrial natriuretic peptide type B receptor	Heart	Doyle *et al.*, 1997
Beta(2)-adrenergic receptors	Heart	Rybin *et al.*, 2000; Xiang *et al.*, 2002
Ca^{2+}-ATPase	Heart and lung	Hammes *et al.*, 1998; Schnitzer *et al.*, 1995b
Caspase-3	Heart	Oxhorn and Buxton, 2003
Epidermal growth factor receptor	Heart	Liu *et al.*, 2003
eNOS	Heart and lung	Feron *et al.*, 1996; Garcia-Cardena *et al.*, 1996a
Extracellular signal-related kinase-2	Lung	Lisanti *et al.*, 1994
Heterotrimeric G-protein subunits	Lung	Lisanti *et al.*, 1994; Oh and Schnitzer, 2001
Inositol triphosphate receptors	Lung	Schnitzer *et al.*, 1995b
Lipid phosphate phosphohydrolase	Lung	Nanjundan and Possmayer, 2001
Muscarinic receptors	Heart	Feron *et al.*, 1997
Na^{+}-Ca^{2+} exchanger	Heart	Bossuyt *et al.*, 2002
Na^{+}-K^{+}-ATPase	Heart	Liu *et al.*, 2003
Protein kinase C	Heart and lung	Lisanti *et al.*, 1994; Rybin *et al.*, 1999
Signal transducer and activator of transcription-3	Lung	Shah *et al.*, 2002
Tissue factor pathway inhibitor	Lung	Lupu *et al.*, 1997
Voltage-gated Na^{+} channels	Heart	Yarbrough *et al.*, 2002

rafts of rat lung tissue, isolated rat alveolar type II cells, and mouse lung epithelial cell lines (Nanjundan and Possmayer, 2001). The presence of the Ca^{2+}-ATPase and inositol triphosphate receptors in caveolae purified from rat lung endothelium also supports the idea that caveolar membrane domains function in lung signal transduction (Schnitzer *et al.*, 1995b). Similarly, the Ca^{2+}-ATPase, the Na^{+}-K^{+}-ATPase, the Na^{+}-Ca^{2+} exchanger, and the voltage-gated Na^{+} channels have been colocalized with Cav-3 in cardiomyocyte caveolae (Hammes *et al.*, 1998; Bossuyt *et al.*, 2002; Yarbrough *et al.*, 2002; Liu *et al.*, 2003).

The caveolar compartmentalization of several receptors further supports the hypothesis that caveolin proteins regulate cardiopulmonary signal transduction. For instance, upon agonist stimulation, muscarinic acetylcholine receptors translocate into caveolae, where they interact with Cav-3 in rat cardiomyocytes (Feron *et al.*, 1997). Conversely, the unstimulated adenosine receptor has been reported to be localized within cardiomyocytes caveolae and to translocate out of caveolae upon agonist stimulation (Lasley *et al.*, 2000).

Beta-adrenergic receptors have been shown to be enriched within the caveolar domains of neonatal rat ventricular myocytes (Ostrom *et al.*, 2000). However, differential targeting of cardiomyocyte beta-adrenergic receptor subtypes was subsequently reported (Rybin *et al.*, 2000; Xiang *et al.*, 2002). Caveolae purified from rat ventricular cardiomyocytes were highly enriched in the beta-2 subtype, whereas the beta-1 subtype was distributed between caveolae and the other cellular fractions (Rybin *et al.*, 2000). Furthermore, administration of filipin, a cholesterol-chelating reagent that disrupts lipid rafts and caveolae structure, selectively affected beta-2 subtype signaling in neonatal murine cardiomyocytes (Xiang *et al.*, 2002). Interestingly, caveolae appear to negatively regulate the activity of the beta-2 adrenergic receptors (Xiang *et al.*, 2002). Indeed, filipin-treated myocytes showed a larger and more prolonged increase in their contraction rate upon isoproterenol stimulation (Xiang *et al.*, 2002). The atrial natriuretic peptide (ANP) and its type B receptor (ANP-RB) have been shown to co-localize with Cav-3 in rat atrial myocytes (Doyle *et al.*, 1997). Similarly, the epidermal growth factor receptor (EGFR) is concentrated within rat cardiac caveolae (Liu *et al.*, 2003).

Purification of lung endothelial caveolae demonstrated the presence of endothelial nitric oxide synthase (eNOS), as well as its interactions with Cav-1 (Garcia-Cardena *et al.*, 1996a,b). Similarly, purification of caveolae from cardiac myocyte lysates demonstrated the presence of eNOS and highlighted its interactions with Cav-3 (Feron *et al.*, 1996). The direct interaction of eNOS with either Cav-1 or Cav-3 was later shown to inhibit eNOS activity (Garcia-Cardena *et al.*, 1996a; Feron *et al.*, 1998). Experiments with Cav-1$^{-/-}$ mice dramatically support the idea that Cav-1 functions as a natural endogenous inhibitor of eNOS activation (Drab *et al.*, 2001; Razani *et al.*, 2001; Schubert *et al.*, 2002). Indeed, Cav-1$^{-/-}$ mice show microvascular hyperpermeability, with increased para-cellular transport and defects in tight junctions morphology (Schubert *et al.*, 2002). Interestingly, treatment of Cav-1 null mice with nitro-L-arginine methyl ester (L-NAME), a NOS inhibitor, successfully reverses the microvascular hyperpermeability phenotype (Schubert *et al.*, 2002). Moreover, an increase in vascular flow and pressure has been shown to rapidly activate lung caveolar eNOS by releasing it from its inhibitory association with Cav-1, allowing it to associate with calmodulin (Rizzo *et al.*, 1998). Heat shock protein 90 (hsp90) was subsequently shown to facilitate the dissociation of eNOS from Cav-1, allowing its activation (Gratton *et al.*, 2000).

Cardiopulmonary caveolae have been implicated in the regulation of the anticoagulant properties of the endothelium, as well as the apoptotic process (Lupu *et al.*, 1997; Oxhorn and Buxton, 2003). Indeed, the tissue factor pathway inhibitor (TFPI), the main regulator of the pro-coagulant activity of tissue factor VIIa complex, has been shown to be enriched within rat lung caveolae (Lupu *et al.*, 1997). Caspase-3, a main apoptotic effector, was reported to be enriched within the caveolar fractions purified from cardiac endothelial cells (Oxhorn and Buxton, 2003).

D. Pathological Roles of Caveolins

The expression of caveolin proteins appears to be dysregulated in numerous cardiac and lung disorders such as hypertrophic cardiomyopathy, ischemic cardiomyopathy, dilated cardiomyopathy, lung fibrosis, pulmonary interstitial edema, pulmonary hypertension, and several lung cancers.

1. Caveolins in Hypertrophic Cardiomyopathy

Caveolin protein levels have been shown to be altered in different models of hypertrophic cardiomyopathy. More specifically, Cav-1 and Cav-3 expression levels were decreased in hearts of dogs with peri-nephritic hypertension-induced hypertrophic cardiomyopathy (Piech *et al.*, 2002). Similarly, Cav-1 and Cav-3 levels were shown to be decreased in the hypertrophic left ventricles of spontaneously hypertensive rats (SHR) (Fujita *et al.*, 2001; Piech *et al.*, 2003a). The decreased expression of Cav-1 and Cav-3, together with increased expression of hsp90, may contribute to the maintenance of NO production in the SHR hypertrophic hearts (Piech *et al.*, 2003b). The decrease in cardiac Cav-3 expression in SHR rats was conversely suggested to contribute to the development of hypertrophy by lowering the inhibition of growth factor signaling, such as via the alpha-1-adrenergic receptor, thus facilitating the development of cardiac hypertrophy (Fujita *et al.*, 2001). A decrease in cardiac Cav-1 expression was also shown to occur in the angiotensin II-induced hypertensive rat model (Kobayashi *et al.*, 2001). Interestingly, administration of an angiotensin type 1 receptor antagonist significantly increased cardiac Cav-1 expression and decreased the left ventricular weight in these hypertensive rats (Kobayashi *et al.*, 2001).

However, the exact mechanism(s) leading to decreased caveolin proteins in cardiac hypertrophy still remains to be elucidated. Elevated catecholamines could play an important role in such a decrease, since they have been associated with the downregulation of cardiac Cav-1 and Cav-3 (Oka *et al.*, 1997). Indeed, chronic beta-adrenergic receptor stimulation with isoproterenol reduces both Cav-1 and Cav-3 protein levels in mouse hearts (Oka *et al.*, 1997). Hypoxia may also regulate caveolin protein expression, as demonstrated in rabbits raised in a hypoxic environment, which show lower cardiac Cav-3 protein levels than the ones raised in a normoxic environment (Shi *et al.*, 2000).

The generation of mice overexpressing the dominant-negative Cav-3 P104L mutant strongly supports the importance of reduced Cav-3 in the development of hypertrophic cardiomyopathy (Ohsawa *et al.*, 2004). These mice lack cardiomyocyte caveolae and show thickening of the interventricular septum and the posterior wall of the left ventricle, as well as myocyte hypertrophy (Ohsawa *et al.*, 2004). Similarly, adenovirus-mediated overexpression of mutant Cav-3 has been shown to increase the hypertrophic response to endothelin-1, as well as the endothelin-induced phosphorylation of p42/44 MAP kinase in neonatal rat cardiomyocytes (Koga *et al.*, 2003). Accordingly, adenovirus-mediated overexpression of human wild-type Cav-3 has been shown to prevent phenylephrine and endothelin-induced increases in cell size, leucine incorporation, sarcomeric reorganization, re-expression of beta-myosin heavy chain, as well as phosphorylation of p42/44 MAP kinase in neonatal rat cardiomyocytes (Koga *et al.*, 2003). Interestingly, a Cav-3 T63S mutation was recently identified in patients with a form of familial hypertrophic cardiomyopathy, characterized by left ventricular concentric hypertrophy and elevated left ventricular end-diastolic pressure (Hayashi *et al.*, 2004). Also, this mutation was shown to reduce the cell surface expression of Cav-3 as well as to alter its biochemical targeting to "lipid raft/caveaole" fractions (Hayashi *et al.*, 2004).

Interestingly, transgenic overexpression of Cav-3 in mouse heart also induces a cardiomyopathic phenotype (Aravamudan *et al.*, 2003). Indeed, hearts of 6- to 10-month-old Cav-3 transgenic mice show cardiac myocyte disorganization, chronic inflammation, interstitial fibrosis, increased QRS duration, reduced NOS activity, decreased dystrophin expression, and reduced fractional shortening (Aravamudan *et al.*, 2003). Thus, it appears that both decreased expression and overexpression of Cav-3 can promote cardiomyopathy. However, the transgenic overexpression of Cav-3 did not result in any cardiac hypertrophy (Aravamudan *et al.*, 2003). Hence, the cardiomyopathic phenotype observed in mice transgenically overexpressing Cav-3 was suggested to be due, at least in part, to the disruption of the normal processing of the dystrophin complex and to the reduction of NO levels caused by the Cav-3-dependent inhibition of NOS activity (Aravamudan *et al.*, 2003).

2. Caveolins in Ischemic and Dilated Cardiomyopathy

The regulation of caveolin proteins has also been investigated in animal models of myocardial infarction (MI). Although the total protein expression remained unchanged, the redistribution of Cav-1 and Cav-3 from caveolae to the cytosol was shown to occur in rat hearts subjected to MI (Ratajczak *et al.*, 2003). The amounts of cytosolic Cav-1–eNOS complexes were also shown to be increased in MI rat hearts (Ratajczak *et al.*, 2003). This process was proposed as a mechanism leading to decreased NO production in heart failure (Ratajczak *et al.*, 2003). However, a discrepancy between the decreased expression of eNOS and the maintenance of NOS activity was later observed in hearts of MI rats (Damy *et al.*, 2003). This discrepancy was subsequently ascribed to the upregulation of nNOS expression within cardiomyocytes, as well as its redistribution to the sarcolemma via its interaction with Cav-3 (Damy *et al.*, 2003).

Interestingly, increased expression of Cav-3 was reported in the failing hearts of patients with ischemic cardiomyopathy (Uray *et al.*, 2003). Furthermore, mechanical unloading of those failing hearts resulted in the increased expression of all caveolin isoforms (Uray *et al.*, 2003). The induction of Cav-3 was suggested to increase the inotropic properties of the failing hearts by reducing NO levels in cardiomyocytes, while the induction of Cav-1 was suggested to increase endothelial NO release through the inhibition of the PKC (Uray *et al.*, 2003). Accordingly, perfusion of a Cav-1-derived peptide in an experimental model of myocardial ischemia-reperfusion also suggested that inhibition of the PKC could enhance endothelial NO release and, thus, prevent polymorphonuclear neutrophil adherence to the endothelium (Young *et al.*, 2001). The increased expression of caveolin proteins in mechanical unloading has been suggested to also improve beta-adrenergic responsiveness (Uray *et al.*, 2003).

Similarly, despite the decreased expression of eNOS, hearts of patients with idiopathic dilated cardiomyopathy show increased expression of nNOS and Cav-3 as well as increased amounts of Cav-3–nNOS complex formation (Damy *et al.*, 2004). An increase in Cav-3 expression and caveolar abundance has been demonstrated in dog hearts with pacing-induced dilated cardiomyopathy (Hare *et al.*, 2000). Interestingly, despite decreased basal NOS activity, the increased abundance of Cav-3 and caveolar domains was shown to be associated with increased inotropic responses to NOS inhibition, thus suggesting a novel caveolae-regulated mechanism by which agonist-stimulated NOS activity was increased in failing hearts (Hare *et al.*, 2000).

3. Caveolins in Lung Fibrosis

A loss of Cav-1 expression in alveolar type I cells has been proposed as an indicator of subcellular alterations during lung fibrogenesis (Kasper *et al.*, 1998). Indeed, immunoblotting of lung homogenates from normal and irradiated rat lungs showed a marked decrease in Cav-1 expression in fibrotic tissues (Kasper *et al.*, 1998). Interestingly, although immunofluorescence of those tissues demonstrated a striking decrease in Cav-1 expression in alveolar type I cells, it conversely showed increased expression of Cav-1 in endothelial cells, thus suggesting that different mechanisms exist for regulating caveolin gene expression in distinct cell types (Kasper *et al.*, 1998). The loss of epithelial Cav-1 was suggested to impair metabolic function and to reduce the signal transduction capacity of alveolar type I cells (Kasper *et al.*, 1998). Conversely, upregulation of endothelial Cav-1 might cause an increase in endothelial caveolar transcytosis (Kasper *et al.*, 1998). Accordingly, a decrease in the mRNA and protein levels of both Cav-1 and Cav-2 has recently been shown in a rat alveolar type I cell line treated with bleomycin (Koslowski *et al.*, 2004). An increase in endothelial Cav-1 expression has also been demonstrated in the $CdCl_2$/TGF-beta1-induced lung fibrogenesis rat model (Kasper *et al.*, 2004).

4. Caveolins in Pulmonary Interstitial Edema

An increase in caveolar domains and Cav-1 expression has been shown in rabbit lung tissues with pulmonary interstitial edema (Palestini *et al.*, 2003). The recruitment of caveolae and the increased expression of Cav-1 at the air-blood barrier plasma membrane have been suggested to be secondary to increased hydraulic interstitial pressure (Palestini *et al.*, 2003). Chronic exposure to shear stress has subsequently been shown to increase Cav-1 expression and caveolar domains at the luminal surface of cultured endothelial cells (Rizzo *et al.*, 2003). The pulmonary endothelial cells could, thus, respond to flow increases by the recruitment of caveolae at the cell surface, where they could mediate mechanotransduction responses.

5. Caveolins in Pulmonary Hypertension (PH)

Different studies investigating alterations in NO signaling in PH showed conflicting results concerning Cav-1 expression (Black *et al.*, 2002; Murata *et al.*, 2002; Fike *et al.*, 2004). A 30% decrease in Cav-1 protein levels was first shown in peripheral tissues of 8-week-old lambs subjected to fetal aortopulmonary vascular graft placement (Black *et al.*, 2002). However, Cav-1 expression was subsequently found to be unchanged in pulmonary arteries of both rats and piglets subjected to 7 and 10 days of hypoxia, respectively (Murata *et al.*, 2002; Fike *et al.*, 2004).

In order to clarify this issue, we recently directly investigated the expression of caveolin proteins in the lungs of rats with MI-induced secondary PH (Jasmin *et al.*, 2004). In this model of pulmonary venous hypertension, we showed marked decreases in both Cav-1 and Cav-2 expression, while Cav-3 expression remained unchanged in the lungs of rats subjected to an MI (Fig. 5) (Jasmin *et al.*, 2004). This study also demonstrated that reduced expression of Cav-1 and Cav-2 was associated with increased tyrosine phosphorylation of STAT3 as well as the upregulation of cyclin D1 and D3 protein expression (Jasmin *et al.*, 2004). Thus, the downmodulation of caveolin proteins expression in rats subjected to MI may represent an initiating mechanism leading to the activation of the STAT3/cyclins pathway and, ultimately, to the development of PH and lung remodeling.

However, the expression of caveolin proteins could be differentially regulated in different forms of PH. The pulmonary hypertension has recently been divided into five main classes, consisting of (1) pulmonary arterial hypertension, (2) pulmonary venous hypertension, (3) PH associated with disorders of the respiratory system and/or hypoxemia, (4) PH due to chronic thrombotic and or embolic disease, and (5) PH due to disorders affecting the pulmonary vasculature (Nauser and Stites, 2001). Based on the studies mentioned above, it seems that Cav-1 expression might be decreased in pulmonary arterial hypertension (shunted lambs) and pulmonary venous hypertension (MI rats), but unchanged in PH associated with hypoxemia.

6. Caveolins in Lung Cancer

Several lines of evidence have suggested a tumor suppressor role for Cav-1. For instance, the Cav-1 gene has been localized to a suspected tumor suppressor locus on human chromosome 7 (7q31.1/D7S522), which is deleted in several forms of cancer (Engelman *et al.*, 1998, 1999). Furthermore, targeted downregulation of Cav-1 in NIH 3T3 cells stimulates anchorage-independent growth, drives the formation of tumors when these cell are injected into immunodeficient mice, and causes the hyperactivation of the p42/44 MAP kinase cascade (Galbiati *et al.*, 1998). Interestingly, the transformation phenotype induced by Cav-1 downregulation can be reversed when Cav-1 protein levels are restored to normal (Galbiati *et al.*, 1998).

Cav-1 has also been shown to negatively regulate cell cycle progression, cyclin D1 gene transcription, and several pro-proliferative signaling molecules, such as Ha-Ras, c-Src, platelet-derived growth factor receptor (PDGF-R), and phosphatidylinositol 3-kinase (PI3-kinase) (Li *et al.*, 1996; Song *et al.*, 1996; Yamamoto *et al.*, 1999; Hulit *et al.*, 2000; Zundel *et al.*, 2000; Galbiati *et al.*, 2001). Furthermore, loss of Cav-1 gene expression in Cav-1$^{-/-}$ mice interbred with tumor-prone transgenic mice (MMTV-PyMT) has been shown to accelerate the development of multifocal dysplastic

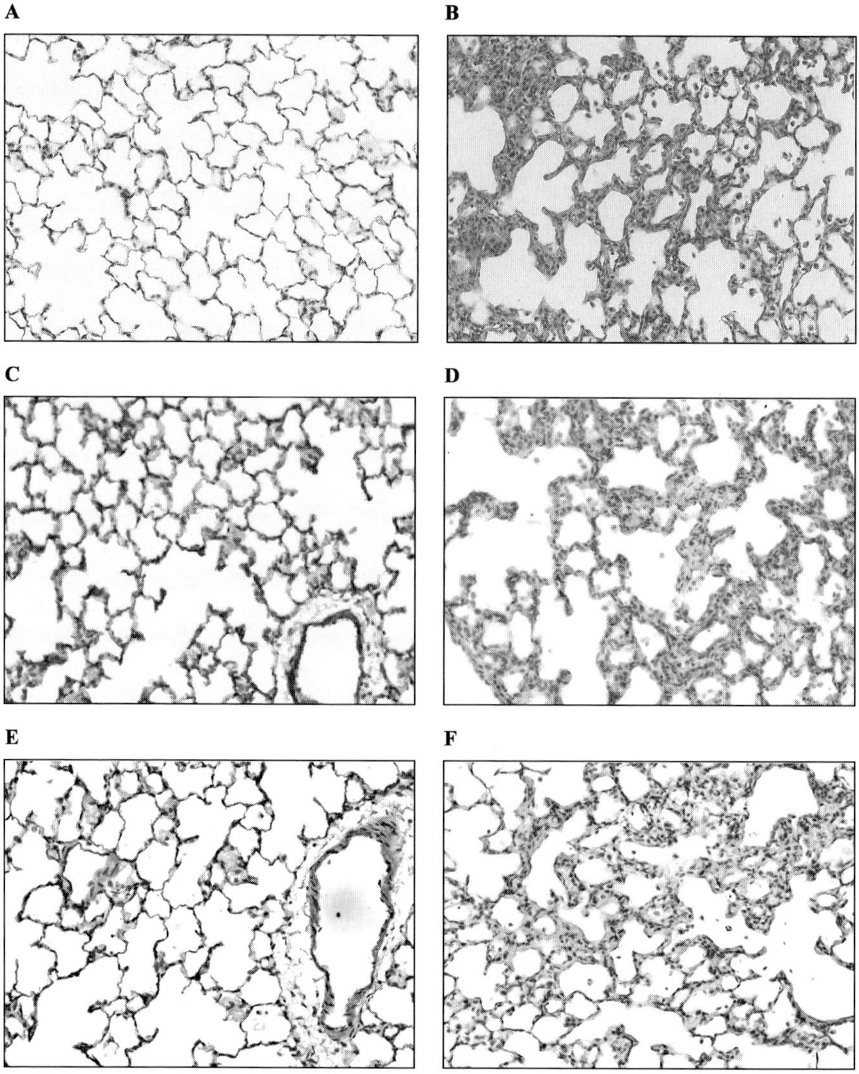

Figure 5. Hematoxylin and eosin (H&E) staining of the lungs of control (A) and MI (B) rats shows lung structural remodeling characterized by hypercellularity and thickened alveolar septa in the MI group. This lung structural remodeling is associated with decreased expression of Cav-1 and Cav-2. (C) and (D) represent immunostaining of Cav-1 in lungs of control and MI rats, respectively. (E) and (F) represent immunostaining of Cav-2 in lungs of control and MI rats, respectively. Modified from Jasmin *et al.* (2004). (See Color Insert.)

mammary lesions and to upregulate cyclin D1 expression (Williams *et al.*, 2003). Accordingly, Cav-1 expression is reduced in a variety of human cancers, such as human breast cancer cells, human colon tumors, human ovarian carcinomas, and human sarcomas (Lee *et al.*, 1998; Bender *et al.*, 2000; Wiechen *et al.*, 2001a,b).

A reduction of Cav-1 expression was reported in lung carcinoma cell lines (Racine *et al.*, 1999). Indeed, with the exception of the Calu-1 cell line, a human squamous cell lung tumor cell line, all mouse and human small cell lung carcinoma and adenocarcinoma cell lines showed decreases in both Cav-1 protein and mRNA levels when compared to a normal bronchial epithelial cell line (Racine *et al.*, 1999). The utilization of a cDNA array technique subsequently revealed the downregulation of the Cav-1 and Cav-2 gene transcriptional activity in human pulmonary adenocarcinoma (Wikman *et al.*, 2002). Furthermore, methylation analysis recently suggested that hypermethylation of the Cav-1 promoter could be implicated in the decreased expression of Cav-1 observed in small cell lung cancer cell lines (Sunaga *et al.*, 2004). Interestingly, although Cav-1 expression was shown to be decreased in 95% of small cell lung cancer cell lines, 76% of non-small cell lung cancer cell lines were conversely shown to retain Cav-1 expression, suggesting different roles for Cav-1 in the development of these two types of lung cancer (Sunaga *et al.*, 2004).

Accordingly, a dual function for Cav-1, both as a tumor suppressor gene and as a metastasis promoting gene, was proposed (Ho *et al.*, 2002). Indeed, although Cav-1 expression was decreased in mildly invasive lung cancer cell lines, it was found to be upregulated in lung adenocarcinoma cell lines with higher invasiveness properties (Ho *et al.*, 2002). It has even been demonstrated that introduction of Cav-1 into a mildly invasive lung cancer cell line increases the cell invasiveness by at least 150% (Ho *et al.*, 2002). Furthermore, evaluation of paraffin-embedded specimens of human lung adenocarcinoma showed increased Cav-1 expression in patients with ipsilateral hilar/peribronchial lymph node metastasis (Ho *et al.*, 2002).

The upregulation of Cav-1 expression has even been identified as an independent prognosis factor in patients with advanced lung adenocarcinoma (Ho *et al.*, 2002). Similarly, increased expression of Cav-1 was shown to correlate with advanced stage disease and to predict mortality in patients with pulmonary squamous cell carcinoma (Yoo *et al.*, 2003). The 5-year survival was significantly decreased, from 82.6% in patients with Cav-1-negative carcinoma to only 58.5% in patients with Cav-1-positive carcinoma (Yoo *et al.*, 2003). Similarly, increased Cav-1 expression has been correlated with tumor grade and metastasis and associated with poor prognosis in many other cancers, such as esophageal squamous cell carcinoma, pancreatic ductal adenocarcinoma, prostate cancer, and clear cell renal cell carcinoma (Yang *et al.*, 1999; Kato *et al.*, 2002; Suzuoki *et al.*, 2002; Joo *et al.*, 2004).

However, the detailed mechanisms underlying the metastasis-promoting role of Cav-1 are still obscure. Inhibition of c-Myc-induced apoptosis by Cav-1 was proposed to promote the progression of prostate cancer (Timme *et al.*, 2000). The upregulation of Cav-1 expression was subsequently proposed to be necessary to induce filopodia formation in lung adenocarcinoma cell lines (Ho *et al.*, 2002). However, overexpression of Cav-1 in MTLn3 cells, a metastatic rat mammary adenocarcinoma cell line, has conversely been shown to inhibit EGF-stimulated lamellipod extension, cell migration, and anchorage-independent growth (Zhang *et al.*, 2000). The overexpression of Cav-1 also accelerates endothelial cell differentiation and capillary tubule formation (Liu *et al.*, 2002). Therefore, Cav-1-mediated angiogenesis could also play an important role in metastasis promotion. Furthermore, Cav-1 expression was recently shown to be required for non-small cell lung cancer cell growth, through the activation of focal adhesion kinase (FAK) (Sunaga *et al.*, 2004).

Caveolin proteins have also been proposed to be implicated in the development of multidrug resistance. Indeed, multidrug-resistant cancer cells express very high Cav-1 and Cav-2 protein levels and show increases in caveolar density (Lavie *et al.*, 1998; Yang *et al.*, 1998). The upregulation of caveolins and caveolar domains in multidrug-resistant cancer cells may accelerate the intracellular cholesterol efflux pathway, thus increasing cytotoxic drug transport to membrane domains capable of clearance or neutralization (Lavie *et al.*, 2001).

A recent study showed that cytotoxic agents themselves were able to increase caveolin expression in drug-sensitive lung cancer cells (Belanger *et al.*, 2003). In fact, application of doxorubicin, paclitaxel, or etoposide to lung adenocarcinoma and small cell lung cancer cell lines upregulated Cav-1 mRNA as well as Cav-1 and Cav-2 protein levels (Belanger *et al.*, 2003). The activation of p53 and p38 MAP kinases by cytotoxic drugs may be a possible mechanism for increased caveolin expression in lung cancer cells (Belanger *et al.*, 2003). Therefore, the increases in Cav-1 and Cav-2 levels reported for several drug-resistant cell lines might simply reflect exposure to the cytotoxic drugs themselves.

7. Caveolins and Other Lung-Related Disorders

The expression of caveolin proteins has also been shown to be dysregulated in the lungs of diabetic rats (Pascariu *et al.*, 2004). Indeed, the alveolar endothelial cells of hyperglycemic rats were shown to contain more caveolae structures and to have increased Cav-1 expression, which may contribute to the observed microvascular hyperpermeability to plasma macromolecules (Pascariu *et al.*, 2004).

E. Caveolins and Aging

Caveolin proteins also appear to be dysregulated during the aging process. Indeed, late passage cells, such as human lung fibroblasts, were found to be enriched in caveolar domains and caveolins (Cav-1 and Cav-2) and showed strong interactions between Cav-1 and the EGF-R when compared to early passage cells (Park *et al.*, 2000). This interaction with Cav-1 was subsequently shown to attenuate the EGF-induced activation of the p42/44 MAP kinase pathway (Park *et al.*, 2000). Interestingly, overexpression of Cav-1 in early passage cells suppressed the activation of p42/44 MAP kinase cascade upon EGF stimulation (Park *et al.*, 2000). Furthermore, reduction of Cav-1 expression in senescent human fibroblasts, using antisense oligonucleotides and small interfering RNA, resulted in the restoration of the normal growth factor response (Cho *et al.*, 2003). Accordingly, increases in Cav-1, Cav-2, and Cav-3 protein levels were observed in the lung homogenates of old rats (27 months of age), as compared to young rats (2 months of age) (Park *et al.*, 2000).

In the heart, other studies showed conflicting results concerning the regulation of the caveolin proteins during the aging process. Although one report showed an increase in the total expression levels of Cav-1 and Cav-3 in the hearts of 27-month-old Wistar rats, another report showed unchanged total expression, but a dissociation of Cav-1 and Cav-3 from the caveolae to the cytosol in 27-month-old Sprague-Dawley rats (Park *et al.*, 2000; Ratajczak *et al.*, 2003).

II. CONCLUSIONS

Caveolin proteins are key regulators of several physiological and pathological processes in the cardiopulmonary system. As mentioned above, caveolins have been implicated in heart and lung development, as well as in cardiopulmonary endocytosis, vesicular trafficking, and signal transduction. Caveolin protein expression also appears to be altered or dysregulated in several cardiopulmonary disorders. A loss of caveolin has been implicated in the development of hypertrophic cardiomyopathy, lung fibrogenesis, pulmonary hypertension, and lung cancer. Conversely, increased caveolin expression has been associated with the development of ischemic and dilated cardiomyopathy, pulmonary interstitial edema, and the appearance of metastasis. The generation and characterization of caveolin-deficient mice strongly support the important roles of the caveolin proteins in the cardiopulmonary system. Indeed, although caveolin-deficient mice are viable, they show multiple cardiopulmonary abnormalities, such as cardiac hypertrophy, pulmonary hypertension, and lung structural remodeling. Thus, tight regulation of caveolin expression appears to be fundamental for normal cardiopulmonary functioning.

Future studies should investigate the cardiopulmonary effects of *in vivo* modulation of caveolin protein expression. Interestingly, *in vivo* delivery of a Cav-1 scaffolding domain peptide has previously been shown to reduce inflammation, block microvascular hyperpermeability, and alter tumor progression in mice (Bucci *et al.*, 2000; Gratton *et al.*, 2003). The perfusion of the Cav-1 scaffolding domain has also been shown to exert cardio-protective effects during myocardial ischemia-reperfusion by reducing polymorphonuclear neutrophils adherence and infiltration (Young *et al.*, 2001).

ACKNOWLEDGMENTS

This work was supported by grants from the National Institutes of Health, and the American Heart Association (to M. P. L.). J. F. J. was supported by a fellowship grant from the "Fonds de la Recherche en Santé du Québec" (FRSQ). P.G.F. was the recipient of a Scientist Development Grant from the American Heart Association.

REFERENCES

Aravamudan, B., Volonte, D., Ramani, R., Gursoy, E., Lisanti, M. P., London, B., and Galbiati, F. (2003). Transgenic overexpression of caveolin-3 in the heart induces a cardiomyopathic phenotype. *Hum. Mol. Genet.* **12,** 2777–2788.

Belanger, M. M., Roussel, E., and Couet, J. (2003). Up-regulation of caveolin expression by cytotoxic agents in drug-sensitive cancer cells. *Anticancer Drugs* **12,** 281–287.

Bender, F. C., Reymond, M. A., Bron, C., and Quest, A. F. (2000). Caveolin-1 levels are down-regulated in human colon tumors, and ectopic expression of caveolin-1 in colon carcinoma cell lines reduces cell tumorigenicity. *Cancer Res.* **60,** 5870–5878.

Biederer, C. H., Ries, S. J., Moser, M., Florio, M., Israel, M. A., McCormick, F., and Buettner, R. (2000). The basic helix-loop-helix transcription factors myogenin and Id2 mediate specific induction of caveolin-3 gene expression during embryonic development. *J. Biol. Chem.* **275,** 26245–26251.

Black, S. M., Bekker, J. M., McMullan, D. M., Parry, A. J., Ovadia, B., Reinhartz, O., Lakshminrushimha, S., Mata-Greenwood, E., Steinhorn, R. H., and Fineman, J. R. (2002). Alterations in nitric oxide production in 8-week-old lambs with increased pulmonary blood flow. *Pediatr. Res.* **52,** 233–244.

Bossuyt, J., Taylor, B. E., James-Kracke, M., and Hale, C. C. (2002). The cardiac sodium-calcium exchanger associates with caveolin-3. *Ann. N Y Acad. Sci.* **976,** 197–204.

Bucci, M., Gratton, J. P., Rudic, R. D., Acevedo, L., Roviezzo, F., Cirino, G., and Sessa, W. C. (2000). *In vivo* delivery of the caveolin-1 scaffolding domain inhibits nitric oxide synthesis and reduces inflammation. *Nat. Med.* **6,** 1362–1367.

Campbell, L., Hollins, A. J., Al-Eid, A., Newman, G. R., von Ruhland, C., and Gumbleton, M. (1999). Caveolin-1 expression and caveolae biogenesis during cell transdifferentiation in lung alveolar epithelial primary cultures. *Biochem. Biophys. Res. Commun.* **262,** 744–751.

Cho, K. A., Ryu, S. J., Park, J. S., Jang, I. S., Ahn, J. S., Kim, K. T., and Park, S. C. (2003). Senescent phenotype can be reversed by reduction of caveolin status. *J. Biol. Chem.* **278,** 27789–27795.

Cohen, A. W., Park, D. S., Woodman, S. E., Williams, T. M., Chandra, M., Shirani, J., Pereira S., Kitsis, R. N., Russell, R. G., Weiss, L. M., Tang, B., Jelicks, L. A., Factor, S. M., Shtutin, V., Tanowitz, H. B., and Lisanti, M. P. (2003). Caveolin-1 null mice develop cardiac hypertrophy with hyperactivation of p42/44 MAP kinase in cardiac fibroblasts. *Am. J. Physiol. Cell Physiol.* **284,** C457–C474.

Damy, T., Ratajczak, P., Robidel, E., Bendall, J. K., Oliviero, P., Boczkowski, J., Ebrahimian, T., Marotte, F., Samuel, J. L., and Heymes, C. (2003). Up-regulation of cardiac nitric oxide synthase 1-derived nitric oxide after myocardial infarction in senescent rats. *FASEB J.* **17,** 1934–1936.

Damy, T., Ratajczak, P., Shah, A. M., Camors, E., Marty, I., Hasenfuss, G., Marotte, F., Samuel, J. L., and Heymes, C. (2004). Increased neuronal nitric oxide synthase-derived NO production in the failing human heart. *Lancet* **363,** 1365–1367.

Doyle, D. D., Ambler, S. K., Upshaw-Earley, J., Bastawrous, A., Goings, G. E., and Page, E. (1997). Type B atrial natriuretic peptide receptor in cardiac myocyte caveolae. *Circ. Res.* **81,** 86–91.

Drab, M., Verkade, P., Elger, M., Kasper, M., Lohn, M., Lauterbach, B., Menne, J., Lindschau, C., Mende, F., Luft, F. C., Schedl, A., Haller, H., and Kurzchalia, T. V. (2001). Loss of caveolae, vascular dysfunction, and pulmonary defects in caveolin-1 gene-disrupted mice. *Science* **293,** 2449–2452.

Engelman, J. A., Zhang, X. L., and Lisanti, M. P. (1998). Genes encoding human caveolin-1 and -2 are co-localized to the D7S522 locus (7q31.1), a known fragile site (FRA7G) that is frequently deleted in human cancers. *FEBS Lett.* **436,** 403–410.

Engelman, J. A., Zhang, X. L., and Lisanti, M. P. (1999). Sequence and detailed organization of the human caveolin-1 and -2 genes located near the D7S522 locus (7q31.1). Methylation of a CpG island in the 5′ promoter region of the caveolin-1 gene in human breast cancer cell lines. *FEBS Lett.* **448,** 221–230.

Feron, O., Belhassen, L., Kobzik, L., Smith, T. W., Kelly, R. A., and Michel, T. (1996). Endothelial nitric oxide synthase targeting to caveolae. Specific interactions with caveolin isoforms in cardiac myocytes and endothelial cells. *J. Biol. Chem.* **271,** 22810–22814.

Feron, O., Dessy, C., Opel, D. J., Arstall, M. A., Kelly, R. A., and Michel, T. (1998). Modulation of the endothelial nitric-oxide synthase-caveolin interaction in cardiac myocytes. Implications for the autonomic regulation of heart rate. *J. Biol. Chem.* **273,** 30249–30254.

Feron, O., Smith, T. W., Michel, T., and Kelly, R. A. (1997). Dynamic targeting of the agonist-stimulated m2 muscarinic acetylcholine receptor to caveolae in cardiac myocytes. *J. Biol. Chem.* **272,** 17744–17748.

Fike, C. D., Aschner, J. L., Zhang, Y., and Kaplowitz, M. R. (2004). Impaired NO signaling in small pulmonary arteries of chronically hypoxic newborn piglets. *Am. J. Physiol. Lung Cell Mol. Physiol.* **286,** L1244–L1254.

Fujita, T., Toya, Y., Iwatsubo, K., Onda, T., Kimura, K., Umemura, S., and Ishikawa, Y. (2001). Accumulation of molecules involved in alpha1-adrenergic signal within caveolae: Caveolin expression and the development of cardiac hypertrophy. *Cardiovasc. Res.* **51,** 709–716.

Galbiati, F., Volonte, D., Engelman, J. A., Watanabe, G., Burk, R., Pestell, R. G., and Lisanti, M. P. (1998). Targeted downregulation of caveolin-1 is sufficient to drive cell transformation and hyperactivate the p42/44 MAP kinase cascade. *EMBO J.* **17,** 6633–6648.

Galbiati, F., Volonte, D., Liu, J., Capozza, F., Frank, P. G., Zhu, L., Pestell, R. G., and Lisanti, M. P. (2001). Caveolin-1 expression negatively regulates cell cycle progression by inducing G0/G1 arrest via a p53/p21 WAF1/Cip1. *Mol. Biol. Cell* **12,** 2229–2244.

Garcia-Cardena, G., Fan, R., Stern, D. F., Liu, J., and Sessa, W. C. (1996a). Endothelial nitric oxide synthase is regulated by tyrosine phosphorylation and interacts with caveolin-1. *J. Biol. Chem.* **271,** 27237–27240.

Garcia-Cardena, G., Oh, P., Liu, J., Schnitzer, J. E., and Sessa, W. C. (1996b). Targeting of nitric oxide synthase to endothelial cell caveolae via palmitoylation: Implications for nitric oxide signaling. *Proc. Natl. Acad. Sci. USA* **93,** 6448–6453.

Ghitescu, L., Fixman, A., Simionescu, M., and Simionescu, N. (1986). Specific binding sites for albumin restricted to plasmalemmal vesicles of continuous capillary endothelium: Receptor-mediated transcytosis. *J. Cell Biol.* **102,** 1304–1311.

Gratton, J. P., Fontana, J., O'Connor, D. S., Garcia-Cardena, G., McCabe, T. J., and Sessa, W. C. (2000). Reconstitution of an endothelial nitric-oxide synthase (eNOS), hsp 90, and caveolin-1 complex *in vitro*. Evidence that hsp90 facilitates calmodulin stimulated displacement of eNOS from caveolin-1. *J. Biol. Chem.* **275,** 22268–22272.

Gratton, J. P., Lin, M. I., Yu, J., Weiss, E. D., Jiang, Z. L., Fairchild, T. A., Iwakiri, Y., Groszmann, R., Claffey, K. P., Cheng, Y. C., and Sessa, W. C. (2003). Selective inhibition of tumor microvascular permeability by cavtratin blocks tumor progression in mice. *Cancer Cell* **4,** 31–39.

Hammes, A., Oberdorf-Maass, S., Rother, T., Nething, K., Gollnick, F., Linz, K. W., Meyer, R., Hu, K., Han, H., Gaudron, P., Ertl, G., Hoffmann, S., Ganten, U., Vetter, R., Schuh, K., Benkwitz, C., Zimmer, H. G., and Neyses, L. (1998). Overexpression of the sarcolemmal calcium pump in the myocardium of transgenic rats. *Circ. Res.* **83,** 877–888.

Hare, J. M., Lofthouse, R. A., Juang, G. J., Colman, L., Ricker, K. M., Kim, B., Senzaki, H., Cao, S., Tunin, R. S., and Kass, D. A. (2000). Contribution of caveolin protein abundance to augmented nitric oxide signaling in conscious dogs with pacing-induced heart failure. *Circ. Res.* **86,** 1085–1092.

Hayashi, T., Arimura, T., Ueda, K., Shibata, H., Hohda, S., Takahashi, M., Hori, H., Koga, Y., Oka, N., Imaizumi, T., Yasunami, M., and Kimura, A. (2004). Identification and functional analysis of a caveolin-3 mutation associated with familial hypertrophic cardiomyopathy. *Biochem. Biophys. Res. Commun.* **313,** 178–184.

Ho, C. C., Huang, P. H., Huang, H. Y., Chen, Y. H., Yang, P. C., and Hsu, S. M. (2002). Up-regulated caveolin-1 accentuates the metastasis capability of lung adenocarcinoma by inducing filopodia formation. *Am. J. Pathol.* **161,** 1647–1656.

Hulit, J., Bash, T., Fu, M., Galbiati, F., Albanese, C., Sage, D. R., Schlegel, A., Zhurinsky, J., Shtutman, M., Ben-Ze'ev, A., Lisanti, M. P., and Pestell, R. G. (2000). The cyclin D1 gene is transcriptionally repressed by caveolin-1. *J. Biol. Chem.* **275,** 21203–21209.

Jasmin, J. F., Mercier, I., Hnasko, R., Cheung, M. W. C., Tanowitz, H. B., Dupuis, J., and Lisanti, M. P. (2004). Lung remodeling and pulmonary hypertension after myocardial infarction: Pathogenic role of reduced caveolin expression. *Cardiovasc. Res.* **63,** 747–755.

Joo, D. J., Oh, D. K., Kim, Y. S., Lee, K. B., and Kim, S. J. (2004). Increased expression of caveolin-1 and microvessel density correlates with metastasis and poor prognosis in clear cell renal cell carcinoma. *BJU Int.* **93,** 291–296.

Kasper, M., Reimann, T., Hempel, U., Wenzel, K. W., Bierhaus, A., Schuh, D., Dimmer, V., Haroske, G., and Muller, M. (1998). Loss of caveolin expression in type I pneumocytes as an indicator of subcellular alterations during lung fibrogenesis. *Histochem. Cell Biol.* **109,** 41–48.

Kasper, M., Seidel, D., Knels, L., Morishima, N., Neisser, A., Bramke, S., and Koslowski, R. (2004). Early signs of lung fibrosis after *in vitro* treatment of rat lung slices with CdCl2 and TGF-beta1. *Histochem. Cell Biol.* **121,** 131–140.

Kathuria, H., Cao, Y., Ramirez, M. I., and Williams, M. C. (2004). Transcription of the caveolin-1 gene is differentially regulated in lung type I epithelial and endothelial cell lines: A role for ETS proteins in epithelial cell expression. *J. Biol. Chem* **279,** 30036–30078.

Kato, K., Hida, Y., Miyamoto, M., Hashida, H., Shinohara, T., Itoh, T., Okushiba, S., Kondo, S., and Katoh, H. (2002). Overexpression of caveolin-1 in esophageal squamous cell carcinoma correlates with lymph node metastasis and pathologic stage. *Cancer* **94,** 929–933.

Kawabe, J. I., Grant, B. S., Yamamoto, M., Schwencke, C., Okumura, S., and Ishikawa, Y. (2001). Changes in caveolin subtype protein expression in aging rat organs. *Mol. Cell Endocrinol.* **176,** 91–95.

Kobayashi, N., Mori, Y., Nakano, S., Tsubokou, Y., Kobayashi, T., Shirataki, H., and Matsuoka, H. (2001). TCV-116 stimulates eNOS and caveolin-1 expression and improves coronary microvascular remodeling in normotensive and angiotensin II-induced hypertensive rats. *Atherosclerosis* **158,** 359–368.

Koga, A., Oka, N., Kikuchi, T., Miyazaki, H., Kato, S., and Imaizumi, T. (2003). Adenovirus-mediated overexpression of caveolin-3 inhibits rat cardiomyocyte hypertrophy. *Hypertension* **42,** 213–219.

Kogo, H., Aiba, T., and Fujimoto, T. (2004). Cell type-specific occurence of caveolin-1alpha and -1 beta in the lung caused by expression of distinct mRNAs. *J. Biol. Chem.* **279,** 25574–25581.

Koslowski, R., Barth, K., Augstein, A., Tschernig, T., Bargsten, G., Aufderheide, M., and Kasper, M. (2004). A new rat type I-like alveolar epithelial cell line R3/1: Bleomycin effects on caveolin expression. *Histochem. Cell Biol* **121,** 509–519.

Lasley, R. D., Narayan, P., Uittenbogaard, A., and Smart, E. J. (2000). Activated cardiac adenosine A(1) receptors translocate out of caveolae. *J. Biol. Chem.* **275,** 4417–4421.

Lavie, Y., Fiucci, G., and Liscovitch, M. (1998). Up-regulation of caveolae and caveolar constituents in multidrug-resistant *Cancer Cells*. *J. Biol. Chem.* **273,** 33766–33773.

Lavie, Y., Fiucci, G., and Liscovitch, M. (2001). Upregulation of caveolin in multidrug resistant *Cancer Cells*: Functional implications. *Adv. Drug Deliv. Rev.* **49,** 317–323.

Lee, S. W., Reimer, C. L., Oh, P., Campbell, D. B., and Schnitzer, J. E. (1998). Tumor cell growth inhibition by caveolin re-expression in human breast cancer cells. *Oncogene* **16,** 1391–1397.

Li, S., Couet, J., and Lisanti, M. P. (1996). Src tyrosine kinases, Galpha subunits, and H-ras share a common membrane-anchored scaffolding protein, caveolin. Caveolin binding negatively regulates the auto activation of Src tyrosine kinases. *J. Biol. Chem.* **271,** 29182–29190.

Lisanti, M. P., Scherer, P. E., Vidugiriene, J., Tang, Z., Hermanowski-Vosatka, A., Tu, Y. H., Cook, R. F., and Sargiacomo, M. (1994). Characterization of caveolin-rich membrane domains isolated from an endothelial-rich source: Implications for human disease. *J. Cell Biol.* **126,** 111–126.

Liu, J., Wang, X. B., Park, D. S., and Lisanti, M. P. (2002). Caveolin-1 expression enhances endothelial capillary tubule formation. *J. Biol. Chem.* **277,** 10661–10668.

Liu, L., Mohammadi, K., Aynafshar, B., Wang, H., Li, D., Liu, J., Ivanov, A. V., Xie, Z., and Askari, A. (2003). Role of caveolae in signal-transducing function of cardiac Na+/K+-ATPase. *Am. J. Physiol. Cell Physiol.* **284,** C1550–C1560.

Lupu, C., Goodwin, C. A., Westmuckett, A. D., Emeis, J. J., Scully, M. F., Kakkar, V. V., and Lupu, F. (1997). Tissue factor pathway inhibitor in endothelial cells colocalizes with glycolipid microdomains/

caveolae. Regulatory mechanism(s) of the anticoagulant properties of the endothelium. *Arterioscler. Thromb. Vasc. Biol.* **17,** 2964–2974.

Murata, T., Sato, K., Hori, M., Ozaki, H., and Karaki, H. (2002). Decreased endothelial nitric-oxide synthase (eNOS) activity resulting from abnormal interaction between eNOS and its regulatory proteins in hypoxia-induced pulmonary hypertension. *J. Biol. Chem.* **277,** 44085–44092.

Nanjundan, M. and Possmayer, F. (2001). Pulmonary lipid phosphate phosphohydrolase in plasma membrane signalling platforms. *Biochem. J.* **358,** 637–646.

Nauser, T. and Stites, S. W. (2001). Diagnosis and treatment of pulmonary hypertension. *Am. Fam. Physician* **63,** 1789–1798.

Newman, G. R., Campbell, L., von Ruhland, C., Jasani, B., and Gumbleton, M. (1999). Caveolin and its cellular and subcellular immunolocalisation in lung alveolar epithelium: Implications for alveolar type I cell function. *Cell Tissue Res.* **295,** 111–120.

Oh, P. and Schnitzer, J. E. (2001). Segregation of heterotrimeric G proteins in cell surface microdomains. Gq binds caveolin to concentrate in caveolae, whereas Gi and Gs target lipid rafts by default. *Mol. Biol. Cell* **12,** 685–698.

Ohsawa, Y., Toko, H., Katsura, M., Morimoto, K., Yamada, H., Ichikawa, Y., Murakami, T., Ohkuma, S., Komuro, I., and Sunada, Y. (2004). Overexpression of P104L mutant caveolin-3 in mice develops hypertrophic cardiomyopathy with enhanced contractility in association with increased endothelial nitric oxide synthase activity. *Hum. Mol. Genet.* **13,** 151–157.

Oka, N., Asai, K., Kudej, R. K., Edwards, J. G., Toya, Y., Schwencke, C., Vatner, D. E., Vatner, S. F., and Ishikawa, Y. (1997). Downregulation of caveolin by chronic beta-adrenergic receptor stimulation in mice. *Am. J. Physiol.* **273,** C1957–C1962.

Ostrom, R. S., Violin, J. D., Coleman, S., and Insel, P. A. (2000). Selective enhancement of beta-adrenergic receptor signaling by overexpression of adenylyl cyclase type 6: Colocalization of receptor and adenylyl cyclase in caveolae of cardiac myocytes. *Mol. Pharmacol.* **57,** 1075–1079.

Oxhorn, B. C. and Buxton, I. L. (2003). Caveolar compartmentation of caspase-3 in cardiac endothelial cells. *Cell Signal* **15,** 489–496.

Page, E., Winterfield, J., Goings, G., Bastawrous, A., and Upshaw-Earley, J. (1998). Water channel proteins in rat cardiac myocyte caveolae: Osmolarity-dependent reversible internalization. *Am. J. Physiol.* **274,** H1988–H2000.

Palestini, P., Calvi, C., Conforti, E., Daffara, R., Botto, L., and Miserocchi, G. (2003). Compositional changes in lipid microdomains of air-blood barrier plasma membranes in pulmonary interstitial edema. *J. Appl. Physiol.* **95,** 1446–1452.

Park, D. S., Woodman, S. E., Schubert, W., Cohen, A. W., Frank, P. G., Chandra, M., Shirani, J., Razani, B., Tang, B., Jelicks, L. A., Factor, S. M., Weiss, L. M., Tanowitz, H. B., and Lisanti, M. P. (2002). Caveolin-1/3 double-knockout mice are viable, but lack both muscle and non-muscle caveolae, and develop a severe cardiomyopathic phenotype. *Am. J. Pathol.* **160,** 2207–2217.

Park, W. Y., Park, J. S., Cho, K. A., Kim, D. I., Ko, Y. G., Seo, J. S., and Park, S. C. (2000). Up-regulation of caveolin attenuates epidermal growth factor signaling in senescent cells. *J. Biol. Chem.* **275,** 20847–20852.

Pascariu, M., Bendayan, M., and Ghitescu, L. (2004). Correlated endothelial caveolin overexpression and increased transcytosis in experimental diabetes. *J. Histochem. Cytochem.* **52,** 65–76.

Piech, A., Dessy, C., Havaux, X., Feron, O., and Balligand, J. L. (2003a). Differential regulation of nitric oxide synthases and their allosteric regulators in heart and vessels of hypertensive rats. *Cardiovasc. Res.* **57,** 456–467.

Piech, A., Dessy, C., Havaux, X., Feron, O., and Balligand, J. L. (2003b). Differential regulation of nitric oxide synthases and their allosteric regulators in heart and vessels of hypertensive rats. *Cardiovasc. Res.* **57,** 456–467.

Piech, A., Massart, P. E., Dessy, C., Feron, O., Havaux, X., Morel, N., Vanoverschelde, J. L., Donckier, J., and Balligand, J. L. (2002). Decreased expression of myocardial eNOS and caveolin in dogs with hypertrophic cardiomyopathy. *Am. J. Physiol. Heart Circ. Physiol.* **282,** H219–H231.

Racine, C., Belanger, M., Hirabayashi, H., Boucher, M., Chakir, J., and Couet, J. (1999). Reduction of caveolin 1 gene expression in lung carcinoma cell lines. *Biochem. Biophys. Res. Commun.* **255,** 580–586.

Ramirez, M. I., Pollack, L., Millien, G., Cao, Y. X., Hinds, A., and Williams, M. C. (2002). The alpha-isoform of caveolin-1 is a marker of vasculogenesis in early lung development. *J. Histochem. Cytochem.* **50,** 33–42.

Ratajczak, P., Damy, T., Heymes, C., Oliviero, P., Marotte, F., Robidel, E., Sercombe, R., Boczkowski, J., Rappaport, L., and Samuel, J. L. (2003). Caveolin-1 and -3 dissociations from caveolae to cytosol in the heart during aging and after myocardial infarction in rat. *Cardiovasc. Res.* **57,** 358–369.

Razani, B., Engelman, J. A., Wang, X. B., Schubert, W., Zhang, X. L., Marks, C. B., Macaluso, F., Russell, R. G., Li, M., Pestell, R. G., Di Vizio, D., Hou, H. J., Kneitz, B., Lagaud, G., Christ, G. H., Edelmann, W., and Lisanti, M. P. (2001). Caveolin-1 null mice are viable but show evidence of hyperproliferative and vascular abnormalities. *J. Biol. Chem.* **276,** 38121–38138.

Razani, B., Wang, X. B., Engelman, J. A., Battista, M., Lagaud, G., Zhang, X. L., Kneitz, B., Hou H., Jr., Christ, G. J., Edelmann, W., and Lisanti, M. P. (2002). Caveolin-2-deficient mice show evidence of severe pulmonary dysfunction without disruption of caveolae. *Mol. Biol. Cell* **22,** 2329–2344.

Rizzo, V., McIntosh, D. P., Oh, P., and Schnitzer, J. E. (1998). *In situ* flow activates endothelial nitric oxide synthase in luminal caveolae of endothelium with rapid caveolin dissociation and calmodulin association. *J. Biol. Chem.* **273,** 34724–34729.

Rizzo, V., Morton, C., DePaola, N., Schnitzer, J. E., and Davies, P. F. (2003). Recruitment of endothelial caveolae into mechanotransduction pathways by flow conditioning *in vitro*. *Am. J. Physiol. Heart Circ. Physiol.* **285,** H1720–H1729.

Rybin, V. O., Xu, X., Lisanti, M. P., and Steinberg, S. F. (2000). Differential targeting of beta-adrenergic receptor subtypes and adenylyl cyclase to cardiomyocyte caveolae. A mechanism to functionally regulate the cAMP signaling pathway. *J. Biol. Chem.* **275,** 41447–41457.

Rybin, V. O., Xu, X., and Steinberg, S. F. (1999). Activated protein kinase C isoforms target to cardiomyocyte caveolae: Stimulation of local protein phosphorylation. *Circ. Res.* **84,** 980–988.

Schnitzer, J. E., Liu, J., and Oh, P. (1995a). Endothelial caveolae have the molecular transport machinery for vesicle budding, docking, and fusion including VAMP, NSF, SNAP, annexins, and GTPases. *J. Biol. Chem.* **270,** 14399–14404.

Schnitzer, J. E. and Oh, P. (1996). Aquaporin-1 in plasma membrane and caveolae provides mercury-sensitive water channels across lung endothelium. *Am. J. Physiol. Heart Circ. Physiol.* **270,** H416–H422.

Schnitzer, J. E., Oh, P., Jacobson, B. S., and Dvorak, A. M. (1995b). Caveolae from luminal plasmalemma of rat lung endothelium: Microdomains enriched in caveolin, Ca^{2+}-ATPase, and inositol trisphosphate receptor. *Proc. Natl. Acad. Sci. USA* **92,** 1759–1763.

Schubert, W., Frank, P. G., Razani, B., Park, D. S., Chow, C. W., and Lisanti, M. P. (2001). Caveolae-deficient endothelial cells show defects in the uptake and transport of albumin *in vivo*. *J. Biol. Chem.* **276,** 48619–48622.

Schubert, W., Frank, P. G., Woodman, S. E., Hyogo, H., Cohen, D. E., Chow, C. W., and Lisanti, M. P. (2002). Microvascular hyperpermeability in caveolin-1 (–/–) knock-out mice. *J. Biol. Chem.* **277,** 40091–40098.

Shah, M., Patel, K., Fried, V. A., and Sehgal, P. B. (2002). Interactions of STAT3 with caveolin-1 and heat shock protein 90 in plasma membrane raft and cytosolic complexes. Preservation of cytokine signaling during fever. *J. Biol. Chem.* **277,** 45662–45669.

Shi, Y., Pritchard, K. A., Jr., Holman, P., Rafiee, P., Griffith, O. W., Kalyanaraman, B., and Baker, J. E. (2000). Chronic myocardial hypoxia increases nitric oxide synthase and decreases caveolin-3. *Free Radic. Biol. Med.* **29,** 695–703.

Song, K. S., Li, S., Okamoto, T., Quilliam, L. A., Sargiacomo, M., and Lisanti, M. P. (1996). Co-purification and direct interaction of Ras with caveolin, an integral membrane protein of caveolae microdomains. Detergent-free purification of caveolae microdomains. *J. Biol. Chem.* **271,** 9690–9697.

Sunaga, N., Miyajima, K., Suzuki, M., Sato, M., White, M. A., Ramirez, R. D., Shay, J. W., Gazdar, A. F., and Minna, J. D. (2004). Different roles for caveolin-1 in the development of non-small cell lung cancer versus small cell lung cancer. *Cancer Res.* **64,** 4277–4285.

Suzuoki, M., Miyamoto, M., Kato, K., Hiraoka, K., Oshikiri, T., Nakakubo, Y., Fukunaga, A., Shichinohe, T., Shinohara, T., Itoh, T., Kondo, S., and Katoh, H. (2002). Impact of caveolin-1 expression on prognosis of pancreatic ductal adenocarcinoma. *Br. J. Cancer* **87,** 1140–1144.

Timme, T. L., Goltsov, A., Tahir, S., Li, L., Wang, J., Ren, C., Johnston, R. N., and Thompson, T. C. (2000). Caveolin-1 is regulated by c-myc and suppresses c-myc-induced apoptosis. *Oncogene* **19,** 3256–3265.

Tiruppathi, C., Song, W., Bergenfeldt, M., Sass, P., and Malik, A. B. (1997). Gp60 activation mediates albumin transcytosis in endothelial cells by tyrosine kinase-dependent pathway. *J. Biol. Chem.* **272,** 25968–25975.

Uray, I. P., Connelly, J. H., Frazier, O. H., Taegtmeyer, H., and Davies, P. J. (2003). Mechanical unloading increases caveolin expression in the failing human heart. *Cardiovasc. Res.* **59,** 57–66.

Wiechen, K., Diatchenko, L., Agoulnik, A., Scharff, K. M., Schober, H., Arlt, K., Zhumabayeva, B., Siebert, P. D., Dietel, M., Schafer, R., and Sers, C. (2001a). Caveolin-1 is down-regulated in human ovarian carcinoma and acts as a candidate tumor suppressor gene. *Am. J. Pathol.* **159,** 1635–1643.

Wiechen, K., Sers, C., Agoulnik, A., Arlt, K., Dietel, M., Schlag, P. M., and Schneider, U. (2001b). Down-regulation of caveolin-**1,** a candidate tumor suppressor gene, in sarcomas. *Am. J. Pathol.* **158,** 833–839.

Wikman, H., Kettunen, E., Seppanen, J. K., Karjalainen, A., Hollmen, J., Anttila, S., and Knuutila, S. (2002). Identification of differentially expressed genes in pulmonary adenocarcinoma by using cDNA array. *Oncogene* **21,** 5804–5813.

Williams, T. M., Cheung, M. W. C., Park, D. S., Razani, B., Cohen, A. W., Muller, W. J., Di Vizio, D., Chopra, N. G., Pestell, R. G., and Lisanti, M. P. (2003). Loss of caveolin-1 gene expression accelerates the development of dysplastic mammary lesions in tumor-prone transgenic mice. *Mol. Biol. Cell* **14,** 1027–1042.

Woodman, S. E., Park, D. S., Cohen, A. W., Cheung, M. W., Chandra, M., Shirani, J., Tang, B., Jelicks, L. A., Kitsis, R. N., Christ, G. J., Factor, S. M., Tanowitz, H. B., and Lisanti, M. P. (2002). Caveolin-3 knock-out mice develop a progressive cardiomyopathy and show hyperactivation of the p42/44 MAPK cascade. *J. Biol. Chem.* **277,** 38988–38997.

Xiang, Y., Rybin, V. O., Steinberg, S. F., and Kobilka, B. (2002). Caveolar localization dictates physiologic signaling of beta 2-adrenoceptors in neonatal cardiac myocytes. *J. Biol. Chem.* **277,** 34280–34286.

Yamamoto, M., Toya, Y., Jensen, R. A., and Ishikawa, Y. (1999). Caveolin is an inhibitor of platelet-derived growth factor receptor signaling. *Exp. Cell Res.* **247,** 380–388.

Yang, C. P. H., Galbiati, F., Volonte, D., Horwitz, S. B., and Lisanti, M. P. (1998). Upregulation of caveolin-1 and caveolae organelles in Taxol-resistant A549 cells. *FEBS Lett.* **439,** 368–372.

Yang, G., Truong, L. D., Wheeler, T. M., and Thompson, T. C. (1999). Caveolin-1 expression in clinically confined human prostate cancer: A novel prognostic marker. *Cancer Res.* **59,** 5719–5723.

Yarbrough, T. L., Lu, T., Lee, H. C., and Shibata, E. F. (2002). Localization of cardiac sodium channels in caveolin-rich membrane domains: Regulation of sodium current amplitude. *Circ. Res.* **90,** 443–449.

Yoo, S. H., Park, Y. S., Kim, H. R., Sung, S. W., Kim, J. H., Shim, Y. S., Lee, S. D., Choi, Y. L., Kim, M. K., and Chung, D. H. (2003). Expression of caveolin-1 is associated with poor prognosis of patients with squamous cell carcinoma of the lung. *Lung Cancer* **42,** 195–202.

Young, L. H., Ikeda, Y., and Lefer, A. M. (2001). Caveolin-1 peptide exerts cardioprotective effects in myocardial ischemia-reperfusion via nitric oxide mechanism. *Am. J. Physiol. Heart Circ. Physiol.* **280,** H2489–H2495.

Zhang, W., Razani, B., Altschuler, Y., Bouzahzah, B., Mostov, K. E., Pestell, R. G., and Lisanti, M. P. (2000). Caveolin-1 inhibits epidermal growth factor-stimulated lamellipod extension and cell migration in metastatic mammary adenocarcinoma cells (MTLn3). *J. Biol. Chem.* **275,** 20717–20725.

Zhao, Y. Y., Liu, Y., Stan, R. V., Fan, L., Gu, Y., Dalton, N., Chu, P. H., Peterson, K., Ross, J., Jr., and Chien, K. R. (2002). Defects in caveolin-1 cause dilated cardiomyopathy and pulmonary hypertension in knockout mice. *Proc. Natl. Acad. Sci. USA* **99,** 11375–11380.

Zundel, W., Swiersz, L. M., and Giaccia, A. (2000). Caveolin 1-mediated regulation of receptor tyrosine kinase-associated phosphatidylinositol 3-kinase activity by ceramide. *Mol. Biol. Cell* **20,** 1507–1514.

Abbreviations and Acronyms

AMF	Autocrine motilitiy factor
APOE	Apolipoprotein E
CaM	Calmodulin
Cav-3	Caveolin-3
CBM	Caveolin-binding motif
CD	Circular dichroism
CK	Creatine kinase
CSD	Caveolin scaffolding domain
CT	Cholera toxin
Cyp	Cyclophilin
CytD	Cytochalasin D
DHE	Dihydroergosterol
DMD	Duchenne muscular dystrophy
DPH	Diphenylhexatriene
E_2	Estradiol
ECFP-Mem	Enhanced cyan fluorescent protein-Mem
EGFR	Epidermal growth factor receptor
eNOS	Endothelial nitric oxide synthase
ER	Endoplasmic reticulum
ER	Estrogen receptor
ERα	Estrogen receptor α
ERα-myc	Epitope-tagged ERα
ERβ	Estrogen receptor β
ERK	Extracellular signal-related kinase
FAK	Focal adhesion kinase
FKRP	Fukutin-related protein
Gα_i	G-protein α_i
GAP	GTPase activating protein
GFP	Green fluorescent protein
GIC	Golgi intermediate compartment
GLUT4	Glucose transporter 4
GM1	Ganglioside M1
GPCR	G-protein-coupled receptor
GSL	Glycosphingolipid
HDL	High-density lipoprotein
Hsp56	Heat shock protein 56
IP	Immunoprecipitation
IRS	Insulin receptor substrate
LDL	Low-density lipoprotein
LGMD	Limb-girdle muscular dystrophy
MMP	Metalloproteinases
NMR	Nuclear magnetic resonance
nNOS	Neuronal nitric oxide synthase
oxLDL	Oxidized low-density lipoprotein
PC	Phosphatidylcholine
PE	Phosphatidylethanolamine

PGDF	Platelet-derived growth factor
P-gp	P-glycoprotein
PH	Pulmonary hypertension
PI	Phosphatidylinositol
PI3	Phosphatidylinositol 3
PIRC	Percussion/pressure-induced rapid muscle contractions
PKC	Protein kinase C
PKCα	Protein kinase Cα
PLA_2	Phospholipase A_2
PM	Plasma membrane
PT	Pertussis toxin
RGS	Regulator of G-protein signaling
RMD	Rippling muscle disease
RTK	Receptor tyrosine kinases
S1P	Sphingosine-1-phosphate
SCP-2	Sterol carrier protein-2
SL	Sphingolipids
SM	Sphingomyelin
SMC	Smooth muscle cells
SRB1	Scavenger receptor class B Type I
SRE	Sterol regulatory element
STAT	Signal tranducer/activator of transcription
SUV	Small unilamellar vesicles
THC	Tetrahydrechrysene
uPAR	Urokinase receptor
VAMP	Vesicle-associated membrane protein
VCAM	Vascular cell adhesion molecule
VEGF	Vascular endothelial growth factor
VSMC	Vascular smooth muscle cells
VVO	Vesiculo-vacuolar organelles

Index

M

N

P

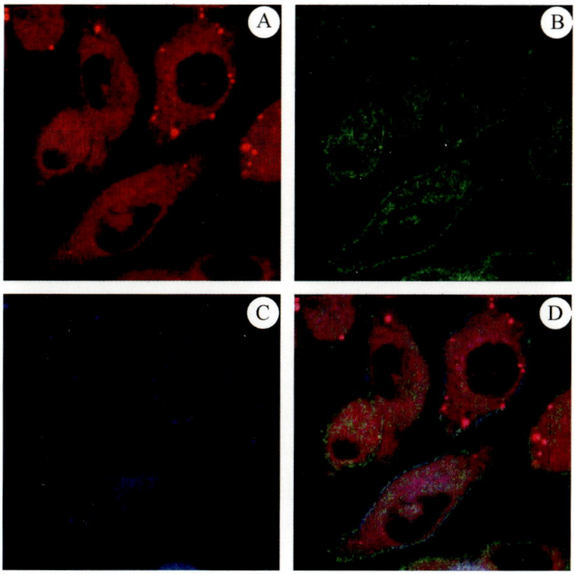

Chapter 1, Figure 4. Multiphoton laser scanning microscopy of fluorescent sterol (DHE), a plasma membrane marker (ECFP-Mem), and a lipid marker (Nile red) in L-cell fibroblasts. The plasmid pECFP-Mem (Clontech) was stably expressed in L-cells (L $arpt^{-}tk^{-}$) by transfection using Superfect (Qiagen) according to the manufacturer's instructions. After G418 selection and PCR verification to select resistant clones, L-cells stably expressing ECFP-Mem were cultured for 2 days on two-well Lab-Tek chambered coverglasses (VWR, Sugarland, TX) with Higuchi medium containing 10% fetal bovine serum and supplemented with a total concentration of 20 μg/ml of dehydroergosterol (DHE) in the form of large unilamellar vesicles. These vesicles were prepared with 1-palmitoyl-2-oleoyl-sn-glycero-3-phosphocholine (POPC) and dehydroergosterol (65:35 POPC:DHE), wherein DHE was almost completely monomeric (McIntosh *et al.*, 2003). Cells were then incubated with 100–400 nM Nile red (Molecular Probes, Eugene, OR) for approximately 30 min. Multiphoton laser scanning microscopy (MPLSM) was performed using a Bio-Rad MRC1024 MP system attached Axiovert 135 (Zeiss Inc., New York, NY) microscope with a Zeiss 63 × Plan-Apochromat (1.4 N.A.) oil immersion objective. All three fluorescent probes (DHE, Nile red, ECFP-Mem) were simultaneously excited using multiphoton excitation at 920 nm with a femtosecond Coherent Mira 900 Ti:Sapphire laser pumped with a Spectra-Physics Millenia X. Fluorescence emission was detected through an external three-channel detector system provided by Dr. Warren Zipfel (Cornell University, Ithaca, NY). Emission of the three fluorescent probes was individually selected by the INDO dichroic filter set provided by Dr. Warren Zipfel. (A) Nile red emission detected at 525–650 nm. (B) Dehydroergosterol emission detected at 360–430 nm. (C) ECFP-Mem emission detected at 485–515 nm. (D) Merged image with Nile red (red), dehydroergosterol (green), and ECFP-Mem (blue).

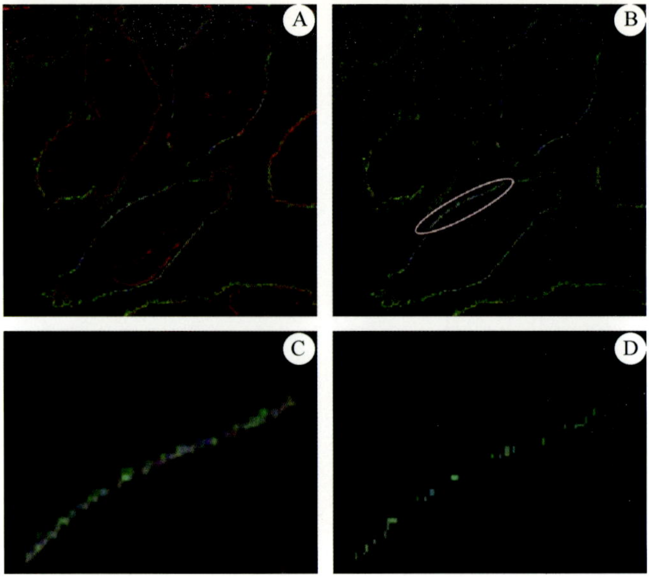

Chapter 1, Figure 5. Statistical segmentation of the plasma membrane of L-cells expressing ECFP-Mem (blue), supplemented with dehydroergosterol (green), and subsequently stained with Nile red (red). (A) Initial segmentation result from the rank-statistic technique produced by a combination of two windows chosen from two different plasma membrane sections. (B) Final segmentation result after binary masking was applied to further remove remaining intracellular regions and enhancing definition of the plasma membrane. A typical segment (circled in white) of the plasma membrane was chosen for statistical image analysis. (C) A blowup of the selected segment (all three channels) showing the discontinuity of the dehydroergosterol along the plasma membrane for the full data set. (D) A blowup of the selected segment (all three channels) after 85th percentile thresholding of the dehydroergosterol intensities (peak DHE intensities) along the plasma membrane.

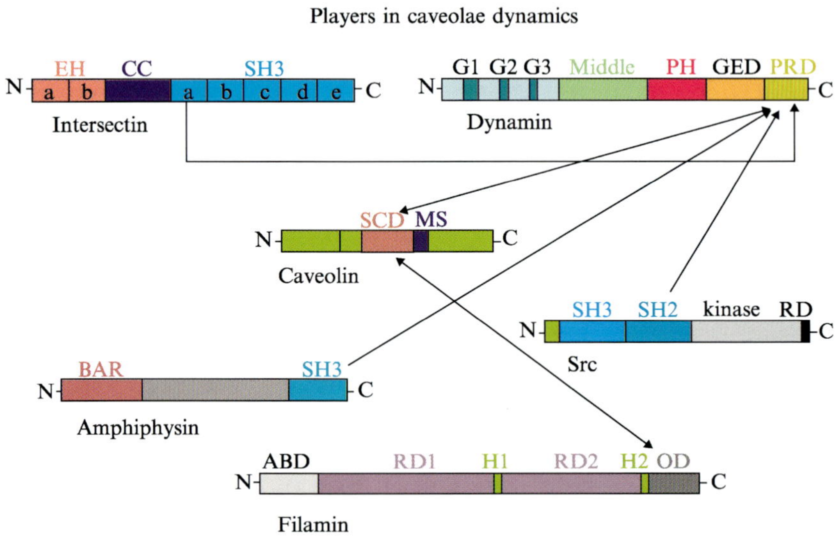

Chapter 3, Figure 1. Multiple structural proteins interact with the caveolins to mediate caveolae formation and fusion. Arrows point to different caveolin-associated proteins and corresponding domains for which direct interactions have been documented. EH, Eps15 homology domain; CC, coiled-coil domain; SH2, SH3, Src homology 2, 3 domains; G1-3, GTPase domains 1-3; Middle, middle domain; PH, plekstrin homology domain; GED, GTPase effector domain; PRD, proline-rich domain; SCD, scaffolding domain; MS, membrane-spanning domain; Kinase, Src kinase catalytic domain; RD, Src kinase regulatory domain; BAR, Bin/Amphiphysin/Rvs domain; ABD, actin-binding domain; RD1-2, rod domain 1, 2; HD1-2, hinge domain 1, 2; OD, oligomerization domain.

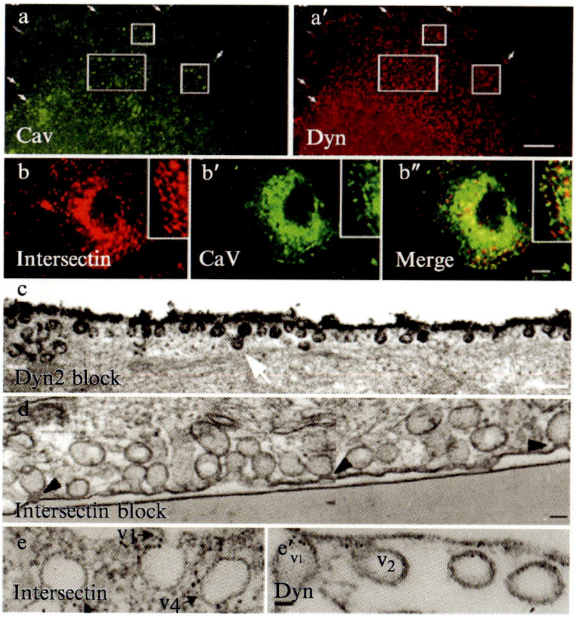

Chapter 3, Figure 2. Dynamin and intersectin in caveolae membrane dynamics. (a, a′) Dynamin localizes to caveolae in cultured hepatocytes. Fluorescence micrographs representing laser scanning confocal microscopy of cultured hepatocytes that were double-labeled with a monoclonal anti-caveolin antibody (a) and the polyclonal anti-Pan61 antibody (a′) to label endogenous dynamin. A significant number of vesicular structures are labeled with both antibodies (arrows and outlined areas), indicating a colocalization of dynamin and caveolin. Scale bars, 8.0 μm. (b–b″) Subcellular distribution of intersectin by double immunofluorescence in cultured endothelial cells (ECs). Intersectin displays a punctate staining pattern both at the plasma membrane and throughout the cytoplasm (b). Caveolin immunostaining in cultured ECs (b′). The merged image (b″) reflects extensive colocalization for both intersectin and caveolin (panel b″ and inset b″). Scale bars, (a–c) 10 μm; (insets) 5 μm. (c) Accumulation of surface attached caveolae in cultured hepatocytes injected with anti-dynamin antibodies. Electron micrographs of dynamin antibody-injected cells that were fixed and stained with ruthenium red. Dark vesicles reveal both surface (small arrows) and deeper (large arrows) membrane invaginations that are continuous with the plasma membrane. Scale bars, 0.15 μm. (d) Caveolae proliferation in ECs overexpressing wt intersectin. Electron micrographs show large numbers of caveolae forming along the cell border (arrows). Note the caveolar profiles displaying staining dense rings (arrowhead). Scale bars, 50 nm. (e, e′) Immuno-EM localization of dynamin (e) and intersectin (e′) to the necks of formed caveolae in cultured ECs. Immunogold labeling shows gold particles (6 nm) that are preferentially associated with the neck region of caveolae open to the EC surface (e, v1). Frequently, more than two gold particles labeled a caveolar profile or its neck. Scale bars, (e) 50 nm. (a, a′), and (c) reprinted with permission from Henley *et al.*, 1998. (b–b″) and (d–e′) reprinted with permission from Predescu *et al.*, 2003.

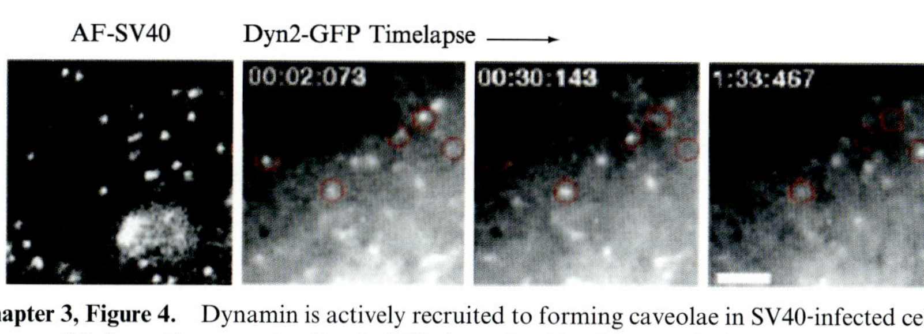

Chapter 3, Figure 4. Dynamin is actively recruited to forming caveolae in SV40-infected cells. Fluorescence images of living cells expressing Dyn2-GFP show that Dyn2 is recruited to caveolae-bound AF-SV40 viral particles. Dyn2 recruitment (red circles) appears as "blinking" at the cell surface as actin and caveolin are recruited, then disappear as caveolar scission occurs. Scale bars, 2.5 μm. Images reprinted with permission from Helenius *et al.*, 2002.

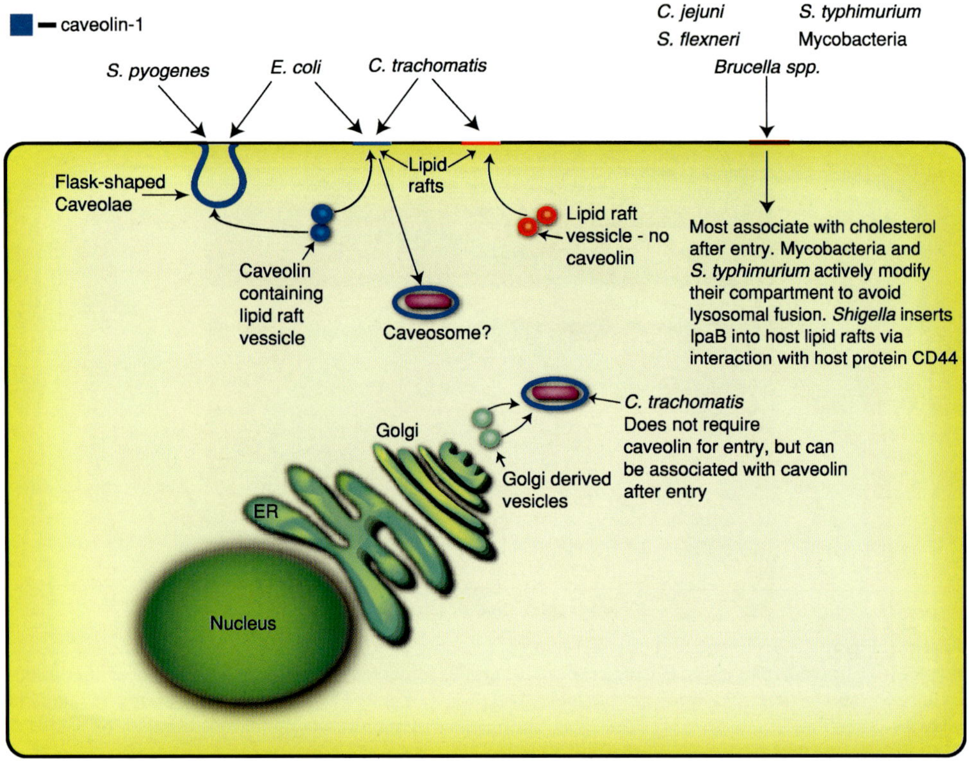

Chapter 4, Figure 1. Model depicting the major defining characteristics of entry of various bacteria into most cells via lipid rafts.

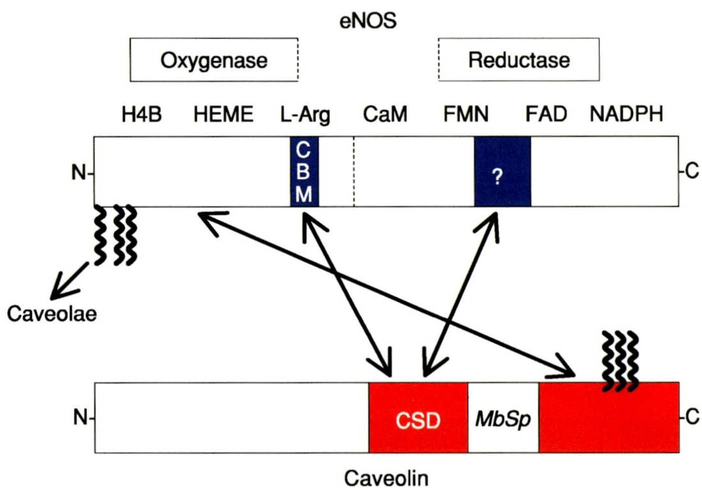

Chapter 5, Figure 1. Scheme of the caveolin–eNOS interactions. Both the oxygenase and the reductase domains of eNOS interact with caveolin. The caveolin-binding motif (CBM) corresponds to the consensus sequence FPAAPFSGW that recognizes the caveolin scaffolding domain (CSD) within the caveolin sequence. At the other side of the membrane-spanning region of caveolin (MBSp), the 40 C-terminal residues of caveolin (involved in caveolin oligomer–oligomer interaction) also interact with the oxygenase domain of eNOS. Both proteins are acylated, and in the case of eNOS, myristoylation on the glycine and palmitoylation on two cysteines are key for facilitating the targeting of eNOS to lipid rafts/caveolae. Note that eNOS is active as a dimmer and that the size of the two proteins is not to scale (human sequences: eNOS, 1205 residues and caveolin-1, 179 residues).

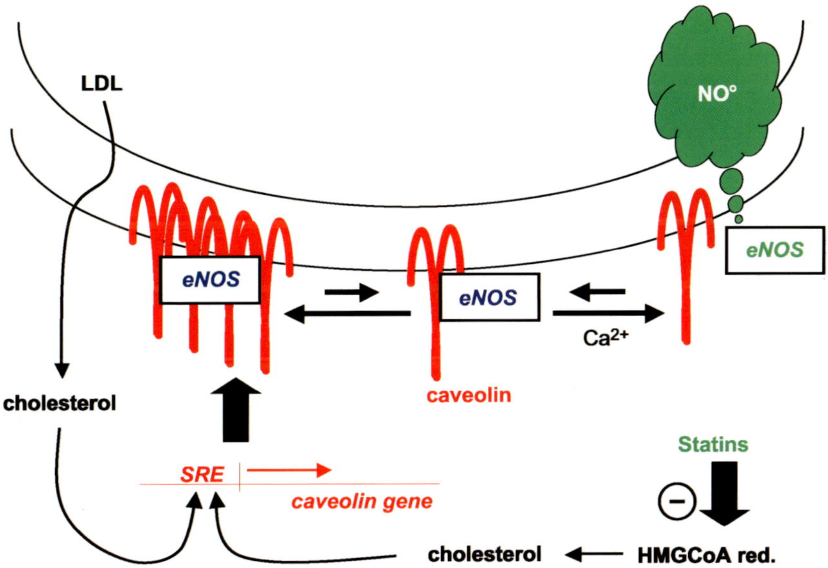

Chapter 5, Figure 2. Regulation of the caveolin–eNOS interaction by cholesterol levels. The presence of sterol regulatory elements (SRE) and other regulatory sequences in the promoter of the caveolin gene accounts for the stimulatory transcription of caveolin in the presence of excess cholesterol, as encountered when endothelial cells are exposed to high levels of LDL cholesterol. The consecutive increase in caveolin blocks the basal and agonist-stimulated eNOS activity in endothelial cells. Statins, by inhibiting the endogenous synthesis of cholesterol in (peripheral) endothelial cells (as well as indirectly by reducing circulatory LDL cholesterol), negatively affect caveolin expression. A reduction in caveolin abundance (or a restoration if first exposed to high LDL levels) facilitates the calcium-calmodulin-mediated disruption of the caveolin–eNOS complex (e.g., displaces the equilibrium toward the caveolin-free eNOS state), thereby leading to NO release.

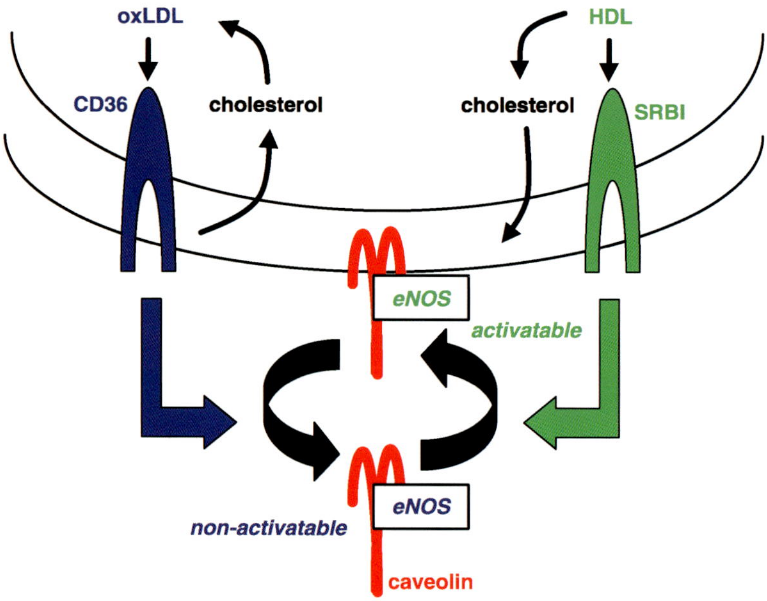

Chapter 5, Figure 3. Regulation of the caveolin–eNOS interaction by oxLDL and HDL. The lipoprotein receptors CD36 and scavenger receptor class B member 1 (SRBI) are both enriched in caveolae. Oxidized LDL binds to CD36 and loads itself with cholesterol, leading to a marked depletion of caveolar cholesterol. Consecutively, caveolin and eNOS are translocated to intracellular compartments wherein basal and agonist-stimulated NO productions are dramatically reduced. HDL binds to SRBI and provisions cholesterol esters to the cell, thereby maintaining the integrity of caveolae and even reversing the deleterious effects of oxLDL. The normal subcellular location of caveolin and eNOS allows eNOS to be activated when needed and NO to exert its major function of vascular homeostasis control.

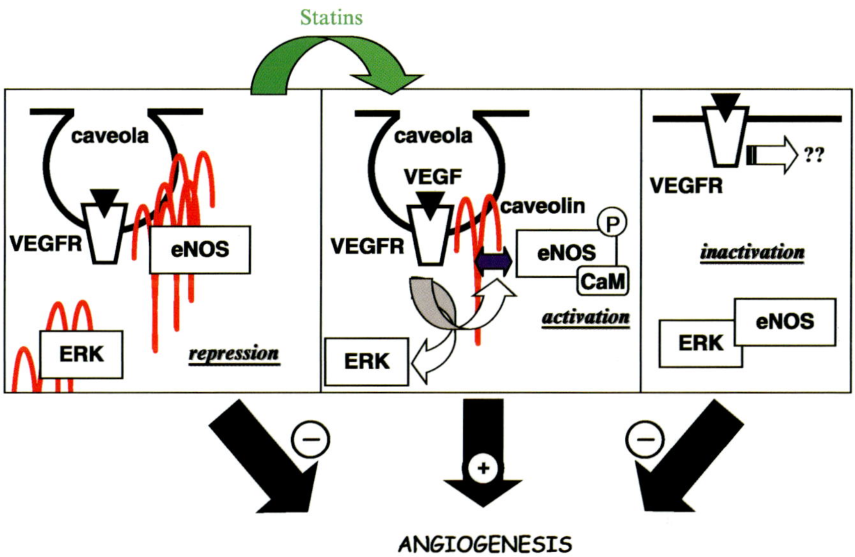

Chapter 5, Figure 4. Model of the apparent paradoxical regulation of NO-mediated angiogenesis by caveolin. This three-panel scheme represents the relationship between the abundance of caveolin and the NO-mediated angiogenesis. Middle panel: the VEGF binding to the VEGFR-2 receptor leads to the activation of eNOS through the local increase in intracellular calcium (the endoplasmic reticulum has been located near to caveolae) and the activation of the PI3K/Akt pathway leading to the phosphorylation of eNOS on Serine 1177. Activation of ERK is also induced upon VEGF exposure and, together with the increase in NO, leads to endothelial cell migration, proliferation, and reorganization in a new vascular network. Left panel: when caveolin is increased in the cells, such as after exposure to high levels of LDL cholesterol or expression of recombinant caveolin, eNOS and ERK are maintained inactivated, preventing full activation on VEGF exposure and leading to a defect in angiogenesis. Such a negative effect may be desired to inhibit tumor neovascularization and thereby block tumor growth. Right panel: in the absence of caveolin, as encountered in mice deficient for the caveolin gene, the compartmentation of the different actors of the VEGF/NO signaling cascade is lacking and leads to a decrease in the efficacy of VEGF to stimulate angiogenesis.

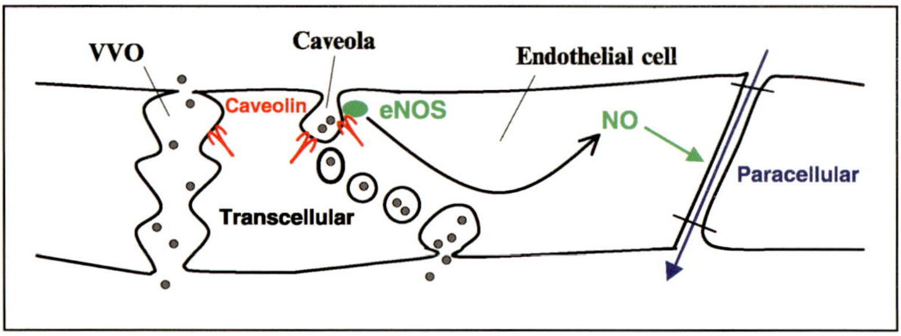

Permeability mode	Transcytosis (CAVEOLAE)	Paracellular (GAPS)
Wild-type	+	+
$eNOS^{-/-}$	+	-
$Cav^{-/-}$	-	++
$Cav^{-/-}$ + L-NAME	-	-

Chapter 5, Figure 5. Implication of caveolin and eNOS in endothelial cell permeability. NO is known to regulate the paracellular permeability (e.g., the transport through the intercellular junctions), but the location of eNOS in caveolae makes the role of NO more complex. Indeed, the other major mode of permeability, transcytosis either through the fission of caveolae and transport from one pole of the cell to the other or through fusion of several caveolae to form vesicular-vacuolar organelles (VVO) or channels connecting each pole of the cell, requires caveolae and caveolin. It is therefore likely that alteration in transcytotic permeability will affect caveolin function and thereby the eNOS locale and regulation. In the bottom of the figure, the phenotypes observed in mice deficient for eNOS or caveolin gene are illustrated. In $eNOS^{-/-}$ mice, transcytosis is the major mode of permeability in endothelial cells, whereas in $Cav^{-/-}$ mice, paracellular transport is mostly involved in macromolecules transport and becomes exquisitely sensitive to the NOS inhibitor L-NAME. The efficacy of this pathway is even increased when compared to wild-type mice (in order to compensate for the lack of caveolae).

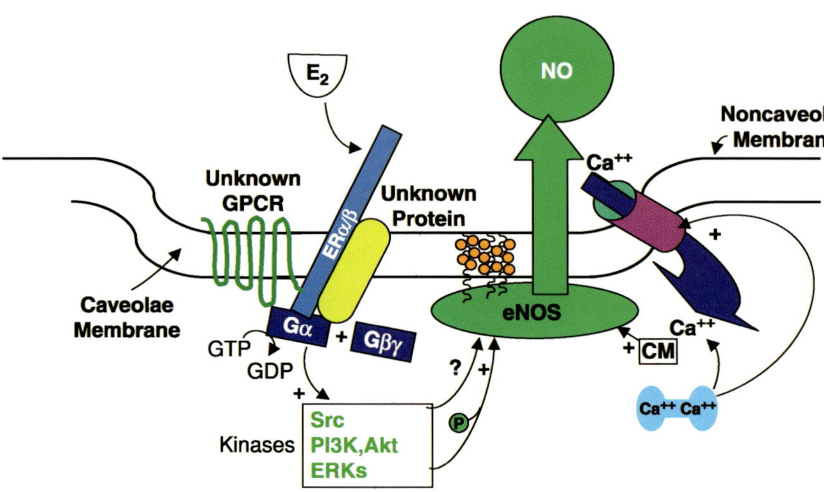

Chapter 6, Figure 7. ER signaling to eNOS in endothelial cell caveolae. eNOS is localized to cholesterol-enriched (orange circles) caveolae by myristoylation and palmitoylation. Both ERα and ERβ have also been localized to endothelial cell caveolae. Ligand binding to ER leads to $G_{\alpha i}$ activation, which mediates downstream events. ER interaction with $G_{\alpha i}$ may be direct, or it may involve a classical *G*-protein-coupled receptor (GPCR) or an alternative unknown protein intermediate. The downstream events include the activation of Src tyrosine kinase, PI3 kinase-Akt kinase signaling causing eNOS phosphorylation, and ERK activation having an unknown impact on eNOS. In addition, there is perturbation of the local calcium environment, involving either the release of intracellular stores, which are likely to be in close proximity to caveolae membranes, or calcium influx, resulting in calmodulin (CM) binding to the enzyme.

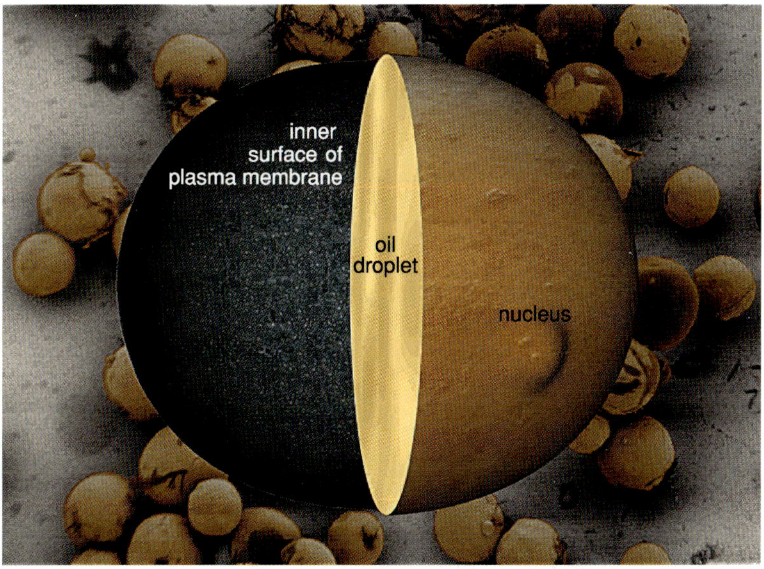

Chapter 8, Figure 1. Scanning electron microscopy reconstruction of an adipocyte. Images from scanning electron microscopy of the surface of intact adipocytes and the inner surface of plasma membranes were used to construct this image, to scale, of an adipocyte. The inner surface of the plasma membrane is seen to be speckled with caveolae of varying size.

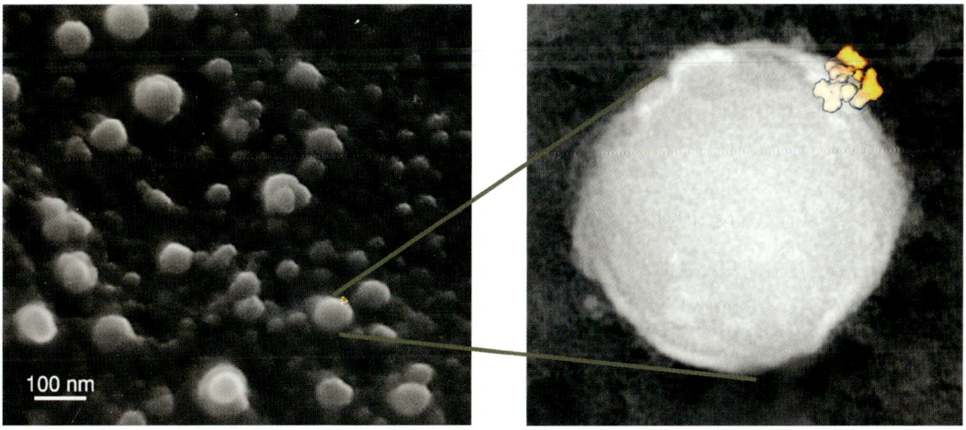

Chapter 8, Figure 2. Scanning electron microscopy of adipocyte plasma membrane inner surface with an insulin receptor model inserted to scale. From Luo *et al.* (1999); Thorn *et al.* (2003).

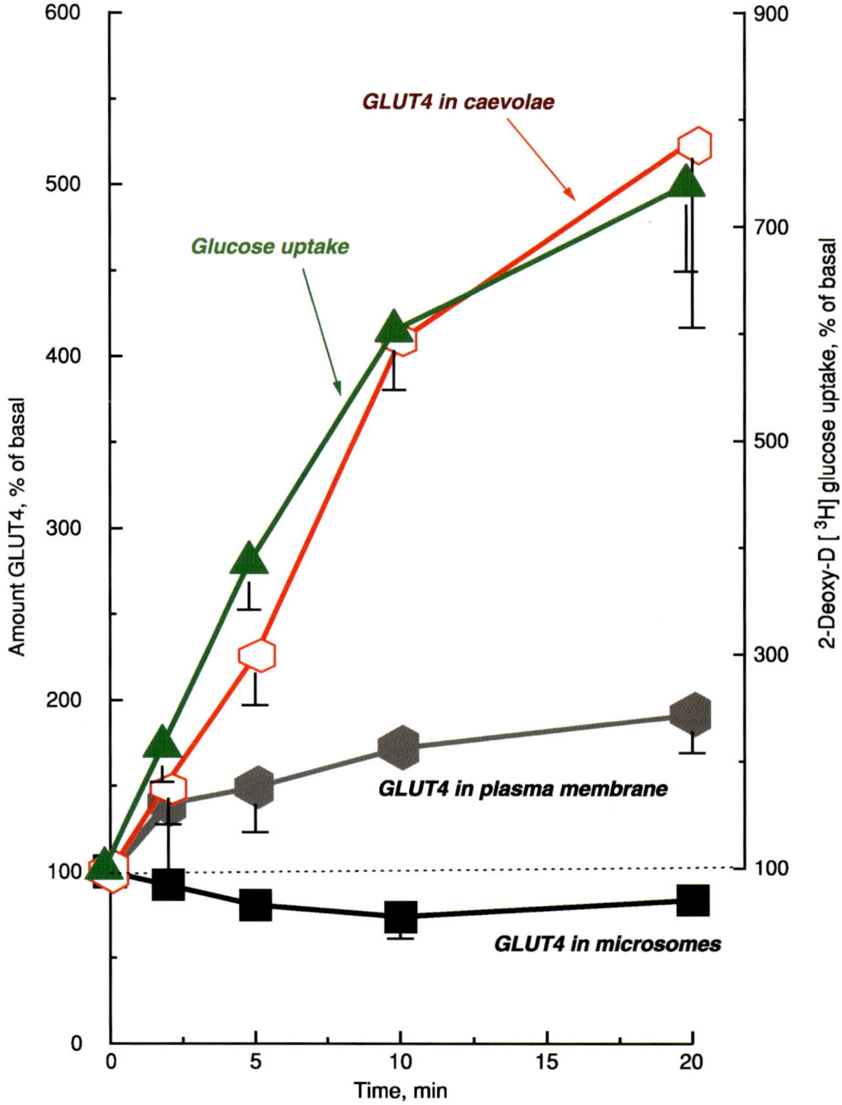

Chapter 8, Figure 4. Translocation of GLUT4 to caveolae and glucose transport in caveolae. Rat adipocytes were incubated with insulin for the indicated time and subjected to subcellular fractionation. Caveolae were prepared from plasma membranes by treatment with non-ionic detergent and collecting the insoluble residue. The amount of GLUT4 protein in the different fractions was determined after SDS-PAGE and immunoblotting. Glucose uptake was measured from the same cell batches in parallel. From Gustavsson *et al.* (1996).

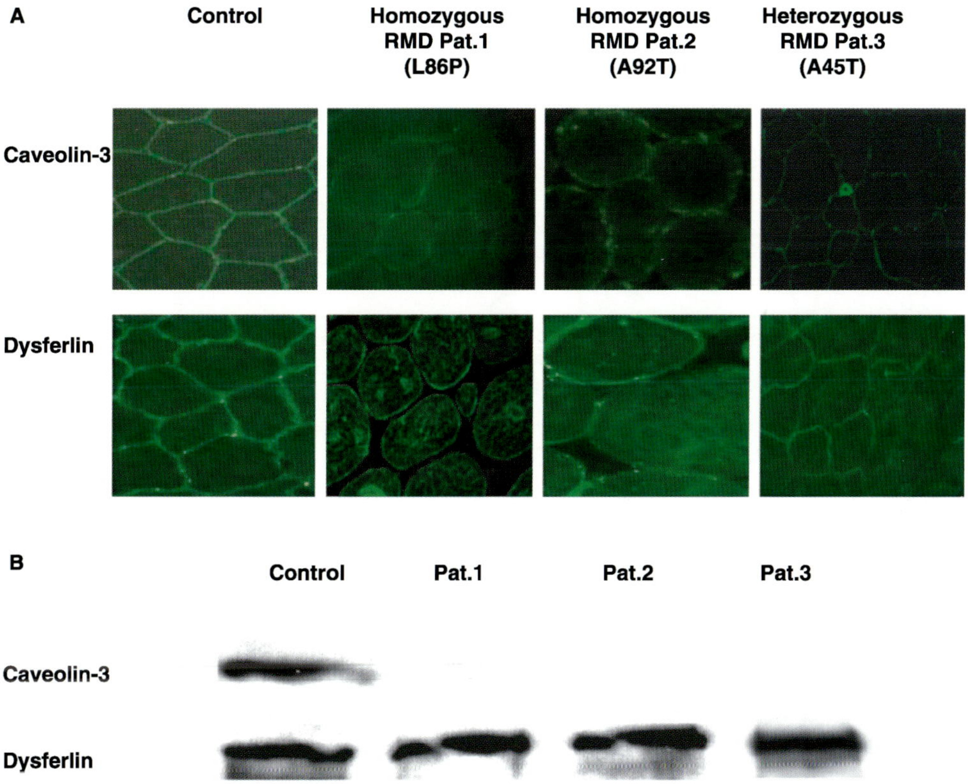

Chapter 9, Figure 2. (A) Immunofluorescence and (B) immunoblot analysis of CAV3 and dysferlin on muscle biopsies of two patients with homozgyous (L86P, A92T) and one heterozygous (A45T) RMD. In the RMD patients there is a drastic reduction in the intensity of CAV3 immunostaining at the sarcolemma and in immunoblotting. Dysferlin is reduced in RMD by immunohistochemistry, but normal in immunoblotting (Kubisch *et al.*, 2003).

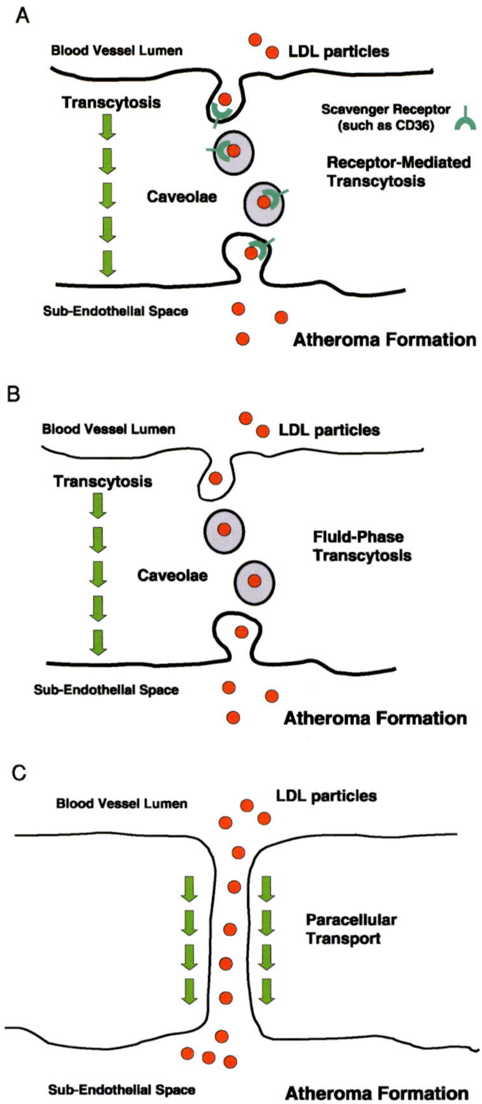

Chapter 10, Figure 2. Models of trans-endothelial transport. (A and B) Caveolae-mediated transcytosis. In this model, LDL particles are transferred from the blood vessel lumen to the intima. Two possible pathways exist for this endothelial transcellular transport. (A) An LDL particle may bind to a specific receptor found in endothelial cell caveolae and induce the endocytosis/transcytosis of caveolae vesicle, and the particle is then transferred to the subendothelial space (receptor-mediated transcytosis). (B) Alternatively, an LDL particle can be transported via a non-specific, or fluid-phase, form of caveolar transcytosis. (C) Para-cellular transport of LDL particles across the endothelial cell barrier. Under certain conditions, inter-endothelial tight junctions may be altered, increasing endothelium permeability. Modified from Frank and Lisanti (2004).

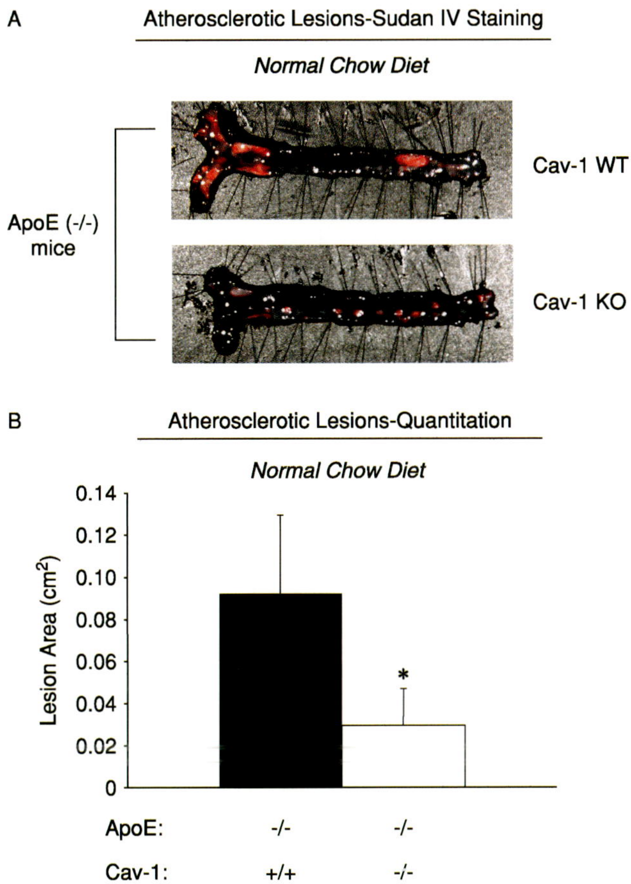

Chapter 10, Figure 3. Caveolin-1 deficiency reduces the extent of atheromatous lesions in ApoE$^{-/-}$ mice. (A) En face visualization of aortas, which were harvested from ApoE$^{-/-}$/Cav-1$^{+/+}$ and ApoE$^{-/-}$ /Cav-1$^{-/-}$ mice, fed a normal chow diet. (B) Quantification of the total area occupied by atherosclerotic lesions per aorta. Note that in animals fed a normal chow diet, a Cav-1 deficiency leads to dramatic reductions (~70%) in the total area occupied by these atherosclerotic lesions. An asterisk (*) indicates statistical significance ($p < 0.05$); $n = 5$ mice for each experimental group. Modified from Frank *et al.* (2004).

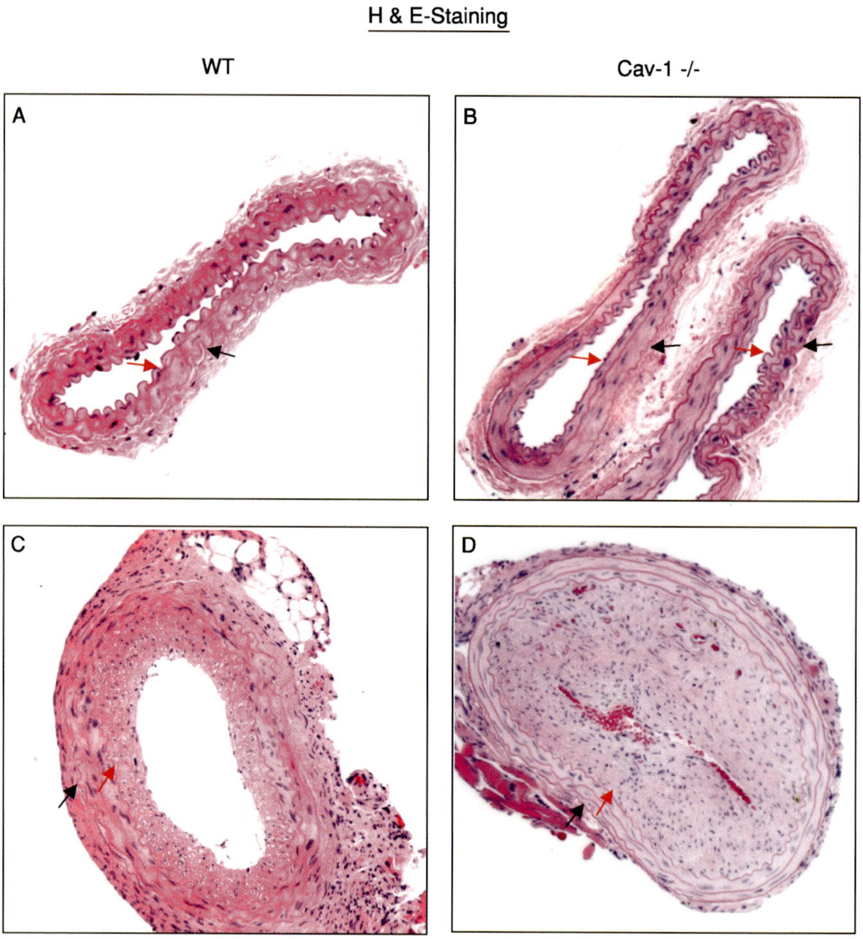

Chapter 10, Figure 4. Histological analysis of wild-type (WT) and Cav-$1^{-/-}$ carotid arteries, retrieved 4 weeks post-ligation. (A and B) H&E staining of control wild-type (A) and control Cav-$1^{-/-}$ (B) non-ligated arteries. Note that both WT and Cav-$1^{-/-}$ carotid arteries appear normal, without any evidence of baseline SMC hyperproliferation or migration. (C and D) H&E staining of ligated wild-type (C) and ligated Cav-$1^{-/-}$ (D) arteries. Note that the carotids of Cav-$1^{-/-}$ mice show significantly more intimal hyperplasia, with subtotal luminal obstruction, as compared to WT mice. In (A)–(D), red arrows denote the internal elastic lamina and black arrows denote the external elastic lamina. Modified from Hassan *et al.* (2004).

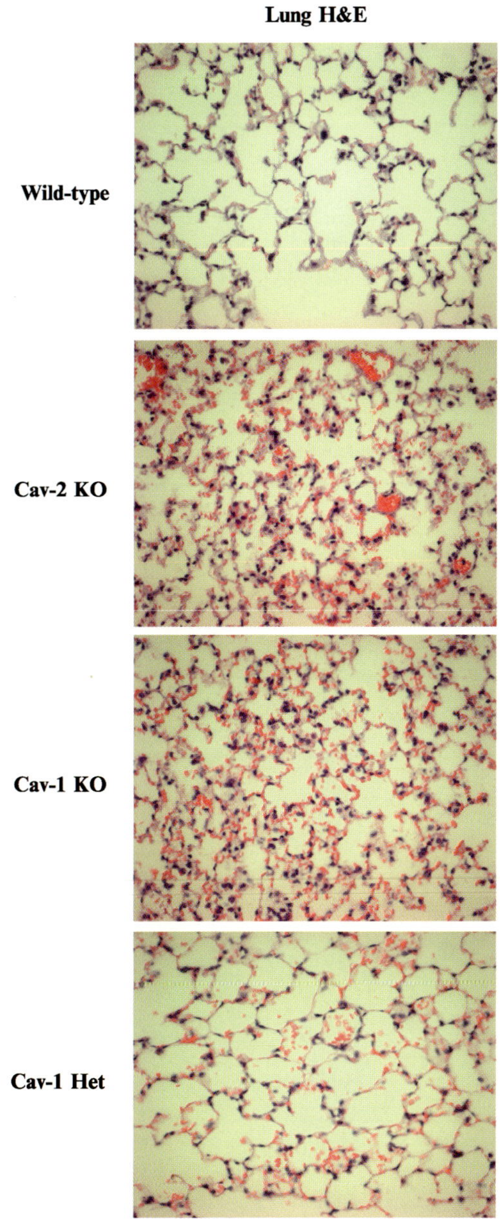

Chapter 11, Figure 4. Cav-1$^{-/-}$ and Cav-2$^{-/-}$ mice show lung abnormalities characterized by hypercellularity and thickening of the alveolar septa, as compared to wild-type and Cav-1$^{-/+}$ heterozygous mice. Modified from Razani *et al.* (2002).

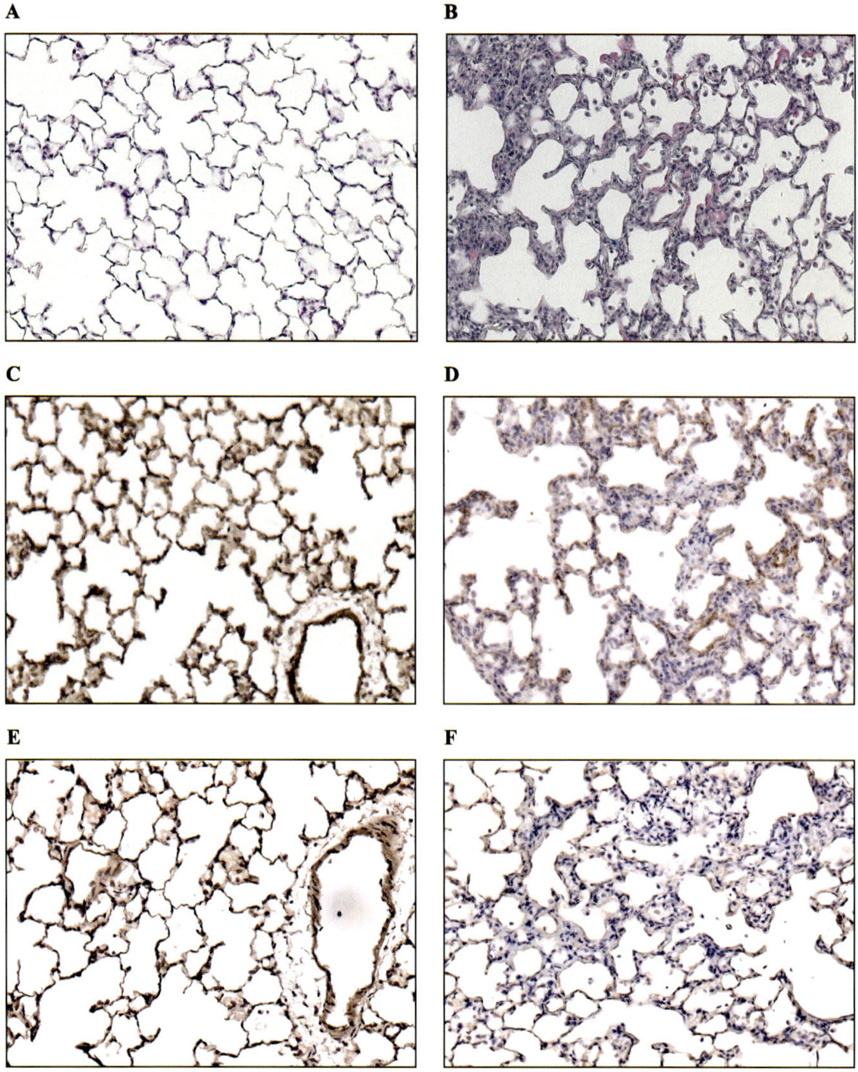

Chapter 11, Figure 5. Hematoxylin and eosin (H&E) staining of the lungs of control (A) and MI (B) rats shows lung structural remodeling characterized by hypercellularity and thickened alveolar septa in the MI group. This lung structural remodeling is associated with decreased expression of Cav-1 and Cav-2. (C) and (D) represent immunostaining of Cav-1 in lungs of control and MI rats, respectively. (E) and (F) represent immunostaining of Cav-2 in lungs of control and MI rats, respectively. Modified from Jasmin *et al.* (2004).